G·L·O·B·A·L S·T·U·D·I·E·S

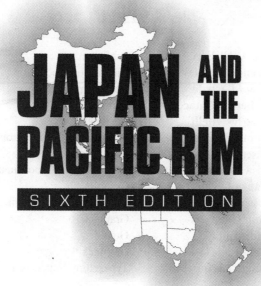

JAPAN AND THE PACIFIC RIM

SIXTH EDITION

Dr. Dean W. Collinwood
University of Utah

OTHER BOOKS IN THE GLOBAL STUDIES SERIES

- Africa
- China
- Europe
- India and South Asia
- Latin America
- The Middle East
- Russia, the Eurasian Republics,
  and Central/Eastern Europe

McGraw-Hill/Dushkin
530 Old Whitfield Street, Guilford, Connecticut 06437
Visit us on the Internet—*http://www.dushkin.com*

STAFF

Ian A. Nielsen	Publisher
Brenda S. Filley	Director of Production
Lisa M. Clyde	Developmental Editor
Roberta Monaco	Editor
Charles Vitelli	Designer
Robin Zarnetske	Permissions Coordinator
Joseph Offredi	Permissions Assistant
Lisa Holmes-Doebrick	Administrative Coordinator
Laura Levine	Graphics
Michael Campbell	Graphics/Cover Design
Tom Goddard	Graphics
Eldis Lima	Graphics
Nancy Norton	Graphics
Juliana Arbo	Typesetting Supervisor

Cataloging in Publication Data
Main entry under title: Global Studies: Japan and the Pacific Rim. 6/E
  1. East Asia—History—20th century–. 2. East Asia—Politics and government—20th century–. I. Title: Japan and the Pacific Rim. II. Collinwood, Dean W., *comp.*
ISBN 0–07–243296–9

Sixth Edition

Printed in the United States of America   1234567890BAHBAH54321   Printed on Recycled Paper

Japan and the Pacific Rim

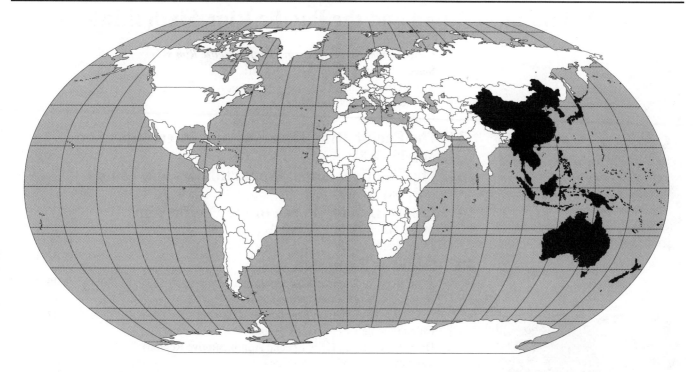

AUTHOR/EDITOR

Dr. Dean W. Collinwood

The author/editor of *Global Studies: Japan and the Pacific Rim* was awarded his Ph.D. from the University of Chicago, his M.Sc. in international relations from the University of London, and his B.A. in political science with a minor in Japanese from Brigham Young University. He was a Fulbright scholar at the University of Tokyo and Tsuda College in Japan and has conducted research throughout Asia and the Pacific. Dr. Collinwood is research professor of management at the University of Utah in Salt Lake City, where he also directs the U.S.–China Center. He also heads an international business consulting firm that includes the U.S.–Japan Center as one of its divisions. He is the past president of the Western Conference of the Association for Asian Studies and former chairman of the Utah Asian Studies Consortium. He is the author of many books and articles on Japan, Korea, and other countries.

SERIES CONSULTANT
H. Thomas Collins
PROJECT LINKS
George Washington University

Contents

Global Studies: Japan and the Pacific Rim, Sixth Edition

Pacific Rim   Page 8

Pacific Islands   Page 16

Japan   Page 33

Australia   *Page 41*

Singapore   *Page 93*

Taiwan   *Page 101*

Using Global Studies: Japan and the Pacific Rim

THE GLOBAL STUDIES SERIES

The Global Studies series was created to help readers acquire a basic knowledge and understanding of the regions and countries in the world. Each volume provides a foundation of information—geographic, cultural, economic, political, historical, artistic, and religious—that will allow readers to better assess the current and future problems within these countries and regions and to comprehend how events there might affect their own well-being. In short, these volumes present the background information necessary to respond to the realities of our global age.

Each of the volumes in the Global Studies series is crafted under the careful direction of an author/editor—an expert in the area under study. The author/editors teach and conduct research and have traveled extensively through the regions about which they are writing.

In *Japan and the Pacific Rim,* the author/editor has written regional essays on the Pacific Rim and the Pacific Islands and country reports for each of the countries covered, including a special report on Japan.

MAJOR FEATURES OF THE GLOBAL STUDIES SERIES

The Global Studies volumes are organized to provide concise information on the regions and countries within those areas under study. The major sections and features of the books are described here.

Regional Essays

For *Global Studies: Japan and the Pacific Rim,* the author/editor has written two essays focusing on the religious, cultural, sociopolitical, and economic differences and similarities of the countries and peoples in the region: "The Pacific Rim: Diversity and Interconnection," and "The Pacific Islands: Opportunities and Limits." Detailed maps accompany each essay.

Country Reports

Concise reports are written for each of the countries within the region under study. These reports are the heart of each Global Studies volume. *Global Studies: Japan and the Pacific Rim, Sixth Edition,* contains 20 country reports, including a lengthy report on Japan.

The country reports are composed of five standard elements. Each report contains a detailed map visually positioning the country among its neighboring states; a summary of statistical information; a current essay providing important historical, geographical, political, cultural, and economic information; a historical timeline, offering a convenient visual survey of a few key historical events; and four "graphic indicators," with summary statements about the country in terms of development, freedom, health/welfare, and achievements.

A Note on the Statistical Reports

The statistical information provided for each country has been drawn from a wide range of sources. (The most frequently referenced are listed on page 217.) Every effort has been made to provide the most current and accurate information available. However, occasionally the information cited by these sources differs to some extent; and, all too often, the most current information available for some countries is dated. Aside from these difficulties, the statistical summary of each country is generally quite complete and up to date. Care should be taken, however, in using these statistics (or, for that matter, any published statistics) in making hard comparisons among countries. We have also provided comparable statistics for the United States and Canada, which can be found on pages viii and ix.

World Press Articles

Within each Global Studies volume is reprinted a number of articles carefully selected by our editorial staff and the author/editor from a broad range of international periodicals and newspapers. The articles have been chosen for currency, interest, and their differing perspectives on the subject countries. There are 30 articles in *Global Studies: Japan and the Pacific Rim, Sixth Edition.*

The articles section is preceded by an annotated table of contents as well as a topic guide. The annotated table of contents offers a brief summary of each article, while the topic guide indicates the main theme(s) of each article. Thus, readers desiring to focus on articles dealing with a particular theme, say, environment, may refer to the topic guide to find those articles.

WWW Sites

An extensive annotated list of selected World Wide Web sites can be found on the facing page (vii) in this edition of *Global Studies: Japan.* In addition, the URL addresses for country-specific Web sites are provided on the statistics page of most countries. All of the Web site addresses were correct and operational at press time. Instructors and students alike are urged to refer to those sites often to enhance their understanding of the region and to keep up with current events.

Glossary, Bibliography, Index

At the back of each Global Studies volume, readers will find a glossary of terms and abbreviations, which provides a quick reference to the specialized vocabulary of the area under study and to the standard abbreviations (NIC, ASEAN, etc.) used throughout the volume.

Following the glossary is a bibliography, which lists general works, national histories, and current-events publications and periodicals that provide regular coverage on Japan and the Pacific Rim.

The index at the end of the volume is an accurate reference to the contents of the volume. Readers seeking specific information and citations should consult this standard index.

Currency and Usefulness

Global Studies: Japan and the Pacific Rim, like other Global Studies volumes, is intended to provide the most current and useful information available necessary to understand the events that are shaping the cultures of the region today.

This volume is revised on a regular basis. The statistics are updated, regional essays and country reports revised, and world press articles replaced. In order to accomplish this task, we turn to our author/editor, our advisory boards, and—hopefully—to you, the users of this volume. Your comments are more than welcome. If you have an idea that you think will make the next edition more useful, an article or bit of information that will make it more current, or a general comment on its organization, content, or features that you would like to share with us, please send it in for serious consideration.

Selected World Wide Web Sites for Japan and the Pacific Rim

All of these Web sites are hot-linked through the *Global Studies* home page:
http://www.dushkin.com/globalstudies **(just click on a book).**

Some Web sites are continually changing their structure and content, so the information listed may not always be available.

GENERAL SITES

1. CNN Online Page—**http://www.cnn.com**—A U.S. 24-hour video news channel. News is updated every few hours.

2. C-SPAN ONLINE—**http://www.c-span.org**—See especially C-SPAN International on the Web for International Programming Highlights and archived C-SPAN programs.

3. I-Trade International Trade Resources & Data Exchange—**http://www.i-trade.com**—Monthly exchange-rate data, U.S. Document Export Market Information (GEMS), U.S. Global Trade Outlook, and the CIA Worldfact Book are available here.

4. Political Science RESOURCES—**http://www.psr.keele.ac.uk/psr.htm**—Dynamic gateway to country sources available via European addresses is presented on this Web site.

5. ReliefWeb—**http://wwwnotes.reliefweb.int**—Access UN's Department of Humanitarian Affairs clearinghouse for international humanitarian emergencies here.

6. Social Science Information Gateway (SOSIG)—**http://sosig. esrc.bris.ac.uk**—This presents the project of the Economic and Social Research Council (ESRC). It catalogs 22 subjects and lists developing-countries' URL addresses.

7. United Nations System—**http://www.unsystem.org**—Access the official Web site for the United Nations system of organizations here.

8. U.S. Agency for International Development (USAID)—**http://www. info.usaid.gov**—Graphically presented U.S. trade statistics related to Japan, China, Taiwan, and other Pacific Rim countries are available at this site.

9. U.S. Central Intelligence Agency Home Page—**http://www. cia.gov**—This site includes publications of the CIA, current Worldfact Book, and maps.

10. U.S. Department of State Home Page—**http://www.state.gov/ index.html**—This site organizes alphabetically: Country Reports, Human Rights, International Organizations, etc.

11. UT International Network Information Systems—**http://inic. utexas.edu**—This Gateway has pointers to international sites, including Japan, China, and Taiwan.

12. World Bank Group—**http://www.worldbank.org**—Find news (i.e., press releases, summary of new projects, speeches), publications, topics in development, countries and regions on this Web site. It links to other financial organizations.

13. World Health Organization (WHO)—**http://www.who.int/**—Maintained by WHO's headquarters in Geneva, Switzerland, it is possible to use the Excite search engine to conduct keyword searches from here.

14. World Trade Organization—**http://www.wto.org**—This Web site's topics include legal frameworks, trade and environmental policies, recent agreements, etc.

15. WWW Virtual Library Database—**http://conbio.net/vl/database/**—Easy search for country-specific sites that provide news, government, and other information is possible from this site.

ASIA

16. Aseanweb—**http://www.asean.or.id**—This site's menu includes "What's New?" and data on economics, politics, security, and print publications on the nations of ASEAN—the Association of Southeast Asian Nations.

17. Asia Gateway—**http://www.asiagateway.com**—Access country profiles, including lifestyles, business, and other data from here. Look in "What's New" for news highlights.

18. Asiatour—**http://asiatour.com/index.htm**—Travel and historical information for many Asian countries is presented on this Web site..

19. Asia-Yahoo—**http://www.yahoo.com/Regional/Regions/Asia/**—This specialized Yahoo search site permits keyword search on Asian events, countries, or topics.

20. NewsDirectory.com—**http://www.newsd.com**—This site, a Guide to English-Language Media Online, lists over 7,000 actively updated papers and magazines.

21. Orientation Asia—**http://as.orientation.com**—Link to specific countries, late-breaking news from this Web site.

22. Signposts to Asia and the Pacific—**http://www-signposts. uts.edu**—Databases, news, key country contacts, articles, and links to other relevant sites are available here.

23. South-East Asia Information—**http://sunsite.nus.edusg/asiasvc. html**—This is an excellent gateway for country-specific research. Information on Internet Providers and Universities in Southeast Asia provide links to Asian online services.

CHINA

24. Chinese Security Home Page—**http://members.aol.com/ mehampton/chinasec.html**—Information is listed under Chinese Military Links, Data Sources on Chinese Security Issues, Key Newspapers and News Services, and Key Scholarly Journals and Magazines.

25. Inside China Today—**http://www.insidechina.com**—The European Information Network is organized under Headline News, Government, and Related Sites, Mainland China, Hong Kong, Macau, and Taiwan.

26. Internet Guide for China Studies—**http://sun.sino.uni-heidelberg. de/igcs/index.html**—Coverage of news media, politics, legal and human rights information, as well as China's economy, philosophy and religion, society, arts, culture, and history may be found here.

JAPAN

27. Japan Ministry of Foreign Affairs—**http://www.mofa.go.jp**—"What's New" lists events, policy statements, press releases. Foreign Policy section has speeches, archive, information under Countries and Region, Friendship.

28. Japan Policy Research Institute (JPRI)—**http://www.jpr.org**—Headings on this site include "What's New" and Publications before 1996.

29. The Japan Times Online—**http://www.japantimes.co.jp**—This daily online newspaper is offered in English and contains late-breaking news.

The United States (United States of America)

GEOGRAPHY

Area in Square Miles (Kilometers):
3,618,770 (9,578,626) (about ½ the size of Russia)

Capital (Population): Washington, D.C. (568,000)

Environmental Concerns: air and water pollution; limited freshwater resources; desertification; loss of habitat; waste disposal

Geographical Features: vast central plain, mountains in the west, hills and low mountains in the east; rugged mountains and broad river valleys in Alaska; volcanic topography in Hawaii.

Climate: mostly temperate

PEOPLE

Population

Total: 276,000,000

Annual Growth Rate: 0.91%

Rural/Urban Population Ratio: 24/76

Major Languages: predominantly English; a sizable Spanish-speaking minority; many others

Ethnic Makeup: 69.1% white; 12.5% Latino; 12.1% black or African American; 3.6% Asian; 0.7% Amerindian

Religions: 56% Protestant; 28% Roman Catholic; 2% Jewish; 4% others; 10% none or unaffiliated

Health

Life Expectancy at Birth: 74 years (male); 80 years (female)

Infant Mortality Rate (Ratio): 6.82/1,000

Physicians Available (Ratio): 1/365

Education

Adult Literacy Rate: 97% (official; estimates vary widely)

Compulsory (Ages): 7–16; free

COMMUNICATION

Telephones: 173,000,000 main lines

Daily Newspaper Circulation: 238 per 1,000 people

Televisions: 776 per 1,000 people

Internet Service Providers: 7,600 (1999 est.)

TRANSPORTATION

Highways in Miles (Kilometers): 3,906,960 (6,261,154)

Railroads in Miles (Kilometers): 149,161 (240,000)

Usable Airfields: 13,387

Motor Vehicles in Use: 206,000,000

GOVERNMENT

Type: federal republic

Independence Date: July 4, 1776

Head of State: President George W. Bush

Political Parties: Democratic Party; Republican Party; others of minor political significance

Suffrage: universal at 18

MILITARY

Military Expenditures (% of GDP): 3.8%

Current Disputes: none

ECONOMY

Per Capita Income/GDP: $33,900/$9.25 trillion

GDP Growth Rate: 4.1%

Inflation Rate: 2.2%

Unemployment Rate: 4.2%

Labor Force: 13,943,000

Natural Resources: minerals; precious metals; petroleum; coal; copper; timber; arable land

Agriculture: food grains; feed crops; fruits and vegetables; oil-bearing crops; livestock; dairy products

Industry: diversified in both capital- and consumer-goods industries

Exports: $663 billion (primary partners Canada, Mexico, Japan)

Imports: $912 billion (primary partners Canada, Japan, Mexico)

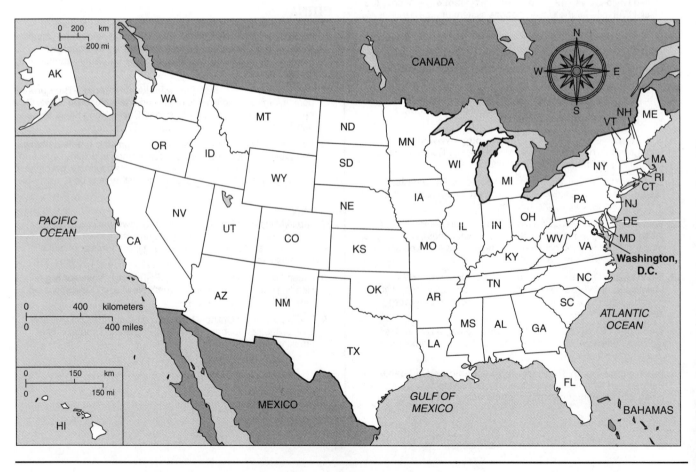

Canada

GEOGRAPHY

Area in Square Miles (Kilometers): 3,850,790 (9,976,140) (slightly larger than the United States)

Capital (Population): Ottawa (1,000,000)

Environmental Concerns: air pollution and resulting acid rain severely affecting lakes and damaging forests; water pollution; industrial damage to agriculture and forest productivity

Geographical Features: permafrost in the north, mountains in the west, central plains, and a maritime culture in the east

Climate: from temperate in south to subarctic and arctic in north

PEOPLE

Population

Total: 31,300,000

Annual Growth Rate: 1.02%

Rural/Urban Population Ratio: 23/77

Major Languages: both English and French are official

Ethnic Makeup: 28% British Isles origin; 23% French origin; 15% other European; 6% others; 2% indigenous; 26% mixed

Religions: 46% Roman Catholic; 16% United Church; 10% Anglican; 28% others

Health

Life Expectancy at Birth: 76 years (male); 83 years (female)

Infant Mortality Rate (Ratio): 5.08/1,000

Physicians Available (Ratio): 1/534

Education

Adult Literacy Rate: 97%

Compulsory (Ages): primary school

COMMUNICATION

Telephones: 18,500,000 main lines

Daily Newspaper Circulation: 215 per 1,000 people

Televisions: 647 per 1,000 people

Internet Service Providers: 750 (1999 est.)

TRANSPORTATION

Highways in Miles (Kilometers): 559,240 (902,000)

Railroads in Miles (Kilometers): 22,320 (36,000)

Usable Airfields: 1,411

Motor Vehicles in Use: 16,800,000

GOVERNMENT

Type: confederation with parliamentary democracy

Independence Date: July 1, 1867

Head of State/Government: Queen Elizabeth II; Prime Minister Jean Chrétien

Political Parties: Progressive Conservative Party; Liberal Party; New Democratic Party; Reform Party; Bloc Québécois

Suffrage: universal at 18

MILITARY

Military Expenditures (% of GDP): 1.2%

Current Disputes: none

ECONOMY

Currency (U.S.$ Equivalent): 1.53 Canadian dollars = $1

Per Capita Income/GDP: $23,300/$722.3 Billion

GDP Growth Rate: 3.6%

Inflation Rate: 1.7%

Labor Force: 15,900,000 million

Natural Resources: petroleum; natural gas; fish; minerals; cement; forestry products; wildlife; hydropower

Agriculture: grains; livestock; dairy products; potatoes; hogs; poultry and eggs; tobacco; fruits and vegetables

Industry: oil production and refining; natural-gas development; fish products; wood and paper products; chemicals; transportation equipment

Exports: $277 billion (primary partners United States, Japan, United Kingdom)

Imports: $259.3 billion (primary partners United States, Japan, United Kingdom)

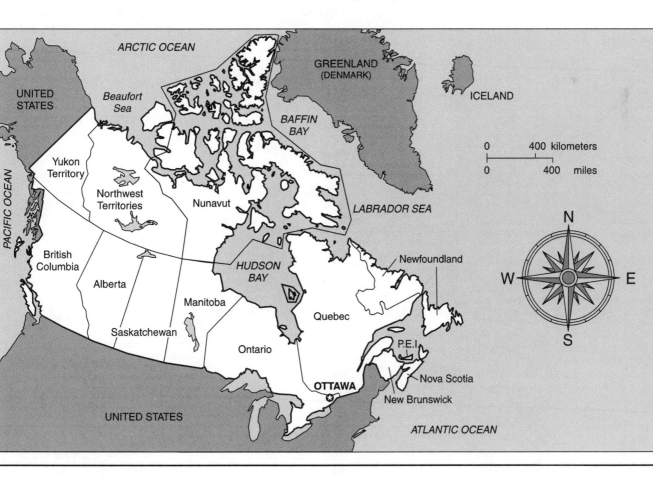

GLOBAL STUDIES

This map is provided to give you a graphic picture of where the countries of the world are located, the relationships they have with their region and neighbors, and their positions relative to economic and political power blocs. We have focused on certain areas to illustrate these crowed regions more clearly.

Scale: 1 to 125,000,000

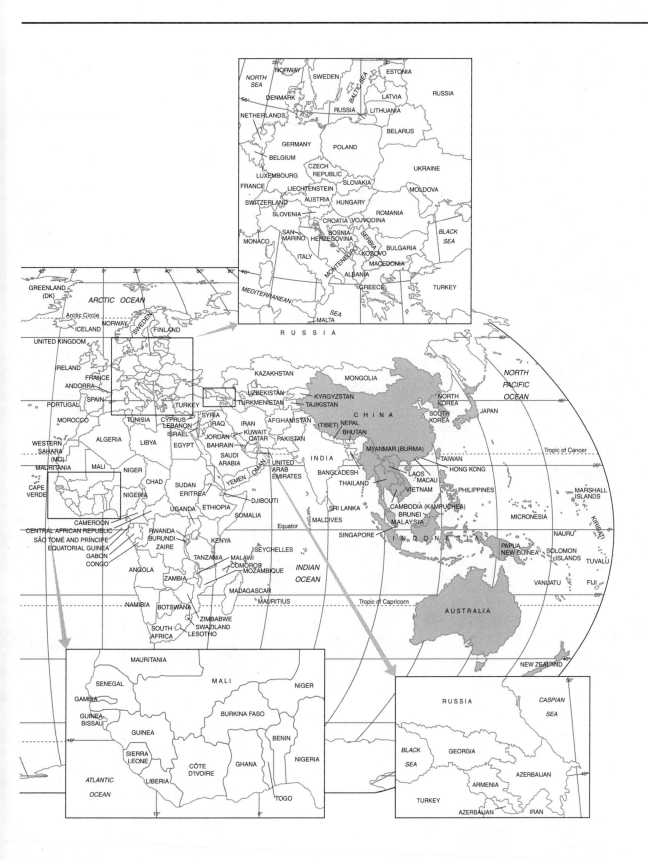

Pacific Rim Map

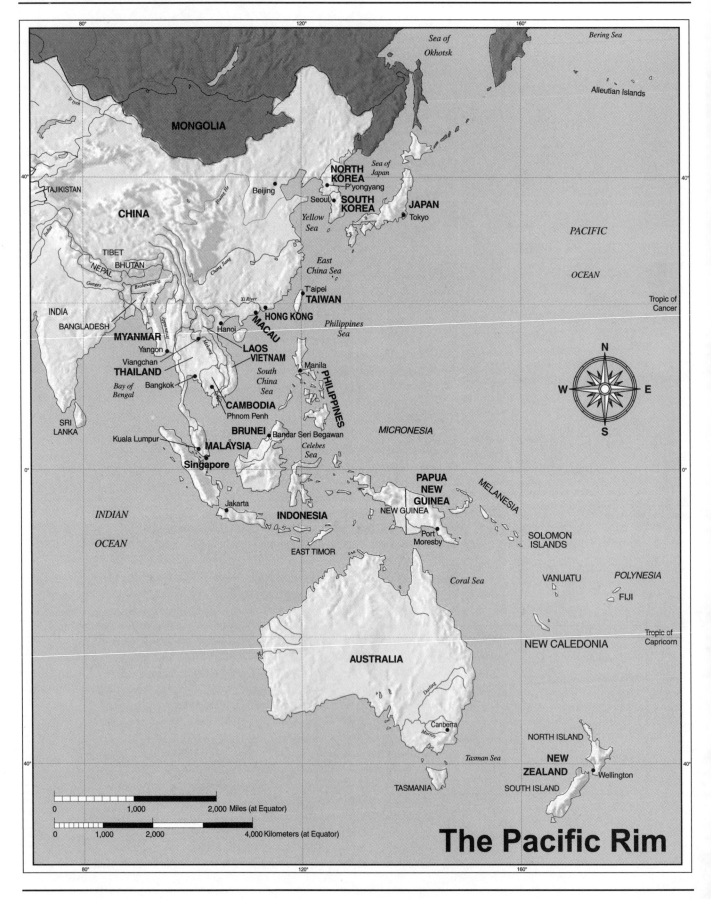

The Pacific Rim

The Pacific Rim: Diversity and Interconnection

WHAT IS THE PACIFIC RIM?

The term *Pacific Rim,* as used in this book, refers to 21 countries or administrative units along or near the Asian side of the Pacific Ocean, plus the numerous islands of the Pacific. Together, they are home to approximately 30 percent of the world's population and produce about 20 percent of the world's gross national product (GNP).

It is not a simple matter to decide which countries to include in a definition of the Pacific Rim. For instance, if we were thinking geographically, we might include Mexico, Chile, Canada, the United States, Russia, and numerous other countries that border the Pacific Ocean, while eliminating Myanmar (Burma) and Laos, since they are not technically on the rim of the Pacific. But our definition, and hence our selected inclusions, stem from fairly recent developments in economic and geopolitical power that have affected the countries of Asia and the Pacific in such a way that these formerly disparate regions are now being referred to by international corporate and political leaders as a single bloc.

Most people living in the region that we have thus defined do not think of themselves as "Pacific Rimmers." In addition, many social scientists, particularly cultural anthropologists and comparative sociologists, would prefer not to apply a single term to such a culturally, politically, and sociologically diverse region. It is true that many of the countries in question share certain cultural influences, such as Buddhism and rice cultivation. But these commonalities have not prevented the region from fracturing into dozens of societies, often very antagonistic toward one another.

However, for more than two decades, something has been occurring in the region that seems to be having the effect of uniting the area in a way it has never been united before. If current trends continue, it is likely that all the countries in the Pacific Rim will one day be operating some version of free-market capitalism and will be espousing similar fundamental values—namely, materialism and consumerism. It is also likely that there will emerge a common awareness of the value of peaceful interdependence, rather than aggression, for realizing improvements in the standard of living for all and for increasing the capacity of the region to supply the basic needs of its inhabitants, for the first time in history.

What are the powerful forces that are fueling these trends? There are many, including nationalism and the rapid advances in global communications. But the one that for the past two decades has stood out as the defining force in the region is the yen—the Japanese currency—and its accompanying Japanese business strategy. For more than 20 years, Japanese money has been flowing throughout the Pacific Rim in the form of aid and investment, while Japan's high-tech, export-oriented approach to making money has been facilitating development and helping other regional countries to create their own engines of economic growth in a way that none of them had experienced before.

It is true that during the 1990s, Japan's economy stagnated and the yen weakened in value. It is also true that Japan's banking sector suffered heavy losses during those years, due to sloppy lending and bad loan management. Those events led some observers to speculate that Japan's role as the economic engine of the Pacific Rim is coming to an end; they predicted that other countries, like China, would assume the lead in the future. Those predictions are probably wrong, and certainly premature; for even after a decade of recession, Japan remains the second-largest economy in the world (after the United States), and its economic output continues to dwarf all other Pacific Rim economies *combined.* Moreover, many Japanese corporations still make large profits, even during these hard times, and the government continues to provide economic aid to many Asian countries (such as the $1.56 billion aid package granted to Indonesia in 2000). Furthermore, the long-term effects of Japan's multi-billion-dollar investments in Asia will remain for years to come.

In the 1960s, when the Japanese economy had completely recovered from the devastation of World War II, the Japanese looked to North America and Europe for markets for their increasingly high-quality products. Japanese business continues to seek out markets and resources globally; but, in the 1980s, in response to the movement toward a truly common European economic community as well as in response to free trade agreements among North American countries, Japan began to invest more heavily in countries nearer its own borders. The Japanese hoped to guarantee themselves market and resource access should they find their products frozen out of the emerging European and North American economic blocs. The unintended, but not unwelcome, consequences of this policy were the revitalization of many Asia–Pacific economies and the solidification of lines of communication between governments and private citizens within the region. Recognizing this interconnection has prompted many people to refer to the countries we treat in this book as a single unit, the Pacific Rim.

TROUBLES IN THE RIM

Twenty years ago, television images of billionaire Japanese businesspeople buying up priceless artworks at auction houses, and filthy-rich Hong Kong Chinese driving around in Rolls-Royces, overshadowed the harsh realities of life for most people in the Rim. For the most part, Pacific Rim countries have not met the needs of their peoples. Whether it is the desire of affluent Japanese for larger homes and two-car garages, or of rice farmers in Myanmar (formerly called Burma) for the right to sell their grain for personal profit, or of Chinese students to speak their minds without repression—in these and many other ways, the Pacific Rim has failed its peoples. In Vietnam, Myanmar, Laos, and Cambodia, for example, life is so difficult that thousands of families have risked their lives to leave their homelands. Some have swum across the wide Mekong River on moonless nights to avoid detection by guards, while others have sailed into the South China Sea on creaky and overcrowded boats (hence the

(Photo by Lisa Clyde)

The small island of Macau was acknowledged by China as a Portuguese settlement in 1557. The Portuguese influence is evident in the architecture of the downtown plaza pictured above. Today Macau is a gambling mecca, drawing an enormous number of avid fans from Hong Kong. This last outpost of European colonial power returned to Chinese control on December 20, 1999, after 442 years under Portugal.

name commonly given such refugees: "boat people"), hoping that people of goodwill, rather than marauding pirates, will find them and transport them to a land of safety. Despite the cut-off of refugee-support funds from the United Nations (UN), thousands of refugees remain unrepatriated, languishing in camps in Thailand, Malaysia, and other countries. Thousands of villagers driven from their homes by the Myanmar Army await return. Meanwhile, the number of defectors from North Korea has been increasing steadily.

Between 1975 and 1994, almost 14,000 refugees reached Japan by boat, along with 3,500 Chinese nationals who posed as refugees in hopes of being allowed to live outside China. In 1998, the Malaysian government, citing its own economic problems, added to the dislocation of many people when it began large-scale deportations of foreign workers, many from Indonesia. Many of these individuals had lived in Malaysia for years. This "Operation Get Out" was expected to affect at least 850,000 people. These examples, and many others not mentioned here, stand as tragic evidence of the social and political instability of many Pacific Rim nations and of the intense ethnic rivalries that divide the people of the Rim.

Warfare

Of all the Rim's troubles, warfare has been the most devastating. In Japan and China alone, an estimated 15.6 million people died as a result of World War II. Not only have there been wars in which foreign powers like Britain, the United States, France, and the former Soviet Union have been involved, but there have been and continue to be numerous battles between local peoples of different tribes, races, and religions.

The potential for serious conflict remains in most regions of the Pacific Rim. Despite international pressure, the military dictators of Myanmar continue to wage war against the Karens and other ethnic groups within its borders. Japan remains locked in a dispute with Russia over ownership of islands to the north of Hokkaido. Taiwan and China still lay claim to each other's territory, as do the two Koreas; and both Taiwan and Japan lay claim to the Senkaku Island chain. The list of disputed borders, lands, islands, and waters in the Pacific Rim is very long; indeed, there are some 30 unresolved disputes involving almost every country of Asia and some of the Pacific Islands.

Of growing concern is a 340,000-square-mile area of the South China Sea. When the likelihood of large oil deposits near the rocks and reefs of the Spratly Islands was announced in the 1970s, China, Taiwan, Vietnam, the Philippines, Malaysia, and Brunei instantly laid claim to the area. By 1974, the Chinese Air Force and Navy were bombing a South Vietnamese settlement on the islands; by 1988, Chinese warships were attacking Vietnamese transport ships in the area. And by 1999, the Philippine Navy was sinking Chinese fishing boats off Mischief Reef. Both China and Vietnam have granted nearby oil-drilling concessions to different U.S. oil companies, so the situation remains tense, especially because China claims sovereignty over almost the entire South China Sea and has been flexing its muscles in the area by stopping, boarding, and sometimes confiscating other nations' ships in the area.

In addition to these national disputes, ethnic tensions—most Asian nations are composed of hundreds of different ethnic groups with their own languages and religions—are sometimes severe. In Fiji, it is the locals versus the immigrant Indians; in Southeast Asia, it is the locals versus the Chinese or the Muslims versus the Christians; in China, it is the Tibetans and most other ethnic groups versus the Han Chinese.

With the end of the Cold War in the late 1980s and early 1990s, many Asian nations found it necessary to seek new military, political, and economic alliances. For example, South Korea made a trade pact with Russia, a nation that, during the Soviet era, would have dealt only with North Korea; and, forced to withdraw from its large naval base in the Philippines, the United States increased its military presence in Singapore. The United States also began encouraging its ally Japan to assume a larger military role in the region. However, the thought of Japan re-arming itself causes considerable fear among Pacific Rim nations, almost all of which suffered defeat at the hands of the Japanese military only six decades ago. Nevertheless, Japan has acted to increase its military preparedness, within the narrow confines of its constitutional prohibition against re-armament. It now has the second-largest military budget in the world (it spends only 1 percent of its budget on defense, but because its economy is so large, the actual absolute expenditure is huge).

In response, China has increased its purchases of military equipment, especially from cash-strapped Russia. As a result, whereas the arms industry is in decline in some other world regions, it is big business in Asia. Four of the nine largest armies in the world are in the Pacific Rim. Thus, the tragedy of warfare, which has characterized the region for so many centuries, could continue unless governments manage conflict very carefully and come to understand the need for mutual cooperation.

In some cases, mutual cooperation is already replacing animosity. Thailand and Vietnam are engaged in sincere efforts to resolve fishing-rights disputes in the Gulf of Thailand and water-rights disputes on the Mekong River; North and South Korea have agreed to allow some cross-border visitation; and even Taiwan and China have amicably settled issues relating to fisheries, immigration, and hijackings. Yet greed and ethnic and national pride are far too often just below the surface; left unchecked, they could catalyze a major confrontation in the region.

Overpopulation

Another serious problem in the Pacific Rim is overpopulation. There are well over 2 billion people living in the region. Of those, approximately 1.2 billion are Chinese. Even though China's government has implemented the strictest family-planning policies in world history, the country's annual growth rate is such that more than 1 million inhabitants are added *every month*. This means that more new Chinese are born each year than make up the entire population of Australia! The World Health Organization (WHO) reports, however, that about 217 million people in East Asia use contraceptives today, as compared to only 18 million in 1965. Couples in some countries, including Japan, Taiwan, and South Korea, have been voluntarily limiting family size. Other states, such as China and Singapore, have promoted family planning through government incentives and punishments. The effort is paying off. The United Nations now estimates that the proportion of the global population living in Asia will remain relatively unchanged between now and the year 2025, and China's share

(UN photo by Shaw McCutcheon)

The number of elderly people in China will triple by the year 2025. Even though–and, ironically, because–it limits most couples to only one child, China will be faced with the increasing need of caring for retirement-age citizens. This group of elders in a village near Chengdu represents just the tip of an enormous problem for the future.

will decline. A drop in birth rates is characteristic of almost the enitre region: Japan and Singapore started to have fewer births in the 1950s; Hong Kong, South Korea, the Philippines, Brunei, Taiwan, Malaysia, Thailand, and China in the 1960s; and Indonesia and Myanmar in the 1970s. In fact, in some countries, especially Japan, South Korea, and Thailand, single-child families and an aging population are creating problems in their own right as the ratio of productive workers to the overall population declines.

Still, so many children have already been born that Pacific Rim governments simply cannot meet their needs. For these new Asians, schools must be built, health facilities provided, houses constructed, and jobs created. This is not an easy challenge for many Rim countries. Moreover, as the population density increases, the quality of life decreases. In crowded New York City, for example, the population is about 1,100 per square mile, and residents, finding the crowding to be too uncomfortable, frequently seek more relaxed lifestyles in the suburbs. Yet in Tokyo, the density is approximately 2,400 per square mile; and in Macau, it is 57,576! Today, many of the world's largest cities are in the Pacific Rim: Shanghai, China, has about 13,600,00 people; Jakarta, Indonesia, has more than 9.1 million; Manila, the Philippines, is home to over 8.5 million (in the wider metropolitan area), while Bangkok, Thailand, has about 6.5 million residents. And migration to the cities continues despite miserable living conditions for many (in some Asian cities, 50 percent of the population live in slum housing). One incredibly rapid-growth country is the Philippines; home to only about 7 million in 1898, when it was acquired by the United States, it is projected to have 130 million people in the year 2020.

Absolute numbers alone do not tell the whole story. In many Rim countries, 40 percent or more of the population are under age 15. Governments must provide schooling and medical care as well as plan for future jobs and housing for all these children. Moreover, as these young people age, they will require increased medical and social care. Scholars point out that, between 1985 and 2025, the numbers of old people will double in Japan, triple in China, and quadruple in Korea. In Japan, where replacement-level fertility was achieved in the 1960s, government officials are already concerned about the ability of the nation to care for the growing number of retirement-age people while paying the higher wages that the increasingly scarce younger workers are demanding.

Political Instability

One consequence of the overwhelming problems of population growth, urbanization, and continual military or ethnic conflict is disillusionment with government. In many countries of the Pacific Rim, people are challenging the very right of their governments to rule or are demanding a complete change in the political philosophy that undergirds governments.

For instance, despite the risk of death, torture, or imprisonment, many college students in Myanmar have demonstrated

---

TYPES OF GOVERNMENT IN SELECTED PACIFIC RIM COUNTRIES

PARLIAMENTARY DEMOCRACIES
Australia*
Fiji
New Zealand*
Papua New Guinea

CONSTITUTIONAL MONARCHIES
Brunei
Japan
Malaysia
Thailand

REPUBLICS
Indonesia
The Philippines
Singapore
South Korea
Taiwan

SOCIALIST REPUBLICS
China
Laos
Myanmar (Burma)
North Korea
Vietnam

OVERSEAS TERRITORIES/COLONIES
French Polynesia
New Caledonia

*Australia and New Zealand have declared their intention of becoming completely independent republics.

---

against the current military dictatorship, and rioting students and workers in Indonesia were successful in bringing down the corrupt government of President Suharto. In some Rim countries, opposition groups armed with sophisticated weapons obtained from foreign nations roam the countryside, capturing towns and military installations. In less than a decade, the government of the Philippines endured six coup attempts; elite military dissidents have wanted to impose a patronage-style government like that of former president Ferdinand Marcos, while armed rural insurgents have wanted to install a Communist government. Thousands of students have been injured or killed protesting the governments of South Korea and China. Thailand has been beset by numerous military coups, the former British colony of Fiji recently endured two coups, and half a million residents of Hong Kong took to the streets to oppose Great Britain's decision to turn over the territory to China in 1997. Military takeovers, political assassinations, and repressive policies have been the norm in most of the countries in the region. Millions have spent their entire lives under governments they have never agreed with, and unrest is bound to continue, because people are showing less and less patience with imposed government.

Part of the reason for political unrest is that the region is so culturally fractured, both between countries and, especially, within them. In some countries, dozens of different languages are spoken, people practice very different religions, families trace their roots back to diverse racial and ethnic origins, and wealth is distributed so unfairly that while some people are well educated and well fed, others nearby remain illiterate and malnourished. Under these conditions, it has been difficult for the peoples of the Rim to agree upon the kinds of government that will best serve them; all are afraid that their particular language, religion, ethnic group, and/or social class will be negatively affected by any leader not of their own background.

Identity Confusion

A related problem is that of confusion about personal and national identity. Many nation-states in the Pacific Rim were created in response to Western pressure. Before Western influences came to be felt, many Asians, particularly Southeast Asians, did not identify themselves with a nation but, rather, with a tribe or an ethnic group. National unity has been difficult in many cases, because of the archipelagic nature of some countries or because political boundaries have changed over the years, leaving ethnic groups from adjacent countries inside the

(UN/DPI Photo by Eskinder Debebe)

In this era of political instability, Pacific Rim countries are increasingly using the forum afforded by regional and international organizations to try to work out their problems. Here, Secretary General Kofi Annan of the United Nations greets President Kim Dae Jung of South Korea at the UN's New York headquarters in June 1998.

neighbor's territory. The impact of colonialism has left many people, especially those in places like Singapore, Hong Kong, and the Pacific islands, unsure as to their roots; are they European or Asian/Pacific, or something else entirely?

Indonesia illustrates this problem. People think of it as an Islamic country, as overall its people are 87 percent Muslim. But in regions like North Sumatra, 30 percent are Protestant; in Bali, 94 percent are Hindu; and in the new country of East Timor, 49 percent are Catholic and 51 percent are animist. Similarly, most people will state that Japan is a Buddhist country, yet few Japanese today claim any actual religious affiliation. The Philippines is another example. With 88 different languages spoken, its people spread out over 12 large islands, and a population explosion (the average age is just 16), it is a classic case of psychological (and economic and political) fragmentation. Coups and countercoups rather than peaceful political transitions seem to be the norm, as people have not yet developed a sense of unified nationalism.

Uneven Economic Development

While millionaires in Singapore, Hong Kong, and Japan wrestle with how best to invest their wealth, far more others worry about how they will obtain their next meal. Such disparity illustrates another major problem afflicting the Pacific Rim: uneven economic development.

Many Asians, especially those in the Northeast Asian countries of Japan, Korea, and China, are finding that rapid economic change seems to render the traditions of the past meaningless. Moreover, economic success has produced a growing Japanese interest in maximizing investment returns, with the result that Japan (and, increasingly, South Korea, Taiwan, Singapore, and Hong Kong) is successfully searching out more ways to make money, while resource-poor regions like the Pacific islands lag behind. Indeed, with China receiving "normal trade relations" status from the United States in late 2000, many smaller countries, such as the Philippines, worry that business will move away from them and toward China, where labor costs are considerably lower.

The *developed nations* are characterized by political stability and long-term industrial success. Their per capita income is comparable to Canada, Northern Europe, and the United States, and they have achieved a level of economic sustainability. These countries are closely linked to North America economically. Japan, for instance, exports one third of its products to the United States.

The *newly industrializing countries* (NICs) are currently capturing world attention because of their rapid growth. Hong Kong, for example, has exported more manufactured products per year for the past decade than did the former Soviet Union and Central/Eastern Europe combined. Taiwan, famous for cameras and calculators, has had the highest average economic growth in the world for the past 20 years. South Korea is tops in shipbuilding and steel manufacturing and is the tenth-largest trading nation in the world.

(UN photo by John Isaac)

Some of the Pacific Rim nations are resource-rich, but development has been curtailed by political instability and a strong traditional culture. This worker is farming as his ancestors did with techniques that have not changed for hundreds of years.

The *resource-rich developing nations* have tremendous natural resources but have been held back economically by political and cultural instability and by insufficient capital to develop a sound economy. An example of a country attempting to overcome these drawbacks is Malaysia. Ruled by a coalition government representing nearly a dozen ethnic-based parties, Malaysia is richly endowed with tropical forests and large oil and natural-gas reserves. Developing these resources has taken years (the oil and gas fields began production as recently as 1978) and has required massive infusions of investment monies from Japan and other countries. By the mid-1990s, more than 3,000 companies were doing business in Malaysia, and the country was moving into the ranks of the world's large exporters.

Command economies lag far behind the rest, not only because of the endemic inefficiency of the system but because military dictatorships and continual warfare have sapped the strength of the people. Yet significant changes in some of these countries are now emerging. China and Vietnam, in particular, are eager to modernize their economies and institute market-based reforms. Historically having directed its trade to North America and Europe, Japan is now finding its Asian/Pacific neighbors—especially the socialist-turning-capitalist ones—to be convenient recipients of its powerful economic and cultural influence.

Many of the *less developed countries* (LDCs) are the small micro-states of the Pacific with limited resources and tiny internal markets. Others, like Papua New Guinea, have only recently achieved political independence and are searching for a comfortable role in the world economy.

Environmental Destruction and Social Ills

Environmental destruction in the Pacific Rim is a problem of mammoth proportions. For more than 20 years, the former Soviet Union dumped nuclear waste into the Sea of Japan; China's use of coal in industrial development has produced acid rain over Korea and Japan; deforestation in Thailand, Myanmar, and other parts of Southeast Asia and China has destroyed many thousands of acres of watershed and wildlife habitat. On the Malaysian island of Sarawak, for example, loggers work through the night, using floodlights, to cut timber to satisfy the demands of international customers, especially Japanese. The forests there are disappearing at a rate of 3 percent a year. Highway and hydroelectric-dam construction in many countries in Asia has seriously altered the natural environment. But environmental damage is perhaps most noticeable in the cities. Mercury pollution in Jakarta Bay has led to brain disorders among children in Indonesia's capital city; air pollution in Manila and Beijing ranks among the world's worst, while not far behind are Bangkok and Seoul; water pollution in Hong Kong has forced the closure of many beaches.

An environmentalist's nightmare came true in 1997 and 1998 in Asia, when thousands of acres of timber went up in a cloud of smoke. Fueled by the worst El Niño–produced drought in 30 years and started by farmers seeking an easy way to clear land for farming, wildfires in Malaysia, Indonesia, and Brunei covered much of Southeast Asia in a thick blanket of smoke for months. Singapore reported its worst pollution-index record ever. All countries in the region complained that Indonesia, in particular, was not doing enough to put out the fires. Foreign-embassy personnel, citing serious health risks, left the region until rains—or the lack of anything more to burn—extinguished the flames. Airports had to close, hundreds of people complaining of respiratory problems sought help at hospitals, and many pedestrians and even those inside buildings donned face masks. With valuable timber becoming more scarce all the time, many people around the world reacted with anger at the callous disregard for the Earth's natural resources.

While conservationists are raising the alarm about the region's polluted air and declining green spaces, medical professionals are expressing dismay at the speed with which serious

(UN/DPI Photo by James Bu)

In this United Nations Development Program project, people dig and move earth into flood-damaged areas in North Korea for future crop planting.

diseases such as AIDS are spreading in Asia. The Thai government now believes that more than 2.4 million Thais are HIV-positive. World Health Organization data suggest that the AIDS epidemic is growing faster in Asia and Africa than anywhere else in the world. Added to these problems are drug and alcohol addictions and the attendant impact on family stability.

GUARDED OPTIMISM

Warfare, overpopulation, political instability, identity confusion, uneven development, and environmental and social ills would seem to be an irresolvable set of problems for the people of the Pacific Rim, but the start of the new millenium also gives reason for guarded optimism. Unification talks between North and South Korea have finally resulted in real breakthroughs. For instance, in 2000, a railway between the demilitarized zone (DMZ) separating the two antagonists was being reconnected, after years of disuse. President Vladimir Putin of Russia agreed to reopen discussion with Japan over the decades-long Northern Territories dispute. Other important issues are also under discussion all over the region, and the UN peacekeeping effort in Cambodia seems to have paid off—at least there is a legally elected government in place, and most belligerents have put down their arms.

Until the Asian financial and currency crises of 1998–1999, the world media carried glowing reports on the burgeoning economic strength of many Pacific Rim countries. Typical was the *CIA World Factbook 1996–1997,* which reported high growth in gross national product (GNP) per capita for most Rim countries: South Korea, 9.0 percent; Hong Kong, 5.0 percent; Indonesia, 7.5 percent; Japan (due to recession), 0.3 percent; Malaysia, 9.5 percent; Singapore, 8.9 percent; and

Thailand, 8.6 percent. By comparison, the U.S. GNP growth rate was 2.1 percent; Great Britain, 2.7 percent; and Canada, 2.1 percent. Other reports on the Rim compared 1990s investment and savings percentages with those of 20 years earlier; in almost every case, there had been a tremendous improvement in the economic capacity of these countries.

Throughout the 1980s and most of the 1990s, the rate of economic growth in the Pacific Rim was indeed astonishing. In 1987, for example, the rate of real gross domestic product (GDP) growth in the United States was 3.5 percent over the previous year. By contrast, in Hong Kong, the rate was 13.5 percent; in Taiwan, 12.4 percent; in Thailand, 10.4 percent; and in South Korea, 11.1 percent. In 1992, economic growth throughout Asia averaged 7 percent, as compared to only 4.8 percent for the rest of the world. But recession in Japan (Japan had suffered zero or even negative growth in 1999 and 2000 and sluggish growth for 10 years before that), near financial

PROJECTED GROWTH RATES IN ASEAN* ECONOMIES, in Percent 1998–2001				
Country	**1998**	**1999**	**2000**	**2001**
Cambodia	5.0	5.0	6.0	7.0
Indonesia	−13.7	2.0	4.0	5.0
Laos	4.0	4.0	4.5	5.0
Malaysia	−7.5	5.6	7.5	6.1
Myanmar	4.0	6.6	4.0	4.0
Philippines	−0.5	3.2	4.5	4.3
Singapore	1.5	5.0	6.0	6.2
Thailand	−8.0	4.1	4.5	5.0
Vietnam	4.0	4.4	5.0	6.0

Source: Asian Development Bank
* Association of Southeast Asian Nations

collapse in Indonesia, and problems in other Asian countries have slowed the growth rates throughout the region. Still, Singapore and many other Pacific Rim economies are expected to grow faster than European and North American economies, and even politically chaotic Indonesia believes that its economy will stabilize very soon.

The significance of high growth rates, in addition to improvements in the standard of living, is the shift in the source of development capital, from North America to Asia. Historically, the economies of North America were regarded as the engine behind Pacific Rim growth; and yet today, growth in the United States and Canada trails many of the Rim economies. This anomaly can be explained, in part, by the hard work and savings ethics of Pacific Rim peoples and by their external-market–oriented development strategies. But, without venture capital and foreign aid, hard work and clever strategies would not have produced the rapid economic improvement that Asia has experienced over the past several decades. Japan's contribution to this improvement, through investments, loans, and donations, and often in much larger amounts than other investor nations such as the United States, cannot be overstated. This is why Japan is considered central to our definition of the Pacific Rim as an identifiable unit.

Some subregions are also emerging. There is, of course, the Association of Southeast Asian Nations (ASEAN) regional trading unit; but the one that is gaining world attention is the informal region that people are calling "Greater China," consisting of the emerging capitalist enclaves of the People's Republic of China, Hong Kong, and Taiwan. Copying Japanese strategy and aided by a common written language and culture, this region has the potential of exceeding even the mammoth U.S. economy in the future. For now, however, and despite recent sluggish growth, Japan will remain the major player in the region.

Japan has been investing in the Asia/Pacific region for several decades. However, growing protectionism in its traditional markets as well as changes in the value of the yen and the need to find cheaper sources of labor (labor costs are 75 percent less in Singapore and 95 percent less in Indonesia) have raised Japan's level of involvement so high as to give it the upper hand in determining the course of development and political stability for the entire region. This heightened level of investment started to gain momentum in the mid-1980s. Between 1984 and 1989, Japan's overseas development assistance to the ASEAN countries amounted to $6.1 billion. In some cases, this assistance translated to more than 4 percent of a nation's annual national budget and nearly 1 percent of GDP. Private Japanese investment in ASEAN countries plus Hong Kong, Taiwan, and South Korea was $8.9 billion between 1987 and 1988. In more recent years, the Japanese government or Japanese business invested $582 million in an auto-assembly plant in Taiwan, $5 billion in an iron and steel complex in China, $2.3 billion in a bullet-train plan for Malaysia, and $530 million in a tunnel under the harbor in

ECONOMIC DEVELOPMENT IN SELECTED PACIFIC RIM COUNTRIES

Economists have divided the Rim into five zones, based on the level of development, as follows:

DEVELOPED NATIONS
Australia
Japan
New Zealand

NEWLY INDUSTRIALIZING COUNTRIES (NICs)
Hong Kong
Singapore
South Korea
Taiwan

RESOURCE-RICH DEVELOPING ECONOMIES
Brunei
Indonesia
Malaysia
The Philippines
Thailand

COMMAND ECONOMIES*
Cambodia
China
Laos
Myanmar (Burma)
North Korea
Vietnam

LESS DEVELOPED COUNTRIES (LDCs)
Papua New Guinea
Pacific Islands

*China, Vietnam, and, to a much lesser degree, North Korea are moving toward free-market economies.

Sydney, Australia. Japan is certainly not the only player in Asian development (Japan has "only" about 20 projects under way in Vietnam, for example, as compared to 80 for Hong Kong and 39 for Taiwan), but the volume of Japanese investment is staggering. In Australia alone, nearly 900 Japanese companies are now doing business. Throughout Asia, Japanese is becoming a major language of business.

Although Japan works very hard at globalizing its markets and its resource suppliers, it has also developed closer ties with its nearby Rim neighbors. In a recent year, out of 20 Rim countries, 13 listed Japan as their first- or second-most-important trading partner, and several more put Japan third. Japan receives 42 percent of Indonesia's exports and 26 percent of Australia's; in return, 23 percent of South Korea's imports, 29 percent of Taiwan's, 30 percent of Thailand's, 24 percent of Malaysia's, and 23 percent of Indonesia's come from Japan. Pacific Rim countries are clearly becoming more interdependent—but simultaneously more dependent on Japan—for their economic success.

JAPANESE INFLUENCE, PAST AND PRESENT

This is certainly not the first time in modern history that Japanese influence has swept over Asia and the Pacific. A major thrust began in 1895, when Japan, like the European powers, started to acquire bits and pieces of the region. By 1942, the Japanese were in control of Taiwan, Korea, Manchuria and most of the populated parts of China, and Hong Kong; what are now Myanmar, Vietnam, Laos, and Cambodia; Thailand; Malaysia; Indonesia; the Philippines; part of New Guinea; and dozens of Pacific islands. In effect, by the 1940s, the Japanese were the dominant force in precisely the area that they are influencing now and that we are calling the Pacific Rim.

The similarities do not end there, for, while many Asians of the 1940s were apprehensive about or openly resistant to Japanese rule, many others welcomed the Japanese invaders and even helped them to take over their countries. This was because they believed that Western influence was out of place in Asia and that Asia should be for Asians. They hoped that the Japanese military would rid them of Western rule—and it did: After the war, very few Western powers were able to regain control of their Asian and Pacific colonies.

Today, many Asians and Pacific islanders are concerned about Japanese financial and industrial influence in their countries, but they welcome Japanese investment anyway because they believe that it is the best and cheapest way to rid their countries of poverty and underdevelopment. So far, they are right—by copying the Japanese model of economic development, and thanks to Japanese trade, foreign aid, and investment, the entire region—some countries excepted—has increased its wealth and positioned itself to be a major player in the world economy for the foreseeable future.

It is important to note, however, that many Rim countries, such as Taiwan, Hong Kong, and South Korea, are strong challengers to Japan's economic dominance; in addition, Japan has not always felt comfortable about its position as head of the pack, for fear of a backlash. For example, Japan's higher regional profile has prompted complaints against the Japanese military's World War II treatment of civilians in Korea and China and forced Japan to pledge $1 billion to various Asian countries as a symbolic act of apology.

Why have the Japanese re-created a modern version of the old Greater East Asian Co-Prosperity Sphere of the imperialistic 1940s? We cannot find the answer in the propaganda of wartime Japan—fierce devotion to emperor and nation and belief in the superiority of Asians over all other races are no longer the propellants in the Japanese economic engine. Rather, Japan courts Asia and the Pacific today to acquire resources to sustain its civilization. Japan is about the size of California, but it has five times as many people and not nearly as much arable land. Much of Japan is mountainous; many other parts are off limits because of active volcanoes (one tenth of all the active volcanoes in the world are in Japan); and, after 2,000-plus years of intensive and uninterrupted habitation, the natural forests are long since consumed

(UN photo by Nichiro Gyogyo)

These men work on a Japanese factory ship, a floating cannery that processes salmon harvested from the Pacific.

(though they have been replanted with new varieties), as are most of the other natural resources—most of which were scarce to begin with.

In short, Japan continues to extract resources from the rest of Asia and the Pacific because it is the same Japan as before—environmentally speaking, that is. Take oil. In the early 1940s, Japan needed oil to keep its industries (as well as its military machine) operating, but the United States wanted to punish Japan for its military expansion in Asia, so it shut off all shipments to Japan of any kind, including oil. That may have seemed politically right to the policymakers of the day, but it did not change Japan's resource environment; Japan still did not have its own oil, and it still needed as much oil as before. So Japan decided to capture a nearby nation that did have natural reserves of oil; in 1941, it attacked Indonesia and obtained by force the resource it had been denied through trade.

Japan has no more domestic resources now than it did half a century ago, and yet its demands—for food, minerals, lumber, paper—are greater. Except for fish, you name it—Japan does not have it. A realistic comparison is to imagine trying to feed half the population of the United States solely from the natural output of the state of Montana. As it happens, however, Japan sits next to the continent of Asia, which is rich in almost all the materials it needs. For lumber, there are the forests of Malaysia; for food, there are the farms and ranches of New Zealand and Australia; and for oil, there are Indonesia and Brunei, the latter of which sells about half of its exports to Japan. The quest for resources is why Japan flooded its neighbors with Japanese yen in recent decades and why it will continue to maintain an active engagement with all Pacific Rim countries well into the future.

Catalyst for Development

In addition to the need for resources, Japan has turned to the Pacific Rim in an attempt to offset the anti-Japanese import or protectionist policies of its historic trading partners. Because so many import tariffs are imposed on products sold directly from Japan, Japanese companies find that they can avoid or minimize tariffs if they cooperate on joint ventures in Rim countries and have products shipped from there. The result is that both Japan and its host countries are prospering as never before. Sony Corporation, for example, assembles parts made in both Japan and Singapore to construct video-cassette recorders at its Malaysian factory, for export to North America, Europe, and other Rim countries. Toyota Corporation intends to assemble its automobile transmissions in the Philippines and its steering-wheel gears in Malaysia, and to assemble the final product in whichever country intends to buy its cars.

So helpful has Japanese investment been in spawning indigenous economic powerhouses that many other Rim countries are now reinvesting in the region. In particular, Hong Kong, Singapore, Taiwan, and South Korea are now in a position to seek cheaper labor markets in Indonesia, Malaysia, the Philippines, and Thailand. In recent years, they have invested billions of dollars in the resource- and labor-rich economies of Southeast Asia, increasing living standards and adding to the growing interconnectivity of the region. An example is a Taiwanese company that has built the largest eel-production facility in the world—in Malaysia—and ships its entire product to Korea and Japan.

Eyed as a big consumer as well as a bottomless source of cheap labor is the People's Republic of China. Many Rim countries, such as South Korea, Taiwan, Hong Kong, and Japan, are working hard to increase their trade with China. In 1990, two-way trade between Taiwan and China was already more than $4 billion; between Hong Kong and China, it was $50 billion. Japan was especially eager to resume economic aid to China in 1990 after temporarily withholding aid to China because of the Tiananmen Square massacre in Beijing. For its part, China is establishing free-enterprise zones that will enable it to participate more fully in the regional and world economy. Indeed, it is most eager to be admitted into the World Trade Organization (WTO), a event more likely now that the United States and others have made private trade agreements with China. Already the Bank of China is the second-largest bank in Hong Kong.

Japan and a handful of other economic powerhouses of the Rim are not the only big players in regional economic development. The United States and Canada are major investors in the Pacific Rim (in computers and automobiles, for example), and Europe maintains its historical linkages with the region (such as oil). But there is no question that Japan has been the main catalyst for development. As a result, Japan itself has become wealthy. The Japanese stock market rivals the New York Stock Exchange, and there is a growing number of Japanese multimillionaires. Loans secured by now-deflated land prices have damaged the Japanese banking industry in recent years, but it is likely that the banks will recover and that Japan will retain its preeminence in the region.

Not everyone is pleased with the way Japan has been giving aid and making loans in the region. Money invested by the Japan International Development Organization (JIDO) has usually been closely connected to the commercial interests of Japanese companies. For instance, commercial-loan agreements have often required that the recipient of low-interest loans purchase Japanese products.

Nevertheless, it is clear that many countries would be a lot worse off without Japanese aid. In a recent year, JIDO aid around the world was $10 billion. Japan is the dominant supplier of foreign aid to the Philippines and a major investor; in Thailand, where U.S. aid recently amounted to $20 million, Japanese aid was close to $100 million. Some of this aid, moreover, gets recycled from one country to another within the Rim. Thailand, for example, receives more aid from Japan than any other country, but in turn, it supplies major amounts of aid to other nearby countries. Thus we can see the growing interconnectivity of the region, a reality now recognized formally by the establishment of the Asia-Pacific Economic Cooperation (APEC).

During the militaristic 1940s, Japanese dominance in the region produced antagonism and resistance. However, it also gave subjugated countries new highways, railways, and other infrastructural improvements. Today, while host countries continue to benefit from infrastructural advances, they also receive quality manufactured products. Once again, Northeast Asian, Southeast Asian, and South Pacific peoples have begun to talk about Japanese domination. The difference is that this time, few seem upset about it; many countries no longer believe that Japan has military aspirations against them, and they regard Japanese investment as a first step toward becoming economically strong themselves. Many people are eager to learn the Japanese language; in some cities, such as Seoul, Japanese has displaced English as the most valuable business language. Nevertheless, to deter negative criticism arising from its prominent position in the Rim, Japan has increased its gift giving, such that now it has surpassed the United States as the world's most generous donor of foreign aid.

ASIAN FINANCIAL CRISIS

All over Asia, but especially in Thailand, Indonesia, Malaysia, South Korea, and the Philippines, business leaders and government economists found themselves scrambling in 1998 and 1999 to minimize the damage from the worst financial crisis in decades, a crisis that exploded in late 1997 with currency devaluations in Southeast Asia. With their banks and major corporations in deep trouble, governments began shutting down some of their expensive overseas consular offices, canceling costly public-works projects, and enduring abuse heaped on them by suddenly unemployed citizens.

For years, Southeast Asian countries copied the Japanese model: They stressed exports, and they allowed their governments to decide which industries to develop. This economic-development approach worked very well for a while, but governments were not eager to let natural markets guide production. So, even when profits from government-supported industries were down, the governments believed that they should continue to maintain these industries through loans from Japan and other sources. But banks can loan only so much, especially if it is "risky" money, and eventually the banks' creditworthiness was called into question. Money became tighter. Currencies were devalued, making it harder still to pay off loans and forcing many companies and banks into bankruptcy. Stock markets nose-dived. Thousands of workers were laid off, and many of the once-booming Asian economies hit hard times.

Japan's own economic sluggishness meant that it was not able to serve as the engine behind Asia's economic growth with the same vigor as it did in the 1980s and early 1990s. One after another, the affected countries requested bailout assistance from the International Monetary Fund (IMF), a pool of money donated by some 183 nations. IMF funding seems to have halted the flow of red ink at least temporarily. But this is not without a cost: Each recipient is required to restructure its economy, which will mean that many unprofitable companies and banks will go under and many workers will have to tighten their belts until the cycle rights itself.

The crisis has revealed how interconnected are the economies of the region and how much Asia has depended on Japan to stimulate growth. It is no wonder why IMF officials say that Japan is the key; it must first fix its own problems before the rest of Asia will rebound.

POLITICAL AND CULTURAL CHANGES

Although economic issues are important to an understanding of the Pacific Rim, political and cultural changes are also crucial. The new, noncombative relationship between the United States and the former Soviet bloc means that special-interest groups and governments in the Rim will be less able to rely on the strength and power of those nations to help advance or uphold their positions. Communist North Korea, for instance, can no longer rely on the Soviet bloc for trade and ideological support. North Korea may begin to look for new ideological neighbors or, more significantly, to consider major modifications in its own approach to organizing society.

Similarly, ideological changes are afoot in Myanmar, where the populace are tiring of life under a military dictatorship. The military can no longer look for guaranteed support from the crumbling socialist world.

In the case of Hong Kong, the British government shied away from extreme political issues and agreed to the peaceful annexation in 1997 of a capitalist bastion by a Communist nation, China. It is highly unlikely that such a decision would have been made had the issue of Hong Kong's political status arisen during the anti-Communist years of the cold war. One must not get the impression, however, that suddenly peace has arrived in the Pacific Rim. But outside support for extreme ideological positions seems to be giving way to a pragmatic search for peaceful solutions. This should have a salutary effect throughout the region.

The growing pragmatism in the political sphere is yielding changes in the cultural sphere. Whereas the Chinese formerly looked upon Western dress and music as decadent, most Chinese now openly seek out these cultural commodities and are finding ways to merge these things with the Communist polity under which they live. It is also increasingly clear to most leaders in the Pacific Rim that international mercantilism has allowed at least one regional country, Japan, to rise to the highest ranks of world society, first economically and now culturally and educationally. The fact that one Asian nation has accomplished this impressive achievement fosters hope that others can do so also.

Rim leaders also see, however, that Japan achieved its position of prominence because it was willing to change traditional mores and customs and accept outside modes of thinking and acting. Religion, family life, gender relations, recreation, and many other facets of Japanese life have altered during Japan's rapid rise to the top. Many other Pacific Rim nations—including Thailand, Singapore, and South Korea—seem determined to follow Japan's lead in this regard. Therefore, we are witnessing in certain high-growth Rim economies significant cultural changes: a reduction in family size, a secularization of religious impulses, a desire for more leisure time and better education, and a move toward acquisition rather than "being" as a determinant of one's worth. That is, more and more people are likely to judge others' value by what they own rather than what they believe or do. Buddhist values of self-denial, Shinto values of respect for nature, and Confucian values of family loyalty are giving way slowly to Western-style individualism and the drive for personal comfort and monetary success. Formerly close-knit communities, such as those in American Samoa, are finding themselves struggling with drug abuse and gang-related violence, just as in the more metropolitan countries. These changes in political and cultural values are at least as important as economic growth in projecting the future of the Pacific Rim.

The Pacific Islands Map

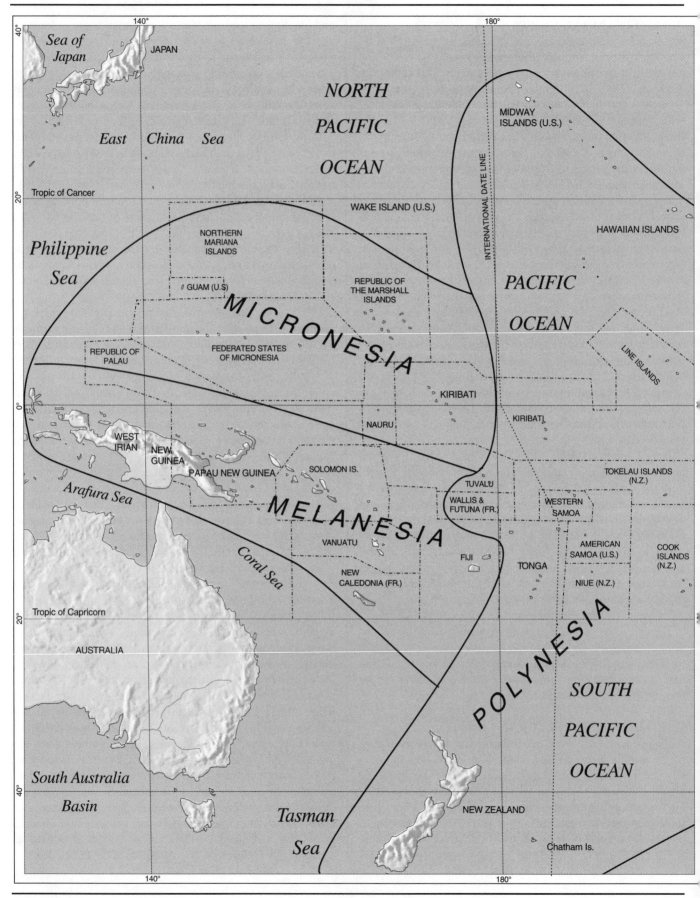

140°   180°

40°

Sea of Japan

JAPAN

NORTH

East China Sea

PACIFIC

OCEAN

MIDWAY ISLANDS (U.S.)

Tropic of Cancer

20°

WAKE ISLAND (U.S.)

HAWAIIAN ISLANDS

Philippine Sea

NORTHERN MARIANA ISLANDS

θ GUAM (U.S)

REPUBLIC OF THE MARSHALL ISLANDS

PACIFIC

OCEAN

MICRONESIA

REPUBLIC OF PALAU

FEDERATED STATES OF MICRONESIA

LINE ISLANDS

KIRIBATI

0°

KIRIBATI

NAURU

WEST IRIAN

NEW GUINEA

PAPAU NEW GUINEA

SOLOMON IS.

TOKELAU ISLANDS (N.Z.)

Arafura Sea

TUVALU

WESTERN SAMOA

MELANESIA

WALLIS & FUTUNA (FR.)

Coral Sea

VANUATU

FIJI

AMERICAN SAMOA (U.S.)

COOK ISLANDS (N.Z.)

NEW CALEDONIA (FR.)

TONGA

NIUE (N.Z.)

Tropic of Capricorn

20°

AUSTRALIA

POLYNESIA

SOUTH

South Australia Basin

PACIFIC

OCEAN

40°

Tasman Sea

NEW ZEALAND

Chatham Is.

140°   180°

INTERNATIONAL DATE LINE

14

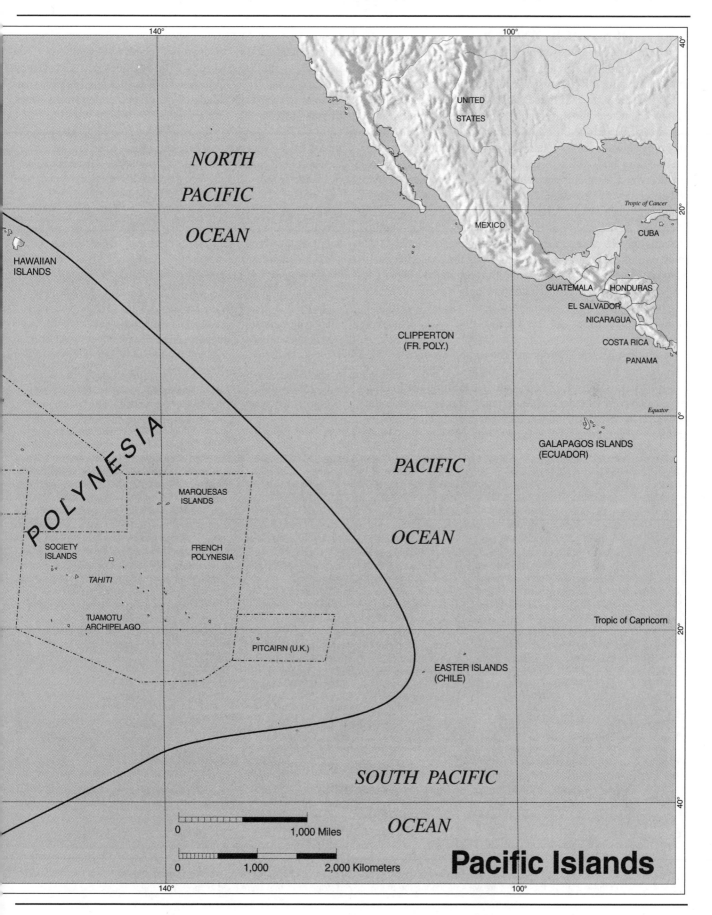

140°   100°   40°

NORTH

PACIFIC

OCEAN   UNITED

                                                     STATES

                                                                        Tropic of Cancer   20°

HAWAIIAN   MEXICO   CUBA
ISLANDS

                                                     GUATEMALA   HONDURAS

                                                     EL SALVADOR

                                      CLIPPERTON   NICARAGUA
                                      (FR. POLY.)
                                                          COSTA RICA

                                                              PANAMA

P O L Y N E S I A
                                      PACIFIC   Equator   0°

                                                     GALAPAGOS ISLANDS
                                                     (ECUADOR)

              MARQUESAS
              ISLANDS
                                      OCEAN

SOCIETY   FRENCH
ISLANDS   POLYNESIA

     TAHITI

TUAMOTU   Tropic of Capricorn   20°
ARCHIPELAGO

              PITCAIRN (U.K.)
                                      EASTER ISLANDS
                                      (CHILE)

                                      SOUTH PACIFIC   40°

0   1,000 Miles   OCEAN

0   1,000   2,000 Kilometers

Pacific Islands

140°   100°

The Pacific Islands: Opportunities and Limits

PLENTY OF SPACE, BUT NO ROOM

There are about 30,000 islands in the Pacific Ocean. Most of them are found in the South Pacific and have been classified into three mammoth regions: Micronesia, composed of some 2,000 islands with such names as Palau, Nauru, and Guam; Melanesia, where 200 different languages are spoken on such islands as Fiji and the Solomon Islands; and Polynesia, comprised of Hawaii, Samoa, Tahiti, and other islands.

Straddling both sides of the equator, these territories are characterized as much by what is *not* there as by what *is*—that is, between every tiny island lie hundreds and often thousands of miles of open ocean. A case in point is the Cook Islands, in Polynesia. Associated with New Zealand, this 15-island group contains only 92 square miles of land but is spread over 714,000 square miles of open sea. So expansive is the space between islands that early explorers from Europe and the Spanish lands of South America often unknowingly bypassed dozens of islands that lay just beyond view in the vastness of the 64 million square miles of the Pacific—the world's largest ocean.

However, once the Europeans found and set foot on the islands, they inaugurated a process that irreversibly changed the history of island life. Their goals in exploring the Pacific were to convert islanders to Christianity and to increase the power and prestige of their homelands (and themselves) by obtaining resources and acquiring territory. They thought of themselves and European civilization as superior to others and often treated the "discovered" peoples with contempt. An example is the "discovery" of the Marquesas Islands (from whence came some of the Hawaiian people) by the Peruvian Spaniard Alvaro de Mendana. Mendana landed in the Marquesas in 1595 with some women and children—and, significantly, 378 soldiers. Within weeks, his entourage had planted three Christian crosses, declared the islands to be the possession of the king of Spain, and killed 200 islanders. Historian Ernest S. Dodge describes the brutality of the first contacts:

> The Spaniards opened fire on the surrounding canoes for no reason at all. To prove himself a good marksman one soldier killed both a Marquesan and the child in his arms with one shot as the man desperately swam for safety.... The persistent Marquesans again attempted to be friendly by bringing fruit and water, but again they were shot down when they attempted to take four Spanish water jars. Magnanimously the Spaniards allowed the Marquesans to stand around and watch while mass was celebrated.... When [the islanders] attempted to take two canoe loads of ... coconuts to the ships half the unarmed natives were killed and three of the bodies hung in the rigging in grim warning. The Spaniards were not only killing under orders, they were killing for target practice.
>
> —*Islands and Empires; Western Impact on the Pacific and East Asia*

(UN photo by Nagata Jr.)

In the South Pacific area of Micronesia, some 2,000 islands are spread over an ocean area of 3 million square miles. There remain many relics of the diverse cultures found on these islands; these boys are walking among the highly prized stone discs that were used as money on the islands of the Yap District.

All over the Pacific, islanders were "pacified" through violence or deception inflicted on them by the conquering nations of France, England, Spain, and others. Rivalries among the European nations were often acted out in the Pacific. For example, the Cook Islands, inhabited by a mixture of Polynesian and Maori peoples, were partly controlled by the Protestant Mission of the London Missionary Society, until the threat of incursions by French Catholics from Tahiti persuaded the British to declare the islands a protectorate of Britain. New Zealand eventually annexed the islands, and it controlled them until 1965.

Business interests frequently took precedence over islanders' sovereignty. In Hawaii, for instance, when Queen Liliuokalani proposed to limit the influence of the business community in island governance, a few dozen American business leaders—without the knowledge of the U.S. president or Congress, yet with the unauthorized help of 160 U.S. Marines—overthrew the Hawaiian monarch, installed Sanford Dole (of Dole Pineapple fame) as president, and petitioned Congress for Hawaii's annexation as a U.S. territory.

Whatever the method of acquisition, once the islands were under European or American control, the colonizing nations insisted that the islanders learn Western languages, wear Western clothing, convert to Christianity, and pay homage to faraway rulers whom they had never seen.

This blatant Eurocentrism ignored the obvious—that the islanders already had rich cultural traditions that both predated European culture and constituted substantial accomplishments in technology, the arts, and social structure. Islanders were skilled in the construction of boats suitable for navigation on the high seas and of homes and religious buildings of varied architecture; they had perfected the arts of weaving and cloth-making, tattooing (the word itself is Tahitian), and dancing. Some cultures organized their political affairs much as had early New Englanders, with village meetings to decide issues by consensus, while others had developed strong chieftainships and kingships with an elaborate variety of rituals and taboos (a Tongan word) associated with the ruling elite. Island trade involving vast distances brought otherwise disparate people together. And, although reading and writing were not known on most islands, some evidence of an ancient writing system has been found.

Despite these cultural attributes and a long history of skill in interisland or intertribal warfare, the islanders could not withstand the superior force of European firearms. Within just a few generations, the entire Pacific had been conquered and colonized by Britain, France, the Netherlands, Germany, the United States, and other nations.

CONTEMPORARY GROUPINGS

The Pacific islands today are classified into three racial/cultural groupings. The first, Micronesia, with a population of approximately 414,000 people, contains seven political entities, four of which are politically independent and three

---

TYPES OF PACIFIC ISLAND GOVERNMENTS

The official names of some of the Pacific island nations indicate the diversity of government structures found there:

Republic of Fiji
Federated States of Micronesia
Republic of Palau
Independent State of Papua New Guinea
Kingdom of Tonga
Republic of Vanuatu
Independent State of Western Samoa

---

of which are affiliated with the United States. Guam is perhaps the best known of these islands. Micronesians share much in common genetically and culturally with Asians. The term *Micronesia* refers to the small size of the islands in this group.

The second grouping, Melanesia, with a population of some 5.5 million (if New Guinea is included), contains six political entities, four of which are independent and two of which are affiliated with colonial powers. The best known of these islands is probably Fiji. The term *Melanesia* refers to the dark skin of the inhabitants, who, despite appearances, apparently have no direct ties with Africa.

Polynesia, the third grouping, with a population of 536,000, contains 12 political entities, three of which are independent, while the remaining are affiliated with colonial powers. *Polynesia* means "many islands," the most prominent grouping of which is probably Hawaii. Most of the cultures in Polynesia have some ancient connections with the Marquesas Islands or Tahiti.

Subtracting the atypically large population of the island of New Guinea leaves about 2.2 million people in the region that we generally think of as the Pacific islands. Although it is possible that some of the islands were peopled by or had contact with ancient civilizations of South America, the overwhelming weight of scholarship places the origins of the Pacific islanders in Southeast Asia, Indonesia, and Australia.

Geologically, the islands may be categorized into the tall, volcanic islands, which have abundant water, flora, and fauna, and are suitable for agriculture; and the dry, flat, coral islands, which have fewer resources (though some are rich in phosphate). It also appears that the farther away an island is from the Asian or Australian continental landmass, the less varied and plentiful are the flora and fauna.

THE PACIFIC COMMUNITY

During the early years of Western contact and colonization, maltreatment of the indigenous peoples and diseases such as measles and influenza greatly reduced their numbers and their cultural strength. Moreover, the carving up of the Pacific by different Western powers superimposed a cultural fragmentation on the region that added to the separateness resulting naturally

THE CASE OF THE DISAPPEARING ISLAND

It wasn't much to begin with, but the way things are going, it won't be *anything* very soon. Nauru, a tiny, $8\frac{1}{2}$-square-mile dot of phosphate dirt in the Pacific, is being gobbled up by the Nauru Phosphate Corporation. Made of bird droppings (guano) mixed with marine sediment, Nauru's high-quality phosphate has a ready market in Australia, New Zealand, Japan, and other Pacific Rim countries, where it is used in industry, medicine, and agriculture.

Many Pacific islanders with few natural resources to sell to the outside world envy the 4,500 Nauruans. The Nauruans pay no taxes; yet thanks to phosphate sales, the government is able to provide them with free health and dental care, bus transportation, newspapers, and schooling (including higher education if they are willing to leave home temporarily for Australia, with the trip paid for by the government). Rent for government-built homes, supplied with telephones and electricity, is only about $5 a month. Nor do Nauruans have to work particularly hard for a living. Most laborers in the phosphate pits are imported from other islands; most managers and other professionals come from Australia, New Zealand, and Great Britain.

Phosphate is Nauru's only export, and yet the country makes so much money from it that technically speaking, Nauru is the richest country per capita in the world. Unable to spend all the export earnings (even though it owns and operates several Boeing 737s, a number of hotels on other islands, and the tallest skyscraper in Melbourne, Australia), the government puts lots of the money away in trust accounts for a rainy day.

It all sounds nice, but the island is literally being mined away. Already there is only just a little fringe of green left along the shore, where everyone lives, and the government is debating what should happen when even the ground under people's homes is mined and shipped away. Some think that topsoil should be brought in to see if the moonlike surface of the excavated areas can be revitalized. Others think that moving away makes sense—with all its money, the government could just buy another island somewhere and move everyone (an idea that Australia suggested years ago, even before Nauru's independence in 1968). Of course, since the government owns the phosphate company, it could just put a halt to any more mining. But if it does, what would Nauru be to anybody? On the other hand, if it doesn't, will Nauru *be* at all?

---

from distance. Today, however, improved medical care is allowing the populations of the islands to rebound, and the withdrawal or realignment of European and American political power under the post–World War II United Nations policy of decolonization has permitted the growth of regional organizations.

First among the postwar regional groups was the South Pacific Commission. Established in 1947, when Western powers were still largely in control, many of its functions have since been augmented or superseded by indigenously created organizations such as the South Pacific Forum, which was organized in 1971 and has since spawned numerous other associations, including the South Pacific Regional Trade and Economic Agency and the South Pacific Islands Fisheries Development Agency. Through an executive body (the South Pacific Bureau for Economic Cooperation), these associations handle such issues as relief funding, the environment, fisheries, trade, and regional shipping. The organizations have produced a variety of duty-free agreements among countries, and have yielded joint decisions about regional transportation and cultural exchanges. As a result, regional art festivals and sports competitions are now a regular feature of island life. And a regional university in New Zealand attracts several thousand island students a year, as do schools in Hawaii.

Some regional associations have been able to deal forcefully with much more powerful countries. For instance, when the regional fisheries association set higher licensing fees for foreign fishing fleets (most fleets are foreign-owned, because island fishermen usually cannot provide capital for such large enterprises), the Japanese protested vehemently. Nevertheless, the association held firm, and many islands terminated their contracts with the Japanese rather than lower their fees.

In 1994, the Cook Islands, the Federated States of Micronesia, Fiji, Kiribati, the Marshall Islands, Nauru, Niue, Papua New Guinea, the Solomon Islands, Tonga, Tuvalu, Vanuatu, and Western Samoa signed an agreement with the United States to establish a joint commercial commission to foster private-sector businesses and to open opportunities for trade, investment, and training. Through this agreement, the people of the islands hoped to increase the attractiveness of their products to the U.S. market.

Increasingly important issues in the Pacific are the testing of nuclear weapons and the disposal of toxic waste. Island leaders, with the occasional support of Australia and the strong support of New Zealand, have spoken out vehemently against the continuation of nuclear testing in the Pacific by the French government (Great Britain and the United States tested hydrogen bombs on coral atolls for years, but have now stopped) and against the burning of nerve-gas stockpiles by the United States on Johnston Atoll. In 1985, the 13 independent or self-governing countries of the South Pacific adopted their first collective agreement on regional security, the South Pacific Nuclear Free Zone Treaty. Encouraged by New Zealand and Australia, the group declared the Pacific a nuclear-free zone and issued a communique criticizing the dumping of nuclear waste in the region. Some island leaders, however, see the storage of nuclear waste as a way of earning income to compensate those who were affected by the nuclear testing on Bikini and Enewetak Islands. The Marshall Islands, for example, are interested in storing nuclear waste on already contaminated islands; however, the nearby Federated States of Micronesia, which were observers at the Nuclear Free Zone Treaty talks, oppose the idea and have asked the Marshalls not to proceed.

World leaders met in Jamaica in 1982 to sign into international law the Law of the Sea. This law, developed under the auspices of the United Nations, gave added power to the tiny Pacific island nations because it extended the territory under their exclusive economic control to 12 miles beyond their shores or 200 miles of undersea continental shelf. This put many islands in undisputed control of large deposits of nickel, copper, magnesium, and other valuable metals. The seabed areas away from continents and islands were declared the world's common heritage, to be mined by an international company, with profits channeled to developing countries. The United States has negotiated for years to increase the role of industrialized nations in mining the seabed areas; if modifications are made to the treaty, the United States will likely sign the document.

COMING OF AGE?

If the peoples of the Pacific islands are finding more reasons to cooperate economically and politically, they are still individually limited by the heritage of cultural fragmentation left them by their colonial pasts. Western Samoa, for example, was first annexed by Germany in 1900, only to be given to New Zealand after Germany's defeat in World War I. Today, the tiny nation of mostly Christian Polynesians, independent since 1962, uses both English and Samoan as official languages and embraces a formal governmental structure copied from Western parliamentary practice. Yet the structure of its hundreds of small villages remains decidedly traditional, with clan chiefs ruling over large extended families, who make their not particularly profitable livings by farming breadfruit, taro, yams, bananas, and copra.

Political independence has not been easy for those islands that have embraced it, nor for those colonial powers that continue to deny it. Anticolonial unrest continues on many islands, especially the French ones. However, concern over economic viability has led most islands to remain in some sort of loose association with their former colonial overseers. After the defeat of Japan in World War II, the Marshall Islands, the Marianas, and the Carolines were assigned by the United Nations to the United States as a trust territory. The French Polynesian islands have remained overseas "departments" (similar to U.S. states) of France. In such places as New Caledonia, however, there has been a growing desire for autonomy, which France has attempted to meet in various ways while still retaining sovereignty. The UN decolonization policy has made it possible for most Pacific islands to achieve independence if they wish, but many are so small that true economic independence in the modern world will never be possible.

In addition to relations with their former colonial masters, the Pacific islands have found themselves in the middle of many serious domestic political problems. For example, Fiji, a former British colony, weathered two military coups in 1987 and a bizarre coup in 2000. In the 2000 episode, native Fijian businessman George Speight captured the prime minister and other government leaders and held them hostage until they resigned their posts. The prime minister, Mahendra Chaudhry, was not a native Fijian but, rather, an Indian. So many immigrants from India had settled in Fiji over the past few decades that they now constituted some 44 percent of the population, enough to wield political power on the island. This situation had frightened the native Fijians into stripping Indians of various civil rights (restored in 1997), but the election of the first Indian prime minister seemed to push some of the native Fijians past their tolerance limit. While not

THIS IS LIBERATION?

In 1994, the people of the U.S. Territory of Guam celebrated the 50th anniversary of their liberation by U.S. Marines and Army Infantry from the occupying troops of the Japanese Army. During the three years that they controlled the tiny, 30-mile-long island, the Japanese massacred some of the Guamanians and subjected many others to forced labor and internment in concentration camps.

Their liberation, therefore, was indeed a cause for celebration. But the United States quickly transformed the island into its military headquarters for the continuing battle against the Japanese. The entire northern part of the island was turned into a base for B-29 bombers, and the Pacific submarine fleet took up residence in the harbor. Admiral Chester W. Nimitz, commander-in-chief of the Pacific, made Guam his headquarters. By 1946, the U.S. military government in Guam had laid claim to nearly 80 percent of the island, displacing entire villages and hundreds of individual property owners.

Since then, some of the land has been returned, and large acreages have been handed over to the local civilian government—which was to have distributed most of it, but has not yet done so. The local government still controls about one third of the land, and the U.S. military controls another third, meaning that only one third of the island is available to the residents for private ownership. Litigation to recover the land has been bitter and costly involving tens of millions of dollars in legal expenses since 1975. The controversy has prompted some local residents to demand a different kind of relationship with the United States, one that would allow for more autonomy. It has also spurred the growth of nativist organizations such as the Chamorru Nation, which promotes the Chamorru language of the original Malayo–Polynesian inhabitants (spelled *Chamorro* by the Spanish) and organizes acts of civil disobedience against both civilian and military authorities.

Guam was first overtaken by Spain in 1565. It has been controlled by the United States since 1898, except for the brief Japanese interlude. Whether the local islanders, who now constitute a fascinating mix of Chamorro, Spanish, Japanese, and American cultures, will be able to gain a larger measure of autonomy after hundrerds of years of colonization by outsiders is difficult to predict, but the ever-present island motto, *Tano Y Chamorro* ("Land of the Chamorros"), certainly spells out the objective of many of those who call Guam home.

(UN photo/atb)

Bartering is still a common practice in Papua New Guinea. Here, coastal and inland dwellers are shown exchanging fish for yams, taro, and other agricultural products.

all native Fijians feel threatened by the rise of the Indian population, when a Suva television station criticized the latest coup and called for racial tolerance, the station was ransacked by Fijian rioters, who also destroyed some 20 other buildings in the capital city.

Similarly, tribal tensions have wracked the Solomon Islands. In the past few years, residents of the island of Malaita have migrated in large numbers to the larger, main island of Guadalcanal. The locals have complained that the newcomers have taken over their land and the government. Tribal violence related to this issue has killed some 50 islanders in the recent past; but in 2000, Malaita's prime minister, a Malaitan, was kidnapped in Honiara, the capital city, and forced to resign. Violence forced 20,000 residents to flee their homes, and foreign nationals were evacuated to avoid the fierce fighting by opposing paramilitary groups. It is clear that in Fiji, the Solomon Islands, and other places, Pacific islanders will have to find creative ways to allow all members of society to participate in the political process. Without political stability, other pressing issues will be neglected.

No amount of political realignment can overcome the economic dilemma of most of the islands. Japan, the single largest purchaser of island products, as well as the United States and others, are good markets for the Pacific economies, but exports are primarily of mineral and agricultural products (coffee, tea, cocoa, sugar, tapioca, coconuts, mother-of-pearl) rather than of the more profitable manufactured or "value-added" items produced by industrial nations. In addition, there will always be the cost of moving products from the vastness of the Pacific to the various mainland markets.

Another problem is that many of the profits from the island's resources do not redound to the benefit of the islanders. Tuna, for example, is an important and profitable fish catch, but most of the profits go to the Taiwanese, Korean,

Japanese, and American fleets that ply the Pacific. Similarly, tourism profits largely end up in the hands of the multinational hotel owners. About 80 percent of visitors to the island of Guam since 1982 have been Japanese (more than half a million people annually)—seemingly a gold mine for local Guamanians, since each traveler typically spends more than $2,000. However, since the tourists tend to purchase their tickets on Japanese airlines and book rooms in Japanese-owned or -managed hotels, much of the money that they spend for their vacations in Guam never reaches the Guamanians.

The poor economies, especially in the outer islands, have prompted many islanders to move to larger cities (about 1 million islanders now live in the Pacific's larger cities) to find work. Indeed, there is currently a tremendous mixing of all of the islands' peoples. Hawaii, for example, is peopled now with Samoans, Filipinos, and many other islanders; pure Hawaiians are a minority, and despite efforts to preserve the Hawaiian language, it is used less and less. Similarly, Fiji, as we have seen, is now populated by nearly as many immigrants from India as by native Fijians. New Caledonians are outnumbered by Indonesians, Vietnamese, French, and others. And, of course, whites have long outnumbered the Maoris on New Zealand. Guam is peopled with islanders from all of Micronesia as well as from Samoa and other islands.

In addition to interisland migration, many islanders emigrate to Australia, New Zealand, the United States, or other countries and then send money back home to sustain their families. Those remittances are important to the economies of the islands, but the absence of parents or adult children for long periods of time does considerable damage to the social fabric. In a few cases, such as in the Cook Islands and American Samoa, more islanders live abroad than remain on the islands. For example, whereas American Samoa has about

65,000 residents, more than 130,000 Samoans live on the U.S. mainland, in such places as Los Angeles and Salt Lake City. Those who leave often find life abroad quite a shock, for the island culture, influenced over the decades by the missionary efforts of Mormons, Methodists, Seventh-day Adventists, and especially the London Missionary Society, is conservative, cautious, and personal. Metropolitan life, by contrast, is considered by some islanders to be wild and impersonal. Some young emigrants respond to the "cold" environment and marginality of big-city life by engaging in deviant behavior themselves, such as selling drugs and joining gangs.

Island society itself, moreover, is not immune from the social problems that plague larger societies. Many islands report an increasing number of crimes and suicides. Young Samoans, for example, are afflicted with many of the same problems—gangs, drugs, and unemployment—as are their U.S. inner-city counterparts. Samoan authorities have reported increases in incidences of rape, robbery, and other socially dysfunctional behaviors. In addition, the South Pacific Commission and the World Health Organization are now reporting an alarming increase in HIV/AIDS and other sexually transmitted diseases.

For decades, and notwithstanding the imposition of foreign ways, islanders have shared a common culture; most people know how to raise bananas, coconuts, and yams, how to roast pigs and fish, and how to make breadfruit, tapioca, and poi. But much of island culture has depended on an identity shaped and preserved by isolation from the rest of the world. Whether the essence of island life—and especially the identity of the people—can be maintained in the face of increasing integration into a much larger world remains to be seen.

Japan

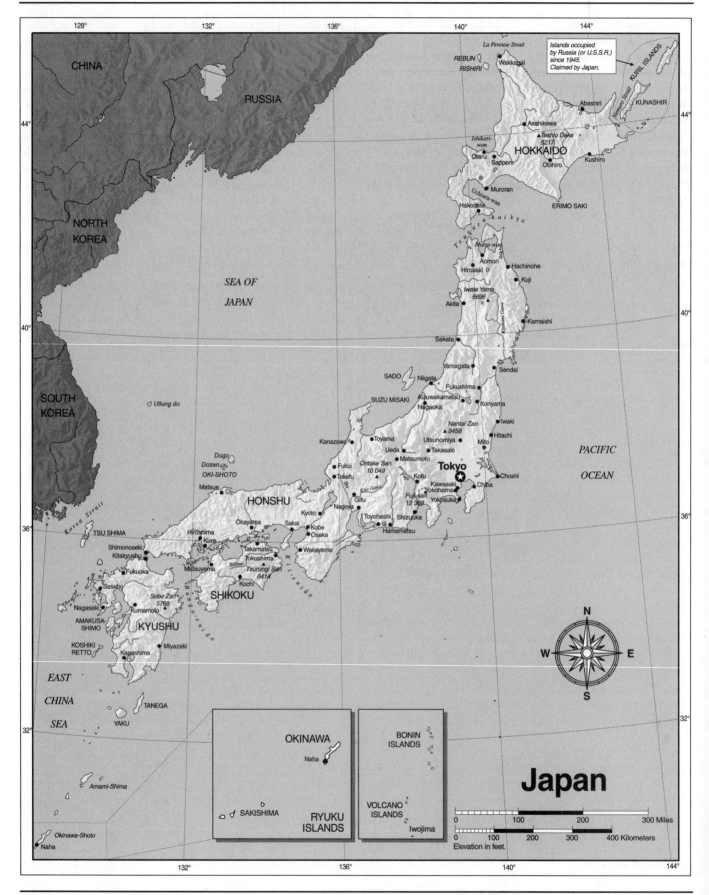

CHINA

RUSSIA

La Perouse Strait

REBUN
RISHIRI

Wakkanai

KURIL ISLANDS

Islands occupied
by Russia (or U.S.S.R.)
since 1945.
Claimed by Japan.

Abashiri

KUNASHIR

Asahikawa

Ishikari-
wan

▲ *Teshio Dake*
5217

HOKKAIDO

NORTH
KOREA

Otaru

Sapporo

Obihiro

Kushiro

Muroran

ERIMO SAKI

Uchiura-wan

SEA OF

JAPAN

Hakodate

Tsugaru-kaikyo

Mutsa-wan

Aomori

Hachinohe

Hirosaki

Kuji

Iwate Yama
6696 ▲

Akita

Kitakami Gawa

Kamaishi

SOUTH
KOREA

⌀ *Ullung do*

Sakata

Yamagata

Sendai

SADO

Niigata

Fukushima

SUZU MISAKI

Aizuwakamatsu

Nagaoka

Koriyama

Iwaki

Hitachi

Dogo
Dozen
OKI-SHOTO

Kanazawa

Toyama

Nantai Zan
▲ *8458*

Utsunomiya

Mito

Ueda

Takasaki

PACIFIC

OCEAN

Fukui

Matsumoto

Takefu

Ibi

Ontake San
10 049 ▲

Matsue

HONSHU

Fuji

Kiso Gawa

Gifu

Nagoya

Tokyo ★

Kofu

Kawasaki
Yokohama

Choshi

Chiba

Fuji-san
12 388

Yokosuka

Kyoto

Sakai

Kobe

Osaka

Toyohashi

Shizuoka

Okayama

Hamamatsu

Hiroshima

Kure

Takamatsu

Wakayama

TSU SHIMA

Shimonoseki

Tokushima

Yoshino

Tsurugi San
6414 ▲

Kii-suido

Kitakyushu

Matsuyama

Fukuoka

Kochi

Bungo-suido

Sasebo

SHIKOKU

Sobo Zan
5768 ▲

Nagasaki

Kumamoto

KYUSHU

AMAKUSA-
SHIMO

KOSHIKI
RETTO

Kagoshima

Miyazaki

EAST

CHINA

SEA

TANEGA

YAKU

Amami-Shima

Okinawa-Shoto

Naha

OKINAWA

Naha

RYUKU
ISLANDS

⌀ SAKISHIMA

BONIN
ISLANDS

VOLCANO
ISLANDS

Iwojima

N

W   E

S

Japan

0   100   200   300 Miles

0   100   200   300   400 Kilometers

Elevation in feet.

Japan

GEOGRAPHY

Area in Square Miles (Kilometers):
145,882 (377,835) (about the size of California)
Capital (Population): Tokyo (7,968,000)
Environmental Concerns: air and water pollution; acidification; depletion of global resources due to Japanese demand
Geographical Features: mostly rugged and mountainous
Climate: tropical in the south to cool temperate in the north

PEOPLE

Population
Total: 126,070,000
Annual Growth Rate: 0.18%
Rural/Urban Population Ratio: 22/78
Ethnic Makeup: 99.4% Japanese; 0.6% others (mostly Korean)
Major Language: Japanese
Religions: primarily Shinto and Buddhist; only 15% claims any formal religious affiliation

Health
Life Expectancy at Birth: 77 years (male); 83 years (female)
Infant Mortality Rate (Ratio): 3.9/1,000
Physicians Available (Ratio): 1/546

Education
Adult Literacy Rate: 99%
Compulsory (Ages): 6–15; free

COMMUNICATION

Telephones: 60,380,900 main lines
Daily Newspaper Circulation: 578 per 1,000 people
Televisions: 619 per 1,000 people
Internet Service Providers: 357 (1999)

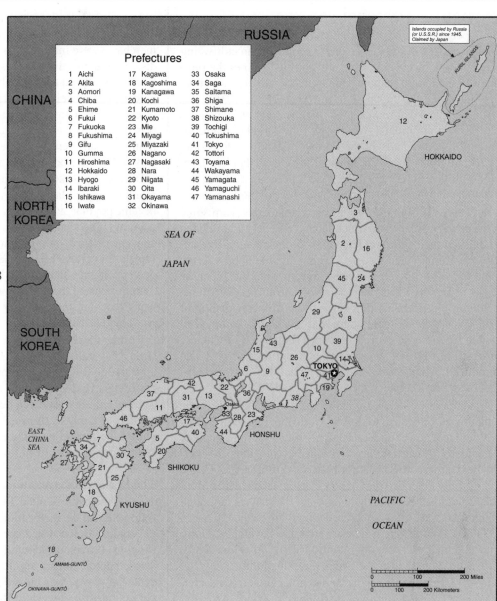

TRANSPORTATION

Highways in Miles (Kilometers): 686,616 (1,144,360)
Railroads in Miles (Kilometers): 14,202 (23,671)
Usable Airfields: 171
Motor Vehicles in Use: 68,030,000

GOVERNMENT

Type: constitutional monarchy
Independence Date: traditional founding 660 B.C.; constitutional monarchy established 1947
Head of State/Government: Emperor Akihito; Prime Minister (at press time the Prime Minister was in transition)
Political Parties: Liberal Democratic Party; Social Democratic Party of Japan; New

Komeito; Democratic Party of Japan; Japan Communist Party; Liberal Party; New Conservative Party
Suffrage: universal at 20

MILITARY

Military Expenditures (% of GDP): 0.9%
Current Disputes: various territorial disputes with Russia, others

ECONOMY

Currency ($ U.S. Equivalent): 122.4 yen = $1
Per Capita Income/GDP: $23,400/$2.95 trillion
GDP Growth Rate: 1% (2000 est.)
Inflation Rate: −0.8%
Unemployment Rate: 4.7%

Labor Force: 67,760,000
Natural Resources: fish
Agriculture: rice; sugar beets; vegetables; fruit; pork; poultry; dairy and eggs; fish
Industry: metallurgy; engineering; electrical and electronics; textiles; chemicals; automobiles; food processing
Exports: $413 billion (primary partners United States, Taiwan, China)
Imports: $306 billion (primary partners United States, China, South Korea)

http://www.cia.gov/cia/publications/factbook/geos/ja.html
http://jin.jcic.or.jp

Japan: Driving Force in the Pacific Rim

HISTORICAL BACKGROUND

The Japanese nation is thought to have begun about 250 B.C., when ancestors of today's Japanese people began cultivating rice, casting objects in bronze, and putting together the rudiments of the Shinto religion. However, humans are thought to have inhabited the Japanese islands as early as 20,000 B.C. Some speculate that remnants of these or other early peoples may be the non-Oriental Ainu people (now largely Japanized) who still occupy parts of the northern island of Hokkaido. Asiatic migrants from China and Korea and islanders from the South Pacific occupied the islands between 250 B.C. and A.D. 400, contributing to the population base of modern Japan.

Between A.D. 300 and 710, military aristocrats from some of the powerful clans into which Japanese society was divided established their rule over large parts of the country. Eventually, the Yamato clan leaders, claiming divine approval, became the most powerful. Under Yamato rule, the Japanese began to import ideas and technology from nearby China, including the Buddhist religion and the Chinese method of writing—which the elite somewhat awkwardly adapted to spoken Japanese, an entirely unrelated language. The Chinese bureaucratic style of government and architecture was also introduced; Japan's first permanent capital was constructed at the city of Nara between the years 710 and 794.

As Chinese influence waned in the period 794–1185, the capital was relocated to Kyoto, with the Fujiwara family wielding real power under the largely symbolic figurehead of the emperor. A warrior class controlled by *shoguns,* or generals, held power at Kamakura between 1185 and 1333 and successfully defended the country from invasion by the Mongols. Buddhism became the religion of the masses, although Shintoism was often embraced simultaneously. Between 1333 and 1568, a very rigid class structure developed, along with a feudalistic economy controlled by *daimyos,* feudal lords who reigned over their own mini-kingdoms.

In 1543, Portuguese sailors landed in Japan, followed a few years later by the Jesuit missionary Francis Xavier. An active trade with Portugal began, and many Japanese (perhaps half a million), including some feudal lords, converted to Christianity. The Portuguese introduced firearms to the Japanese and perhaps taught them Western-style techniques of building castles with moats and stone walls. Wealthier feudal lords were able to utilize these innovations to defeat weaker rivals; by 1600, the country was unified under a military bureaucracy, although feudal lords still retained substantial sovereignty over their fiefs. During this time, the general Hideyoshi attempted an unsuccessful invasion of nearby Korea.

The Tokugawa Era

In the period 1600 to 1868, called the Tokugawa Era, the social, political, and economic foundations of modern Japan were put in place. The capital was moved to Tokyo, cities began to grow in size, and a merchant class arose that was powerful enough to challenge the hegemony of the centuries-old warrior class. Strict rules of dress and behavior for each of the four social classes (samurai, farmer, craftsman, and merchant) were imposed, and the Japanese people learned to discipline themselves to these codes. Western ideas came to be seen as a threat to the established ruling class. The military elite expelled foreigners and put the nation into $2\frac{1}{2}$ centuries of extreme isolation from the rest of the world. Christianity was banned, as was most trade with the West. Even Japanese living abroad were forbidden from returning, for fear that they might have been contaminated with foreign ideas.

During the Tokugawa Era, indigenous culture expanded rapidly. Puppet plays and a new form of drama called *kabuki* became popular, as did *haiku* poetry and Japanese pottery and painting. The samurai code, called *bushido,* along with the concept of *giri,* or obligation to one's superiors, suffused Japanese society. Literacy among males rose to about 40 percent, higher than most European countries of the day. Samurai busied themselves with the education of the young, using teaching methods that included strict discipline, hard work, and self-denial.

During the decades of isolation, Japan grew culturally strong but militarily weak. In 1853, a U.S. naval squadron appeared in Tokyo Bay to insist that Japan open up its ports to foreign vessels needing supplies and desiring to trade. Similar requests had been denied in the past, but the sophistication of the U.S. ships and their advanced weaponry convinced the Japanese military rulers that they no longer could keep Japan isolated from the outside.

The Era of Modernization: The Meiji Restoration

Treaties with the United States and other Western nations followed, and the dislocations associated with the opening of the country to the world soon brought discredit to the ruling shoguns. Provincial samurai took control of the government. The emperor, long a figurehead in Kyoto, away from the center of power, was moved to Tokyo in 1868, beginning the period known as the Meiji Restoration.

Although the Meiji leaders came to power with the intention of ousting all the foreigners and returning Japan to its former state of domestic tranquillity, they quickly realized that the nations of the West were determined to defend their newly won access to the ports of Japan. To defeat the foreigners, they reasoned, Japan must first acquire their knowledge and technology.

Thus, beginning in 1868, the Japanese leaders launched a major campaign to modernize the nation. Ambassadors and scholars were sent abroad to learn about Western-style government, education, and warfare. Implementing these ideas resulted in the abolition of the feudal system and the division of Japan into 43 prefectures, or states, and other administrative districts under the direct control of the Tokyo government. Legal codes that established the formal separation of society into social classes were abolished; and Western-style dress, music, and education were embraced. The old samurai

(Japan National Tourist Organization)

The Japanese emperor has long been a figurehead in Japan. In 1926, Hirohito, pictured above, became emperor and ushered in the era named *Showa.* He died on January 7, 1989, having seen Japan through World War II and witnessed its rise to the economic world power it is today. He was succeeded by his son, Akihito, who named his reign *Heisei,* meaning "Achieving Peace."

class turned its attention from warfare to leadership in the government, in schools, and in business. Factories and railroads were constructed, and public education was expanded. By 1900, Japan's literacy rate was 90 percent, the highest in all of Asia. Parliamentary rule was established along the lines of the government in Prussia, agricultural techniques were imported from the United States, and banking methods were adopted from Great Britain.

Japan's rapid modernization soon convinced its leaders that the nation was strong enough to begin doing what other advanced nations were doing: acquiring empires. Japan went to war with China, acquiring the Chinese island of Taiwan in 1895. In 1904, Japan attacked Russia and successfully acquired Korea and access to Manchuria (both areas having been in the sphere of influence of Russia). Siding against Germany in World War I, Japan was able to acquire Germany's Pacific empire—the Marshall, Caroline, and Mariana Islands. Western nations were surprised at Japan's rapid empire-building but did little to stop it.

The Great Depression of the 1930s caused serious hardships in Japan because, being resource-poor yet heavily populated, the country had come to rely on trade to supply its basic needs. Many Japanese advocated the forced annexation of Manchuria as a way of providing needed resources. This was accomplished easily, albeit with much brutality, in 1931. With militarism on the rise, the Japanese nation began moving away from democracy and toward a military dictatorship. Political parties were eventually banned, and opposition leaders were jailed and tortured.

WORLD WAR II AND THE JAPANESE EMPIRE
The battles of World War II in Europe, initially won by Germany, promised to substantially re-align the colonial empires of France and other European powers in Asia. The military elite of Japan declared its intention of creating a Greater East Asia Co-Prosperity Sphere—in effect, a Japanese empire created out of the ashes of the European empires in Asia that were then dissolving. In 1941, under the guidance of General Hideki Tojo and with the tacit approval of the emperor, Japan captured the former French colony of Indochina (Vietnam, Laos, and Cambodia), bombed Pearl Harbor in Hawaii, and captured oil-rich Indonesia. These victories were followed by others: Japan captured all of Southeast Asia, including Burma (now called Myanmar), Thailand, Malaya, the Philippines, and parts of New Guinea; and expanded its hold in China and in the islands of the South Pacific. Many of these conquered peoples, lured by the Japanese slogan of "Asia for the Asians," were initially supportive of the Japanese, believing that Japan would rid their countries of European colonial rule. It soon became apparent, however, that Japan had no intention of relinquishing control of these territories and would go to brutal lengths to subjugate the local peoples. Japan soon dominated a vast empire, the constituents of which were virtually the same as those making up what we call the Pacific Rim today.

In 1941, the United States launched a counteroffensive against the powerful Japanese military. (American history books refer to this offensive as the Pacific Theater of World War II, but the Japanese call it the *Pacific War.* We use the term *World War II* in this text, for reasons of clarity and consistency.) By 1944, the U.S. and other Allied troops, at the cost of tens of thousands of lives, had ousted the Japanese from most of their conquered lands and were beginning to attack the home islands themselves. Massive firebombing of Tokyo and other cities, combined with the dropping of two atomic bombs on Hiroshima and Nagasaki, convinced the Japanese military rulers that they had no choice but to surrender.

This was the first time in Japanese history that Japan had been conquered, and the Japanese were shocked to hear their emperor, Hirohito—whose voice had never been heard on radio—announce on August 14, 1945, that Japan was defeated. The emperor cited the suffering of the people—almost 2 million Japanese had been killed—devastation of

(U.S. Navy)

On December 7, 1941, Japan entered World War II as a result of its bombing of Pearl Harbor in Hawaii. This photograph, taken from an attacking Japanese plane, shows Pearl Harbor and a line of American battleships.

the cities brought about by the use of a "new and most cruel bomb," and the possibility that, without surrender, Japan as a nation might be completely "obliterated." Emperor Hirohito then encouraged his people to look to the future, to keep pace with progress, and to help build world peace by accepting the surrender ("enduring the unendurable and suffering what is insufferable").

This attitude smoothed the way for the American Occupation of Japan, led by General Douglas MacArthur. Defeat seemed to inspire the Japanese people to adopt the ways of their more powerful conquerors and to eschew militarism. Under the Occupation forces, the Japanese Constitution was rewritten in a form that mimicked that of the United States. Industry was restructured, labor unions encouraged, land reform accomplished, and the nation as a whole demilitarized. Economic aid from the United States, as well as the prosperity in Japan that was occasioned by the Korean War in 1953, allowed Japanese industry to begin to recover from the devastation of war. The United States returned the governance of Japan back to the Japanese people by treaty in 1951 (although some 60,000 troops still remain in Japan as part of an agreement to defend Japan from foreign attack).

By the late 1960s, the Japanese economy was more than self-sustaining and the United States was Japan's primary trading partner (it remains so today, with about a third of

Japanese exports purchased by Americans and similarly a substantial portion of Japanese food imports coming from the United States). Japan's trade with its former Asian empire, however, was minimal, because of lingering resentment against Japan for its wartime brutalities. (In the late 1970s, for example, anti-Japanese riots and demonstrations occurred in Indonesia upon the visit of the Japanese prime minister to that country, and the Chinese government raises the alarm each time Japan effects a modernization of its military.)

Nevertheless, between the 1960s and early 1990s, Japan experienced an era of unprecedented economic prosperity. Annual economic growth was three times as much as in other industrialized nations. Japanese couples voluntarily limited their family size so that each child born could enjoy the best of medical care and social and educational opportunities. The fascination with the West continued, but eventually, rather than "modernization" or "Americanization," the Japanese began to speak of "internationalization," reflecting both their capacity for and their actual membership in the world community, politically, culturally, and economically (but not militarily, because Japan's Constitution forbids Japan from engaging in war).

The Japanese government as well as private industry began to accelerate the drive for diversified markets and resources in the mid-1980s. This was partly in response to protectionist trends in countries in North America and Europe with which

Japan had accumulated huge trade surpluses, but it was also due to changes in Japan's own internal social and economic conditions. Japan's recent resurgence of interest in its neighboring countries and the origin of the bloc of nations we are calling the Pacific Rim can be explained by both external protectionism and internal changes. This time, however, Japanese influence—no longer linked with militarism—is being welcomed by virtually all nations in the region.

DOMESTIC CHANGE

What internal conditions are causing Japan's renewed interest in Asia and the Pacific? One change involves wage structure. For several decades, Japanese exports were less expensive than competitors' because Japanese workers were not paid as well as workers in North America and Europe. Today, however, the situation is reversed: Average manufacturing wages in Japan are now higher than those paid to workers in the United States. Schoolteachers, college professors, and many white-collar workers are also better off in Japan. These wage differentials are the result of successful union activity and demographic changes (although the high cost of living reduces the actual domestic buying power of Japanese wages to about 70 percent that of the United States).

Whereas prewar Japanese families—especially those in the rural areas—were large, today's modern household typically consists of a couple and only one or two children. As Japan's low birth rate began to affect the supply of labor, companies were forced to entice workers with higher wages. An example is McDonald's, increasingly popular in Japan as a fast-food outlet. Whereas young people working at McDonald's outlets in the United States are paid at or slightly above the legal minimum wage of $5.15 an hour, McDonald's employees in Japan are paid more than $7.00 an hour, simply because there are fewer youths available (many schools prohibit students from working during the school year). The cost of land, homes, food—even Japanese-grown rice—is so much higher in Japan than in most of its neighbor countries that employees in Japan expect high wages (household income in Japan is higher even than in the United States).

Given conditions like these, many Japanese companies have found that they cannot be competitive in world markets unless they move their operations to countries like the Philippines or Singapore, where an abundance of laborers keeps wage costs 75 to 95 percent lower than in Japan. Abundant, cheap labor (as well as a desire to avoid import tariffs) is also the reason why so many Japanese companies have been constructed in the economically depressed areas of the U.S. Midwest and South.

Another internal condition that is spurring Japanese interest in the Pacific Rim is a growing public concern for the domestic environment. Beginning in the 1970s, the Japanese courts handed down several landmark decisions in which Japanese companies were held liable for damages to people caused by chemical and other industrial wastes. Japanese industry, real-izing that it no longer had a carte blanche to make profits at the expense of the environment, began moving some of its smokestack industries to new locations in developing-world countries, just as other industrialized nations had done. This has turned out to be a wise move economically for many companies, as it has put their operations closer to their raw materials. This, in combination with cheaper labor costs, has allowed them to remain globally competitive. It also has been a tremendous benefit to the host countries, although environmental groups in many Rim countries are also now becoming active, and industry in the future may be forced to effect actual improvements in their operations rather than move polluting technologies to "safe" areas.

Attitudes toward work are also changing in Japan. Although the average Japanese worker still works about six hours more per week than the typical North American, the new generation of workers—those born and raised since World War II—are not so eager to sacrifice as much for their companies as were their parents. Recent policies have eliminated weekend work in virtually all government offices and many industries, and sports and other recreational activities are becoming increasingly popular. Given these conditions, Japanese corporate leaders are finding it more cost effective to move operations abroad to countries like South Korea, where labor legislation is weaker and long work hours remain the norm.

MYTH AND REALITY OF THE ECONOMIC MIRACLE

The Japanese economy, like any other economy, must respond to market as well as social and political changes to stay vibrant. It just so happened that for several decades, Japan's attempt to keep its economic boom alive created the conditions that, in turn, furthered the economies of all the countries in the Asia/Pacific region. That a regional "Yen Bloc" (so called because of the dominance of the Japanese currency, the yen) had been created was revealed in the late 1990s when sluggishness in the Japanese economy contributed to dramatic downturns in the economies of surrounding countries.

For many years, world business leaders were of the impression that whatever Japan did—whether targeting a certain market or reorienting its economy toward regional trade—turned to gold, as if the Japanese possessed some secret that no one else understood. But when other countries in Asia began to copy the Japanese model, their economies also improved—until, that is, lack of moderation and an inflexible application of the model produced a major correction in 1997 and 1998. Japanese success in business, education, and other fields has been the result of, among other things, hard work, advance planning, persistence, and outside financial help.

However, even with those ingredients in place, Japanese enterprises often fall short. In many industries, for example, Japanese workers are less efficient than are workers in other countries. Japan's national railway system was once found to have 277,000 more employees on its payroll than it needed. At one point, investigators revealed that the system had been so poorly managed for so many years that it had accumulated a public debt of $257 billion. Multimillion-dollar train stations had been built in out-of-the-way towns, for no other reason than that a member of the *Diet* (the Japanese Parliament) happened to live there and had pork-barreled the project. Both government and industry have been plagued by bribery and corruption, as occurred in the Recruit Scandal of the late 1980s, which caused many implicated government leaders, including the prime minister, to resign.

Nor is the Japanese economy impervious to global market conditions. Values of stocks traded on the Tokyo Stock Exchange took a serious drop in 1992; investors lost millions of dollars, and many had to declare bankruptcy. Moreover, the tenacious recession that hit Japan in the early 1990s and that is continuing into the 2000s has seriously damaged many corporations. Large corporations such as Yamaichi Securities and Sogo (a department-store chain) have had to declare bankruptcy, as have many banks. Indeed, at one time, the majority of the top 10 banks in the world were Japanese; but by 2000, only two Japanese banks remained on that prestigious list. Still, the rise of Japan from utter devastation in 1945 to the second-largest economy in the world has been nothing short of phenomenal. It will be helpful to review in detail some of the reasons for that success. We might call these the 10 commandments of Japan's economic success.

THE 10 COMMANDMENTS OF JAPAN'S ECONOMIC SUCCESS

1. Some of Japan's entrenched business conglomerates, called *zaibatsu*, were broken up by order of the U.S. Occupation commander after World War II; this allowed competing businesses to get a start. Similarly, the physical infrastructure—roads, factories—was destroyed during the war. This was a blessing in disguise, for it paved the way for newer equipment and technologies to be put in place quickly.

2. The United States, seeing the need for an economically strong Japan in order to offset the growing attraction of Communist ideology in Asia, provided substantial reconstruction aid. For instance, Sony Corporation got started with help from the Agency for International Development (AID)—an organization to which the United States is a major contributor. Mazda Motors got its start by making Jeeps for U.S. forces during the Korean War. (Other Rim countries that are now doing well can also thank U.S. generosity: Taiwan received $5.6 billion and South Korea received $13 billion in aid during the period 1945–1978.)

3. Japanese industry looked upon government as a facilitator and received useful economic advice as well as political and financial assistance from government planners. (In this regard, it is important to note that many of Japan's civil servants are the best graduates of Japan's colleges and universities.)

Also, the advice and help coming from the government were fairly consistent over time, because the same political party, the Liberal Democratic Party, remained in power for almost the entire postwar period.

4. Japanese businesses selected an export-oriented strategy that stressed building market share over immediate profit.

5. Except in certain professions, such as teaching, labor unions in Japan were not as powerful as in Europe and the United States. This is not to suggest that unions were not effective in gaining benefits for workers, but the structure of the union movement—individual company unions rather than industry-wide unions—moderated the demands for improved wages and benefits.

6. Company managers stressed employee teamwork and group spirit and implemented policies such as "lifetime employment" and quality-control circles, which contributed to group morale. In this they were aided by the tendency of Japanese workers to grant to the company some of the same level of loyalty traditionally reserved for families. In certain ways, the gap between workers and management was minimized.

7. Companies benefited from the Japanese ethic of working hard and saving much. For most of Japan's postwar history, workers labored six days a week, arriving early and leaving late. The paychecks were carefully managed to include a substantial savings component—generally between 15 and 25 percent. This guaranteed that there were always enough cash reserves for banks to offer company expansion loans at low interest.

8. The government spent relatively little of its tax revenues on social-welfare programs or military defense, preferring instead to invest public funds in private industry.

9. A relatively stable family structure (i.e., few divorces and substantial family support for young people, many of whom remained at home until marriage at about age 27), produced employees who were reliable and psychologically stable.

10. The government as well as private individuals invested enormous amounts of money and energy into education, on the assumption that, in a resource-poor country, the mental energies of the people would need to be exploited to their fullest.

Some of these conditions for success are now part of immutable history; but others, such as the emphasis on education, are open to change as the conditions of Japanese life change. A relevant example is the practice of lifetime employment. Useful as a management tool when companies were small and *skilled* laborers were difficult to find, it is now giving way to a freer labor-market system. In some Japanese industries, as many as 30 percent of new hires quit after two years on the job. In other words, the aforementioned conditions for success were relevant to one particular era of Japanese and world history and may not be as effective in other countries or other times. Selecting the right strategy for the right era has perhaps been the single most important condition for Japanese economic success.

CULTURAL CHARACTERISTICS

All these conditions notwithstanding, Japan would never have achieved economic success without its people possessing certain social and psychological characteristics, many of which can be traced to the various religious/ethical philosophies that have suffused Japan's 2,000-year history. Shintoism, Buddhism, Confucianism, Christianity, and other philosophies of living have shaped the modern Japanese mind. This is not to suggest that Japanese are tradition-bound; nothing could be further from the truth. Even though many Westerners think "tradition" when they think Japan, it is more accurate to think of Japanese people as imitative, preventive, pragmatic, obligative, and inquisitive rather than traditional. These characteristics are discussed in this section.

Imitative

The capacity to imitate one's superiors is a strength of the Japanese people; rather than representing an inability to think creatively, it constitutes one reason for Japan's legendary success. It makes sense to the Japanese to copy success, whether it is a successful boss, a company in the West, or an educational curriculum in Europe. It is true that imitation can produce conformity; but, in Japan's case, it is often conformity based on respect for the superior qualities of someone or something rather than simple, blind mimicry.

Once Japanese people have mastered the skills of their superiors, they believe that they have the moral right to a style of their own. Misunderstandings on this point arise often when East meets West. One American schoolteacher, for example, was sent to Japan to teach Western art to elementary-school children. Considering her an expert, the children did their best to copy her work to the smallest detail. Misunderstanding that this was at once a compliment and the first step toward creativity, the teacher removed all of her art samples from the classroom in order to force the students to paint something from their own imaginations. Because the students found this to be a violation of their approach to creativity, they did not perform well, and the teacher left Japan believing that Japanese education teaches conformity and compliance rather than creativity and spontaneity.

This episode is instructive about predicting the future role of Japan vis-à-vis the West. After decades of imitating the West, Japanese people are now beginning to feel that they have the skills and the moral right to create styles of their own. We can expect to see, therefore, an explosion of Japanese creativity in the near future. Some observers have noted, for example, that the global fashion industry seems to be gaining more inspiration from designers in Tokyo than from those in Milan, Paris, or New York. And the Japanese have often registered more new patents with the U.S. Patent Office than any other nation except the United States. The Japanese are also now winning more Nobel prizes than in the past, including the prize for chemistry in 2000.

Preventive

Japanese individuals, families, companies, and the government generally prefer long-range over short-range planning, and they greatly prefer foreknowledge over postmortem analysis. Assembly-line workers test and retest every product to prevent customers from receiving defective products. Some store clerks plug in and check electronic devices in front of a customer in order to prevent bad merchandise from sullying the good reputation of the store, and commuter trains in Japan have three times as many "Watch your step" and similar notices as do trains in the United States and Europe. Insurance companies do a brisk business in Japan; even though all Japanese citizens are covered by the government's national health plan, many people buy additional coverage—for example, cancer insurance—just to be safe.

This concern with prevention trickles down to the smallest details. At train stations, multiple recorded warnings are given of an approaching train to commuters standing on the platform. Parent–teacher associations send teams of mothers around the neighborhood to determine which streets are the safest for the children. They then post signs designating certain roads as "school passage roads" and instruct children to take those routes even if it takes longer to walk to school. The Japanese think that it is better to avoid an accident than to have an emergency team ready when a child is hurt. Whereas Americans say, "If it ain't broke, don't fix it," the Japanese say, "Why wait 'til it breaks to fix it?"

Pragmatic

Rather than pursue a plan because it ideologically fits some preordained philosophy, the Japanese try to be pragmatic on

(Sony Corporation of America)

A contributing factor in the modern economic development of Japan was investment from the Agency for International Development. The Sony Corporation is an example of just how successful this assistance could be. These workers are assembling products that will be sold all over the world.

most points. Take drugs as an example. Many nations say that drug abuse is an insurmountable problem that will, at best, be contained but probably never eradicated, because to do so would violate civil liberties. But, as a headline in the *Asahi Evening News* proclaimed a few years ago, "Japan Doesn't Have a Drug Problem and Means to Keep It That Way." Reliable statistics support this claim, but that is not the whole story. In 1954, Japan had a serious drug problem, with 53,000 drug arrests in one year. At the time, the authorities concluded that they had a big problem on their hands and must do whatever was required to solve it. The government passed a series of tough laws restricting the production, use, exchange, and possession of all manner of drugs, and it gave the police the power to arrest all violators. Users were arrested as well as dealers: It was reasoned that if the addicts were not left to buy the drugs, the dealers would be out of business. Their goal at the time was to arrest all addicts, even if it meant that certain liberties were briefly circumscribed. The plan, based on a do-what-it-takes pragmatism, worked; today, Japan is the only industrialized country without a widespread drug problem. In this case, to pragmatism was added the Japanese tendency to work for the common rather than the individual good.

This approach to life is so much a part of the Japanese mind-set that many Japanese cannot understand why the United States and other industrialized nations have so many unresolved social and economic problems. For instance, when it comes to availability of money for loans to start up businesses or purchase homes, it is clear that one of the West's most serious problems is a low personal savings rate (about three cents saved for every dollar in the United States, comparted to about 15 cents in Japan). This makes money scarce and interest rates on borrowed money relatively high. Knowing that this is a problem in the United States, the Japanese wonder why Americans simply do not start saving more. They think, "We did it in the past with a poorer economy than yours; why can't you?"

Obligative

The Japanese have a great sense of duty toward those around them. Thousands of Japanese workers work late without pay to improve their job skills so that they will not let their fellow workers down. Good deeds done by one generation are remembered and repaid by the next, and lifelong friendships are maintained by exchanging appropriate gifts and letters. North Americans and Europeans are often considered untrustworthy friends because they do not keep up the level of close, personal communications that the Japanese expect of their own friends; nor do the Westerners have as strong a sense of place, station, or position.

Duty to the group is closely linked to respect for superior authority. Every group—indeed, every relationship—is seen as a mixture of people with inferior and superior resources. These differences must be acknowledged, and no one is disparaged for bringing less to the group than someone else. However, equality is assumed when it comes to basic commitment to or effort expended for a task. Slackers are not welcome. Obligation to the group along with respect for superiors motivated Japanese pilots to fly suicide missions during World War II, and it now causes workers to go the extra mile for the good of the company.

That said, it is also true that changes in the intensity of commitment are becoming increasingly apparent. More Japanese than ever before are beginning to feel that their own personal goals are more important than those of their companies or extended families. This is no doubt a result of the Westernization of the culture since the Meiji Restoration, in the late 1800s, and especially of the experiences of the growing number of Japanese—approximately half a million in a given year—who live abroad and then take their newly acquired values back to Japan. (About half of these "away Japanese" live in North America and Western Europe.)

There is no doubt that the pace of "individualization" of the Japanese psyche is increasing and that, more and more, the Japanese attitude toward work is approaching that of the West. Many Japanese companies are now allowing employees to set their own "flex-time" work schedules, and some companies have even asked employees to stop addressing superiors with their hierarical titles and instead refer to everyone as *san,* or Mr. or Ms.

Inquisitive

The image of dozens of Japanese businesspeople struggling to read a book or newspaper while standing inside a packed commuter train is one not easily forgotten, symbolizing as it does the intense desire among the Japanese for knowledge, especially knowledge of foreign cultures. Nearly 6 million Japanese travel abroad each year (many to pursue higher education), and for those who do not, the government and private radio and television stations provide a continuous stream of programming about everything from Caribbean cuisine to French ballet. The Japanese have a yen for foreign styles of dress, foreign cooking, and foreign languages. The Japanese study languages with great intensity. Every student is required to study English; many also study Chinese, Greek, Latin, Russian, Arabic, and other languages, with French being the most popular after English.

Observers inside and outside of Japan are beginning to comment that the Japanese are recklessly discarding Japanese culture in favor of foreign ideas and habits, even when they make no sense in the Japanese context. A tremendous intellectual debate, called *Nihonjin-ron,* is now taking place in Japan over the meaning of being Japanese and the Japanese role in the world. There is certainly value in these concerns, but, as was noted previously, the secret about Japanese traditions is that they are not traditional. That is, the Japanese seem to know that, in order to succeed, they must learn what they need to know for the era in which they live, even if it means modifying or eliminating the past. This is probably the reason why the Japanese nation has endured for more than 2,000 years while many other empires have fallen. In this sense, the Japanese are very forward-looking people and, in their thirst for new modes of thinking and acting, they are, perhaps, revealing their most basic and useful national personality characteristic: inquisitiveness. Given this attitude toward learning, it should come as no surprise that formal schooling in Japan is a very serious business to the government and to families. It is to that topic that we now turn.

SCHOOLING

Probably most of the things that the West has heard about Japanese schools are distortions or outright falsehoods. We hear that Japanese children are highly disciplined, for example; yet in reality, Japanese schools at the elementary and junior high levels are rather noisy, unstructured places, with children racing around the halls during breaks and getting into fights with classmates on the way home. Japan actually has a far lower percentage of its college-age population enrolled in higher education than is the case in the United States—35 percent as compared to 50 percent. Moreover, the Japanese government does not require young people to attend high school (they must attend only until age 15), although 94 percent do anyway. Given these and other realities of school life in Japan, how can we explain the consistently high scores of Japanese on international tests and the general agreement that Japanese high school graduates know almost as much as college graduates in North America?

(AP Wirephoto by Elaine Kurtenbach)

The Japanese take education very seriously. Half of the children start kindergarten at the age of three, and early on they are instilled with respect for authority.

Structurally, schools in Japan are similar to those in many other countries: There are kindergartens, elementary schools, junior high schools, and high schools. Passage into elementary and junior high is automatic, regardless of student performance level. But admission to high school and college is based on test scores from entrance examinations. Preparing for these examinations occupies the full attention of students in their final year of both junior high and high school, respectively. Both parents and school authorities insist that studying for the tests be the primary focus of a student's life at those times. For instance, members of a junior high soccer team may be allowed to play on the team only for their first two years; during their last year, they are expected to be studying for their high school entrance examinations. School policy reminds students that they are in school to learn and to graduate to the next level, not to play sports. Many students even attend after-hours "cram schools" (*juku*) several nights a week to prepare for the exams.

Time for recreational and other nonschool activities is restricted, because Japanese students attend school 240 days out of the year, as compared to about 180 in U.S. schools, including some Saturday mornings (although Saturday classes are being phased out). Summer vacation is only about six weeks long, and students often attend school activities during most of that period. Japanese youths are expected to treat schooling as their top priority over part-time jobs (usually prohibited by school policy during the school year, except for the needy), sports, dating, and even family time.

Children who do well in school are generally thought to be fulfilling their obligations to the family. The reason for this focus is that parents realize that only through education can Japanese youths find their place in society. Joining Japan's relatively small military is generally not an option, opportunities for farming are limited because of land scarcity, and most major companies will not hire a new employee who has not graduated from college or a respectable high school. Thus, the Japanese find it important to focus on education—to do one thing and do it well.

Teachers are held in high regard in Japan, partly because when mass education was introduced, many of the high-status samurai took up teaching to replace their martial activities. In addition, in modern times, the Japan Teacher's Union has been active in agitating for higher pay for teachers. As a group, teachers are the highest-paid civil servants in Japan. They take their jobs very seriously. Public-school teachers, for example, visit the home of each student each year to merge the authority of the home with that of the school, and they insist that parents (usually mothers) play active supporting roles in the school.

Some Japanese youths dislike the system, and discussions are currently under way among Japanese educators on how to improve the quality of life for students. Occasionally the pressure of taking examinations (called "exam hell") produces such stress that a desperate student will commit suicide rather than try and fail. Stress also appears to be the cause of

(Reuters/Bettmann)

Doing well in school is seen by Japanese students as fulfilling their obligation to their families. Education is held in high regard and is seen as a critical element in achieving a better life; it is supported very strongly by parents.

ijime, or bullying of weaker students by stronger peers. In recent years, the Ministry of Education has worked hard to help students deal with school stress, with the result that Japan's youth suicide rate has dropped dramatically, far lower than the youth rate in the United States (although suicide among adults, especially those affected by the downturn in the economy, has risen to its highest level since 1947). Despite these and other problems, most Japanese youths enjoy school and value the time they have to be with their friends, whether in class, walking home, or attending cram school. Some of those who fail their college entrance exams continue to study privately, some for many years, and take the exam each year until they pass. Others travel abroad and enroll in foreign universities that do not have such rigid entrance requirements. Still others enroll in vocational training schools. But everyone in Japan realizes that education—not money, name, or luck—is the key to success.

Parents whose children are admitted to the prestigious national universities—such as Tokyo and Kyoto Universities—consider that they have much to brag about. Other parents are willing to pay as much as $35,000 on average for four years of college at the private (but usually not as prestigious) universities. Once admitted, students find that life slows down a bit. For one thing, parents typically pay more than 65 percent of the costs, and approximately 3 percent is covered by scholarships. This leaves only about 30 percent to be earned by the students; this usually comes from tutoring high school students who are studying for the entrance exams.

Contemporary parents are also willing to pay the cost of a son's or daughter's traveling to and spending a few months in North America or Europe either before college begins or during summer breaks—a practice that is becoming de rigueur for Japanese students, much as taking a "grand tour" of Europe was once expected of young, upper-class Americans and Canadians.

College students may take 15 or 16 courses at a time, but classes usually meet only once or twice a week, and sporadic attendance is the norm. Straight lecturing rather than class discussion is the typical learning format, and there is very little homework beyond studying for the final exam. Students generally do not challenge the professors' statements in class, but some students develop rather close, avuncular-type relationships with their professors outside of class. Hobbies, sports, and club activities (things the students did not have time to do while in public school) occupy the center of life for many college students. Japanese professors visiting universities in North America and Europe are often surprised at how diligently students in those places study during their college years. By contrast, Japanese students spend a lot of time making friendships that will last a lifetime and be useful in one's career and private life.

THE JAPANESE BUSINESS WORLD

Successful college graduates begin their work careers in April, when most large companies do their hiring (although this practice is slowly giving way to individual hiring throughout the year). They may have to take an examination to determine how much they know about electronics or stocks and bonds, and they may have to complete a detailed personality profile. Finally, they will have to submit to a very serious interview with company management. During interviews, the managers will watch their every move; the applicants will be careful to avoid saying anything that will give them "minus points."

Once hired, individuals attend training sessions in which they learn the company song and other rituals as well as company policy on numerous matters. They may be housed in company apartments (or may continue to live at home), permitted to use a company car or van, and advised to shop at company grocery stores. Almost never are employees married at this time, and so they are expected to live a rather spartan life for the first few years.

Employees are expected to show considerable deference to their section bosses, even though, on the surface, bosses do not appear to be very different from other employees. Bosses' desks are out in the open, near the employees; they wear the same uniform; they socialize with the employees after work; even in a factory, they are often on the shop floor rather than sequestered away in private offices. Long-term employees often come to see the section leader as an uncle figure (bosses are usually male) who will give them advice about life, be the best man at their weddings, and provide informal marital and family counseling as needed.

Although there are cases of abuse or unfair treatment of employees, Japanese company life can generally be described as somewhat like a large family rather than a military squad; employees (sometimes called associates) often obey their superiors out of genuine respect rather than forced compliance. Moreover, competition between workers is reduced because everyone hired at the same time receives more or less the same pay and most workers receive promotions at about the same time. Only later in one's career are individualistic promotions given. That said, it is important to note that, under pressure to develop a more inventive workforce to compete against new ideas and products from the West, many Japanese companies are now experimenting with new pay and promotion systems that reward performance, not longevity.

Employees are expected to work hard, for not only are Japanese companies in competition with foreign businesses, but they also must survive the fiercely competitive business climate at home. Indeed, the Japanese skill in international business was developed at home. There are, for example, hundreds of electronics companies and thousands of textile enterprises competing for customers in Japan. And whereas the United States has only four automobile-manufacturing companies, Japan has nine. All these companies entice customers with deep price cuts or unusual services, hoping to edge out unprepared or weak competitors. Many compa-

In the Japanese business world, one's job is taken very seriously and is often seen as a lifelong commitment. These workers have jobs that, in many ways, may be more a part of their lives than are their families.

(UN photo by Jan Corash)

In Japan, not unlike in many other parts of the world, economic well-being often requires two incomes. Still, there is strong social pressure on women to stop working once they have a baby. All generations of family members take part in childrearing.

nies fail. There were once, for instance, almost 40 companies in Japan that manufactured calculators, but today only half a dozen remain, the rest victims of tough internal Japanese competition.

At about age 27, after several years of working and saving money for an apartment, a car, and a honeymoon, the typical Japanese male worker marries. The average bride, about age 25, may have taken private lessons in flower arranging, the tea ceremony, sewing, cooking, and perhaps a musical instrument like the *koto,* the Japanese harp. She probably will not have graduated from college, although she may have attended a specialty college for a while. If she is working, she likely is paid much less than her husband, even if she has an identical position (despite equal-pay laws). She may spend her time in the company preparing and serving tea for clients and employees, dusting the office, running errands, and answering telephones. When she has a baby, she will be expected to quit—although more women today are choosing to remain on the job, and some are advancing into management or are leaving to start their own companies.

Because the wife is expected to serve as the primary caregiver for the children, the husband is expected always to make his time available for the company. He may be asked to work weekends, to stay out late most of the week (about four out of seven nights), or even to be transferred to another branch in Japan or abroad without his family. This loyalty is rewarded in numerous ways: Unless the company goes bankrupt or the employee is unusually inept, he may be permitted to work for the company until he retires, usually at about age 55 or 60, even if the company no longer really needs his

services; he and his wife will be taken on company sightseeing trips; the company will pay most of his health-insurance costs (the government pays the rest); and he will have the peace of mind that comes from being surrounded by lifelong friends and workmates. His association with company employees will be his main social outlet, even after retirement; upon his death, it will be his former workmates who organize and direct his funeral services. Employees who work for companies that are experimenting with new, more individualistic work structures find the insecurity of the arrangements to be discomfiting.

THE FAMILY

The loyalty once given to the traditional Japanese extended family, called the *ie,* has been transferred to the modern company. This is logical from a historical perspective, since the modern company once began as a family business and was gradually expanded to include more workers, or "siblings." Thus, whereas the family is seen as the backbone of most societies, it might be more accurate to argue that the *kaisha,* or company, is the basis of modern Japanese society. As one Japanese commentator explained, "In the West, the home is the cornerstone of people's lives. In Tokyo, home is just a place to sleep at night. . . . Each family member—husband, wife, and children—has his own community centered outside the home."

Thus, the common image that Westerners hold of the centrality of the family to Japanese culture may be inaccurate. For instance, father absence is epidemic in Japan. It is an unusual father who eats more than one meal a day with his family. He may go shopping or to a park with his family when he has free time from work, but he is more likely to go golfing with a workmate. Schooling occupies the bulk of the children's time, even on weekends. And with fewer children than in earlier generations and with appliance-equipped apartments, many Japanese women rejoin the workforce after their children are self-maintaining.

Japan's divorce rate, while rising, is still considerably lower than in other industrialized nations, a fact that may seem incongruent with the conditions described above. Yet, as explained by one Japanese sociologist, Japanese couples "do not expect much emotional closeness; there is less pressure on us to meet each other's emotional needs. If we become close, that is a nice dividend, but if we do not, it is not a problem because we did not expect it in the first place."

Despite these modifications to the common Western image of the Japanese family, Japanese families have significant roles to play in society. Support for education is one of the most important. Families, especially mothers, support the schools by being actively involved in the parent–teacher association, by insisting that children be given plenty of homework, and by saving for college so that the money for tuition is available without the college student having to work.

Another important function of the family is mate selection. Somewhat less than half of current Japanese marriages are arranged by the family or have occurred as a result of far more family involvement than in North America. Families sometimes ask a go-between (an uncle, a boss, or another trusted person) to compile a list of marriageable candidates. Criteria such as social class, blood type, and occupation are considered. Photos of prospective candidates are presented to the unmarried son or daughter, who has the option to veto any of them or to date those he or she finds acceptable. Young people, however, increasingly select their mates with little or no input from parents.

Finally, families in Japan, even those in which the children are married and living away from home, continue to gather for the purpose of honoring the memory of deceased family members or to enjoy one another's company for New Year's Day, Children's Day, and other celebrations.

WOMEN IN JAPAN

Ancient Confucian values held that women were legally and socially inferior to men. This produced a culture in feudal Japan in which the woman was expected to walk several steps behind her husband when in public, to eat meals only after the husband had eaten, to forgo formal education, and to serve the husband and male members of the family whenever possible. A "good woman" was said to be one who would endure these conditions without complaint. This pronounced gender difference (though minimized substantially over the centuries since Confucius) can still be seen today in myriad ways, including in the preponderance of males in positions of leadership in business and politics, in the smaller percentage of women college graduates, and in the pay differential between women and men.

Given the Confucian values noted above, one would expect that all top leaders would be males. However, women's roles are also subject to the complexity of both ancient and modern cultures. Between A.D. 592 and 770, for instance, of the 12 reigning emperors, half were women. In rural areas today, women take an active decision-making role in farm associations. In the urban workplace, some women occupy typically pink-collar positions (nurses, clerks, and so on), but many women are also doctors and business executives; 28,000 are company presidents.

Thus, it is clear that within the general framework of gender inequality imposed by Confucian values, Japanese culture, especially at certain times, has been rather lenient in its application of those values. There is still considerable social pressure on women to stop working once they marry, and particularly after they have a baby, but it is clear that many women are resisting that pressure: one out of every three employees in Japan is female, and nearly 60 percent of the female workforce are married. An equal-pay law was enacted in 1989 that makes it illegal to pay women less for doing comparable work (although it may take years for companies to comply fully). And the Ministry of Education has mandated that home economics and shop classes now be required for both boys and girls; that is, both girls and boys will learn to cook and sew as well as construct things out of wood and metal.

In certain respects, Japanese women seem more assertive than women in the West. For example, in a recent national election, a wife challenged her husband for his seat in the House of Representatives (something that has not been done in the United States, where male candidates usually expect their wives to stump for them). Significantly, too, the former head of the Japan Socialist Party was an unmarried woman, and in 1999 Osaka voters elected a woman as mayor for the first time. Women have been elected to the powerful Tokyo Metropolitan Council and awarded professorships at prestigious universities such as Tokyo University. And, while women continue to be used as sexual objects in pornography and prostitution, certain kinds of misogynistic behavior, such as rape and serial killing, are less frequent in Japan than in Western societies. New laws against child pornography may reduce the abuse of young girls, but a spate of bizarre killings by youths and mafia in recent years is causing the sense of personal safety in Japan to dissipate. Signs in train stations warn of pickpockets, and signs on infrequently traveled paths warn of molesters.

Recent studies show that many Japanese women believe that their lives are easier than those of most Westerners. With their husbands working long hours and their one or two children in school all day, Japanese women find they have more leisure time than Western women. Gender-based social divisions remain apparent throughout Japanese culture, but modern Japanese women have learned to blend these divisions with the realities and opportunities of the contemporary workplace and home.

(The Bettmann Archive)

Religion in Japan, while not having a large active affiliation, is still an intricate part of the texture and history of the culture. This temple in Kyoto was founded in the twelfth century.

RELIGION/ETHICS

There are many holidays in Japan, most of which have a religious origin. This fact, as well as the existence of numerous shrines and temples, may leave the impression that Japan is a rather religious country. This is not true, however. Most Japanese people do not claim any active religious affiliation, but many will stop by a shrine occasionally to ask for divine help in passing an exam, finding a mate, or recovering from an illness.

Nevertheless, modern Japanese culture sprang from a rich religious heritage. The first influence on Japanese culture came from the animistic Shinto religion, from whence modern Japanese acquired their respect for the beauty of nature. Confucianism brought a respect for hierarchy and education. Taoism stressed introspection, and Buddhism taught the need for good behavior now in order to acquire a better life in the future.

Shinto was selected in the 1930s as the state religion and was used as a divine justification for Japan's military exploits of that era, but most Japanese today will say that Japan is, culturally, a Buddhist nation. Some new Buddhist denomina-

tions have attracted thousands of followers. The rudiments of Christianity are also a part of the modern Japanese consciousness, but few Japanese have actually joined Christian churches. Sociologically, Japan, with its social divisions and hierarchy, is probably more of a Confucian society than it is Buddhist or any other philosophy.

Most Japanese regard morality as springing from within the group rather than pronounced from above. That is, a Japanese person may refrain from stealing so as not to offend the owner of an object or bring shame upon the family, rather than because of a divine prohibition against stealing. Thus we find in Japan a relatively small rate of violent—that is, public—crimes, and a much larger rate of white-collar crimes such as embezzlement, in which offenders believe that they can get away with something without creating a public scandal for their families.

THE GOVERNMENT

The Constitution of postwar Japan became effective in 1947 and firmly established the Japanese people as the ultimate

36

source of sovereignty, with the emperor as the symbol of the nation. The national Parliament, or *Diet,* is empowered to pass legislation. The Diet is divided into two houses: the House of Representatives, with 500 members (to be reduced to 480 in the next election) elected for four-year terms; and the House of Councillors, with 252 members elected for six-year terms from each of the 47 prefectures (states) of Japan as well as nationally. The prime minister, assisted by a cabinet, is also the leader of the party with the most seats in the Diet. Prefectures are governed by an elected governor and an assembly, and cities and towns are governed by elected mayors and town councils. The Supreme Court, consisting of a chief judge and 14 other judges, is independent of the legislative branch of government.

Japan's Constitution forbids Japan from engaging in war or from having military capability that would allow it to attack another country. Japan does maintain a well-equipped self-defense force, but it relies on a security treaty with the United States in case of serious aggression against it. In recent years, the United States has been encouraging Japan to assume more of the burden of the military security of the Asian region, and Japan has increased its expenditures in absolute terms. But until the Constitution is amended, Japan is not likely to initiate any major upgrading of its military capability. This is in line with the general wishes of the Japanese people, who, since the devastation of Hiroshima and Nagasaki, have become firmly committed to a pacifist foreign policy. Moreover, Japanese leaders fear that any significant increase in military capability would re-ignite dormant fears about Japanese intentions within the increasingly vital Pacific Rim area.

This tendency toward not wanting to get involved militarily is reflected in one of Japan's most recent performances on the world stage. The Japanese were slow to play any significant part in supporting military expenditures for the Persian Gulf War, even when the outcome had a direct potential effect on their economy. The Iraqi invasion of Kuwait in August 1990 brought on the wrath—against Japan—of a coalition of countries led by the United States in January 1991, but it generated an initial commitment from Japan of only $2 billion (later increased to $9 billion, still a small fraction of the cost) and no personnel of any kind. This meager support was criticized by some foreign observers, who pointed out that Japan relies heavily on Gulf oil.

In 1992, the Japanese government announced its intention of building its own F-16–type jet-fighter planes; and subsequently, amid protests from the public, the Diet voted to send as many as 1,800 Japanese soldiers—the first to go abroad since World War II—to Cambodia to assist in the UN–supervised peacekeeping effort. Countries that had experienced the full force of Japanese domination in the past, such as China and Korea, expressed dismay at these evidences of Japan's modern military capability, but the United States welcomed the moves as an indication of Japan's willingness to share the costs of providing military security to Asia.

The Japanese have formed numerous political parties to represent their views in government. Among these have been the Japan Communist Party, the Social Democratic Party, and the New Frontier Party. For nearly 40 years, however, the most powerful party was the Liberal Democratic Party (LDP). Formed in 1955, it guided Japan to its current position of economic strength, but a series of sex and bribery scandals caused it to lose control of the government in 1993. A shaky coalition of eight parties took control for about a year but was replaced by an even more unlikely coalition of the LDP and the Japan Socialists—historic enemies who were unable to agree on most policies. Eventually, the LDP was able to regain some of its lost political clout; but with some half a dozen changes in the prime ministership in the 1990s and continuous party realignments, it would be an understatement to say that Japan's government is in flux.

Part of the reason for this instability can be explained by Japan's party faction system. Party politics in Japan has always been a mixture of Western-style democratic practice and feudalistic personal relationships. Japanese parties are really several parties rolled into one. That is, parties are divided into several factions, each comprised of a group of loyal younger members headed by a powerful member of the Diet. The senior member has a duty to pave the way for the younger members politically, but they, in turn, are obligated to support the senior member in votes and in other ways. The faction leader's role in gathering financial support for faction members is particularly important, because Diet members are expected by the electorate to be patrons of numerous causes, from charity drives to the opening of a constituent's fast-food business. Because parliamentary salaries are inadequate to the task, outside funds, and thus the faction, are crucial. The size and power of the various factions are often the critical elements in deciding who will assume the office of prime minister and who will occupy which cabinet seats. The role of these intraparty factions is so central to Japanese politics that attempts to ban them have never been successful.

The factional nature of Japanese party politics means that cabinet and other political positions are frequently rotated. This would yield considerable instability in governance were it not for the stabilizing influence of the Japanese bureaucracy. Large and powerful, the career bureaucracy is responsible for drafting more than 80 percent of the bills submitted to the Diet. Many of the bureaucrats are graduates from the finest universities in Japan, particularly Tokyo University, which provides some 80 percent of the senior officials in the more than 20 national ministries. Many of them consider their role in long-range forecasting, drafting legislation, and implementing policies to be superior to that of the elected officials under whom they work. They reason that, whereas the politicians are bound to the whims of the people they represent, bureaucrats are committed to the nation of Japan—to, as it were, the *idea* of Japan. So superior did bureaucrats feel to their elected officials (and bosses) that

Prepottery, paleolithic culture 20,000–4,500 B.C.	Jomon culture with distinctive pottery 4,500–250 B.C.	Yayoi culture with rice agriculture, Shinto religion, and Japanese language 250 B.C.–A.D. 300	The Yamato period; warrior clans import Chinese culture A.D. 300–700	The Nara period; Chinese-style bureaucratic government at the capital at Nara 710–794	The Heian period; the capital is at Kyoto 794–1185	The Kamakura period; feudalism and shoguns; Buddhism is popularized 1185–1333	The Muromachi period; Western missionaries and traders arrive; feudal lords control their own domains 1333–1568	The Momoyama period; feudal lords become subject to one central leader; attempted invasion of Korea 1568–1600

until recently, elected officials were not questioned directly in the Diet; career bureaucrats represented their bosses to the people. Generally speaking, government service is considered a higher calling than are careers in private business, law, and other fields.

In addition to the bureaucracy, Japanese politicians have leaned heavily on big business to support their policies of postwar reconstruction, economic growth, and social reform. Business has accepted heavy taxation so that social-welfare programs such as the national health plan are feasible, and they have provided political candidates with substantial financial help. In turn, the government has seen its role as that of facilitating the growth of private industry (some critics claim that the relationship between government and business is so close that Japan is best described not as a nation but as "Japan, Inc."). Consider, for example, the powerful Ministry of International Trade and Industry (MITI). Over the years, it has worked closely with business, particularly the Federation of Economic Organizations (Keidanren) to forecast potential market shifts, develop strategies for market control, and generally pave the way for Japanese businesses to succeed in the international marketplace. The close working relationship between big business and the national government is an established fact of life in Japan, and, despite criticism from countries with a more laissez-faire approach to business, it will undoubtedly continue into the future, because it has served Japan well.

The prolonged recession in Japan has uncovered inconsistencies in the structure of Japanese society. For instance, we have seen that the Japanese are an inquisitive, change-oriented people, yet the administrative structure of the government and of the business community seems impervious to change. Decisions take an agonizingly long time to make, and outsiders as well as the Japanese public find the government's mode of operation to be far from open and transparent. For instance, in late 1999, one of Japan's 51 nuclear reactors leaked radiation and endangered the lives of many people. At first the government decided not to announce the leak at all. Then, when it finally did, it claimed that only a few people had been affected. In fact, more than 400 people had been affected, and the government seemed incapable of telling the whole truth about the matter. Many blamed bureaucratic rigidity. Similarly, in 1999–2000, the government injected more than $62 billion into the economy (the ninth such stimulus package since the recession started in 1989), yet the economy remained sluggish. Again, many blamed the snail-paced decision-making process as a factor that makes it difficult for

companies to quickly take advantage of new opportunities in the marketplace. Whether Japan can make the structural changes necessary to remain competitive in the new global economy remains to be seen.

THE FUTURE

In the postwar years of political stability, the Japanese have accomplished more than anyone, including themselves, thought possible. Japan's literacy rate is 99 percent, 99 percent of Japanese households have telephones, 99 percent have color televisions, and 75 percent own automobiles. Nationalized health care covers every Japanese citizen, and the Japanese have the longest life expectancy in the world. With only half the population of the United States, a land area about the size of Great Britain, and extremely limited natural resources (it has to import 99.6 percent of its oil, 99.8 percent of its iron, and 86.7 percent of its coal), Japan has nevertheless created the second-largest economy in the world. Where does it go from here?

When the Spanish were establishing hegemony over large parts of the globe, they were driven in part by the desire to bring Christianity to the "heathen." The British, for their part, believed that they were taking "civilization" to the "savages" of the world. China and the former Soviet Union were once strongly committed to the ideals of communism, while the United States has felt that its mission is that of expanding democracy and capitalism.

What about Japan? For what reason do Japanese businesses buy up hotels in New Zealand and skyscrapers in New York? What role does Japan have to play in the world in addition to spawning economic development? What values will guide and perhaps temper Japan's drive for economic dominance?

These are questions that the Japanese people themselves are attempting to answer; but, finding no ready answers, they are beginning to encounter more and more difficulties with the world around them and within their own society. Animosity over the persistent trade imbalance in Japan's favor continues to simmer in Europe and North America as well as in some countries of the Pacific Rim. To deflect these criticisms, Japan has substantially increased its gift-giving to foreign governments, including allocating money for the stabilization or growth of democracy in Central/Eastern Europe and for easing the foreign debt burden of Mexico and other countries.

What Japan has been loathe to do, however, is remove the "structural impediments" that make it difficult for foreign companies to do business in Japan. For example, 50 percent of the automobiles sold in Iceland are Japanese, which means less profit for the American and European manufacturers who

The Tokugawa Era; self-imposed isolation from the West
1600–1868

The Meiji Restoration; modernization; Taiwan and Korea are under Japanese control
1868–1912

The Taisho and Showa periods; militarization leads to war and Japan's defeat
1912–1945

Japan surrenders; the U.S. Occupation imposes major changes in the organization of society
1945

Sovereignty is returned to the Japanese people by treaty
1951

The ruling party is hit by scandals but retains control of the government; Emperor Hirohito dies; Emperor Akihito succeeds
1980s

After years of a slow economy, Japan officially admits it is in a recession; a devastating earthquake in Kobe kills more than 5,000 people
1990s

2000s

A U.S Navy submarine accidentally rams into and sinks a Japanese trawler; nine Japanese are missing and presumed dead; relations with the United States are strained

Experts predict that the worker/retiree ratio will drop from 6:1 to 2:1 by 2020, as the aging population increases

Japanese business undergoes a shift toward more openness and flexibility

Weaknesses in the Japanese economy are a concern all over the world

used to dominate car sales there. Yet, because of high tariffs and other regulations, very few American and European cars have been sold in Japan. Beginning in the mid-1980s, Japan reluctantly began to dismantle many of these trade barriers, and the process has been so successful that Japan now has a lower overall average tariff on nonagricultural products than the United States—its severest critic in this arena.

But Japanese people worry that further opening of their markets may destroy some fundamentals of Japanese life. Rice, for instance, costs much more in Japan than it should, because the Japanese government protects rice farmers with subsidies and limits most rice imports from abroad. The Japanese would prefer to pay less for rice at the supermarket, but they also argue that foreign competition would prove the undoing of many small rice farmers, whose land would then be sold to housing developers. This, in turn, would destroy more of Japan's scarce arable land and weaken the already shaky traditions of the Japanese countryside—the heart of traditional Japanese culture and values.

Today, thousands of foreign firms do business in Japan; some of them, like Polaroid and Schick, control the Japanese market in their products. Foreign investment in Japan has grown about 16 percent annually since 1980. In the case of the United States, the profit made by American firms doing business in Japan (nearly 800 of them) in a single year is just about equal to the amount of the trade imbalance between Japan and the United States. Japanese supermarkets are filled with foreign foodstuffs, and the radio and television airwaves are filled with the sounds and sights of Western music and dress. Japanese youths are as likely to eat at McDonald's or Kentucky Fried Chicken outlets as at traditional Japanese restaurants, and many Japanese have never worn a kimono nor learned to play a Japanese musical instrument. It is clear to many observers that, culturally, Japan already imports much more from the West than the West does from Japan.

Given this overwhelming Westernization of Japan as well as Japan's current capacity to continue imbibing Western culture, even the change-oriented Japanese are beginning to ask where they, as a nation, are going. Will national wealth, as it slowly trickles down to individuals, produce a generation of hedonistic youths who do not appreciate the sacrifices of those before them? Will wealthy Japanese people be satisfied with the small homes and tiny yards that their forebears had to accept? Will there ever be a time when, strapped for resources, the Japanese will once again seek hegemony over other nations? What future role should Japan assume in the international arena, apart from economic development? If these questions remain to be answered, circumstances of international trade have at least provided an answer to the question of Japan's role in the Pacific Rim countries: It is clear that, for the next several decades, Japan will continue to shape the pace and nature of economic development, and thus the political environment, of the entire Pacific Rim.

DEVELOPMENT

Japan has entered a post-smokestack era in which primary industries are being moved abroad, producing a hollowing effect inside Japan and increasing the likelihood of rising unemployment. Nevertheless, prospects for continued growth are excellent, despite the current economic woes.

FREEDOM

Japanese citizens enjoy full civil liberties, and opposition parties and ideologies are seen as natural and useful components of democracy. Certain people, however, such as those of Korean ancestry, have been subject to both social and official discrimination—an issue that is gaining the attention of the Japanese.

HEALTH/WELFARE

The Japanese live longer on average than any other people on earth. Every citizen is provided with inexpensive medical care under a national health-care system, but many people still prefer to save substantial portions of their income for health emergencies and old age.

ACHIEVEMENTS

Japan has achieved virtually complete literacy. Although there are poor areas, there are no slums inhabited by a permanent underclass. The gaps between the social classes appear to be less pronounced than in many other societies. The country seems to be entering an era of remarkable educational and technological achievement.

Australia (Commonwealth of Australia)

GEOGRAPHY

Area in Square Miles (Kilometers):
2,867,896 (7,686,850)
(slightly smaller than the
United States)

Capital (Population): Canberra
(310,000)

Environmental Concerns: depletion
of the ozone layer; pollution;
soil erosion and excessive
salinity; desertification; wildlife
habitat loss; degradation of
Great Barrier Reef; limited
freshwater resources

Geographical Features: mostly
low plateau with deserts;
fertile plain in southeast

Climate: generally arid to
semiarid; temperate to tropical

PEOPLE

Population

Total: 19,169,100
Annual Growth Rate: 1.02%
Rural/Urban Population (Ratio):
15/85
Major Languages: English;
indigenous languages
Ethnic Makeup: 92% Caucasian;
7% Asian; 1% Aboriginal and
others
Religions: 26% Anglican; 26%
Roman Catholic; 24% other
Christian; 24% unaffiliated or
other

Health

Life Expectancy at Birth: 77
years (male); 83 years (female)
Infant Mortality Rate (Ratio):
5.04/1,000
Average Caloric Intake: 118%
of FAO minimum
Physicians Available (Ratio): 1/400

Education

Adult Literacy Rate: 100%
Compulsory (Ages): 6–15; free

COMMUNICATION

Telephones: 9,628,000 main lines
Daily Newspaper Circulation: 258 per
1,000 people
Televisions: 641 per 1,000 people
Internet Service Providers: 709 (1999)

TRANSPORTATION

Highways in Miles (Kilometers): 566,060
(913,000)
Railroads in Miles (Kilometers): 20,968
(33,819)
Usable Airfields: 408
Motor Vehicles in Use: 10,705,000

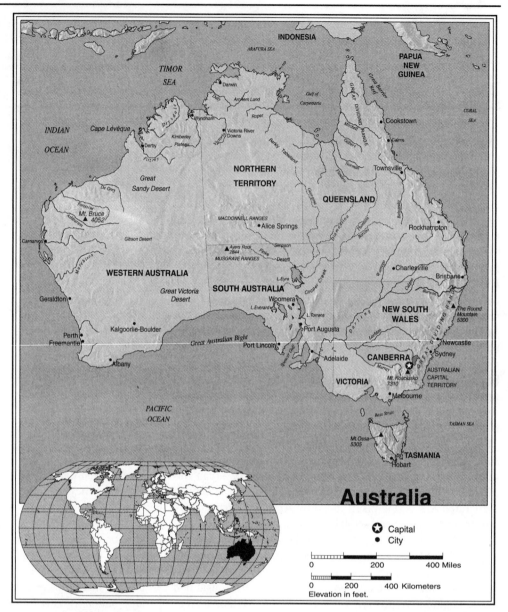

Australia

🟊 Capital
● City

0   200   400 Miles

0   200   400 Kilometers
Elevation in feet.

GOVERNMENT

Type: federal parliamentary state
Independence Date: January 1, 1901
(federation of U.K. colonies)
Head of State/Government: Queen Eliza-
beth II; Prime Minister John Howard
Political Parties: Liberal Party; National
Party; Australian Labour Party;
Australian Democratic Party; Green
Party; One Nation Party
Suffrage: universal and compulsory at 18

MILITARY

Military Expenditures (% of GDP): 1.9$
Current Disputes: territorial claim in
Antarctica

ECONOMY

Currency ($ U.S. Equivalent): 2.03
Australian dollars = $1

Per Capita Income/GDP: $22,200/$416.2
billion
GDP Growth Rate: 4.3%
Inflation Rate: 1.8%
Unemployment Rate: 7.5%
Labor Force: 8,900,000
Natural Resources: bauxite; diamonds;
coal; copper; iron ore; oil; natural gas;
other minerals
Agriculture: wheat; barley; sugarcane;
fruit; livestock
Industry: mining; industrial and transporta-
tion equipment; food processing;
chemicals; steel
Exports: $58 billion (primary partners
Japan, Europe, ASEAN)
Imports: $67 billion (primary partners
Europe, United States, Japan)

THE LAND THAT NO ONE WANTED

Despite its out-of-the-way location, far south of the main trading routes between Europe and Asia, seafarers from England, Spain, and the Netherlands began exploring parts of the continent of Australia in the seventeenth century. The French later made some forays along the coast, but it was the British who first found something to do with a land that others had disparaged as useless: They decided to send their prisoners there. The British had long believed that the easiest solution to prison overcrowding was expulsion from Britain. Convicts had been sent to the American colonies for many years, but after American independence was declared in 1776, Britain began to send prisoners to Australia.

Australia seemed like the ideal spot for a penal colony: It was isolated from the centers of civilization; it had some good harbors; and, although much of the continent was a flat, dry, riverless desert with only sparse vegetation, the coastal fringes were well suited to human habitation. Indeed, although the British did not know it in the 1700s, they had come across a huge continent endowed with abundant natural resources. Along the northern coast (just south of present-day Indonesia and New Guinea) was a tropical zone with heavy rainfall and tropical forests. The eastern coast was wooded with valuable pine trees, while the western coast was dotted with eucalyptus and acacia trees. Minerals, especially coal, gold, nickel, petroleum, iron, and bauxite, were plentiful, as were the many species of unique animals: kangaroos, platypus, and koalas, to name a few.

Today, grazing and agricultural activities generally take place in the central basin of the country, which consists of thousands of miles of rolling plains. The bulk of the population resides along the coast, where good harbors can be found, although they are relatively few in number. Many of Australia's lakes are saltwater lakes left over from an ancient inland sea, much like the Great Salt Lake in North America. Much of the interior is watered by deep artesian wells that are drilled by local ranchers, since Australia's mountain ranges are generally not high enough to proved substantial water supplies.

The British chose to build their first penal colony alongside a good harbor that they called Sydney. By the 1850s, when the practice of transporting convicts stopped, more than 150,000 prisoners, including hundreds of women, had been sent there and to other colonies. Most of them were illiterate English and Irish from the lower socioeconomic classes. Once they completed their sentences, they were set free to settle on the continent. These individuals, their guards, and gold prospectors constituted the beginning of modern Australian society. Today, despite its 19.1 million inhabitants, Australia remains a sparsely populated continent.

RACE RELATIONS

Convicts certainly did not constitute the beginning of human habitation on the continent. Tens of thousands of Aborigines (literally, "first inhabitants") inhabited Australia and the nearby island of Tasmania when Europeans first made contact. Living in scattered tribes and speaking hundreds of entirely unrelated languages, the Aborigines, whose origin is unknown (various scholars see connections to Africa, the Indian subcontinent, or the Melanesian Islands), survived by fishing and nomadic hunting. Succumbing to Euro-

(Australian Information Service photo)

Most Aborigines eventually adapted to the Europeans' customs, but some continue to live in their traditional ways on tribal reservations.

pean diseases, violence, forced removal from their lands, and, finally, neglect, thousands of Aborigines died during the first centuries of contact. Indeed, the entire Tasmanian grouping of people (originally numbering 5,000) is now extinct. Today's Aborigines continue to suffer discrimination. A 1997 bill that would have liberalized land rights for Aborigines failed in Parliament, and even after 10 years of effort, the Council on Aboriginal Reconciliation was not able to persuade the government to issue a formal apology for past abuses (some 62 percent of the population were opposed to a formal apology). Moreover, Parliament has only one Aboriginal member.

Most Aborigines eventually adopted European ways, including Christianity. Today, many live in the cities or work for cattle and sheep ranchers. Others reside on reserves (tribal reservations) in the central and northern parts of Australia. Yet modernization has affected even the reservation Aborigines—some have telephones, and some dispersed tribes in the Northern Territories communicate with one another by satellite-linked video conferencing—but in the main, they continue to live as they have always done, organizing their religion around plant or animal sacred symbols, or totems, and initiating youth into adulthood through lengthy and sometimes painful rituals.

Whereas the United States began with 13 founding colonies, Australia started with six, none of which felt a compelling need to unite into a single British nation until the 1880s, when other European powers began taking an interest in settling the continent. It was not until 1901 that Australians formally separated from Britain (while remaining within the British Commonwealth, with the Queen of England as head of state). Populated almost entirely by whites from Britain or Europe (people of European descent still constitute about 95 percent of Australia's population), Australia has maintained close cultural and diplomatic links with Britain and the West, at the expense of ties with the geographically closer nations of Asia.

Reaction against Polynesians, Chinese, and other Asian immigrants in the late 1800s produced an official "White Australia" policy, which remained intact until the 1960s and effectively excluded nonwhites from settling in Australia. During the 1960s, however, the government made an effort to relax these restrictions and to restore land and some measure of self-determination to Aborigines. In the 1990s, Aborigines successfully persuaded the federal government to block a dam project on Aboriginal land that would have de-

stroyed sacred sites. The federal government sided with the Aborigines against white developers and local government officials. In 1993, despite some public resistance, the government passed laws protecting the land claims of Aborigines and set up a fund to assist Aborigines with land purchases. Evidence of continued racism can be found, however, in such graffiti painted on walls of high-rise buildings as "Go home Japs!" (in this case, the term *Jap,* or, alternatively, *wog,* refers to any Asian, regardless of nationality). The unemployment rate of Aborigines is four times that of the nation as a whole, and there are substantially higher rates of chronic health problems and death by infectious diseases among this population.

ECONOMIC PRESSURES

Despite lingering discriminatory attitudes against nonwhites, events since World War II have forced Australians to reconsider their position, at least economically, vis-à-vis Asia and Southeast Asia. Australia has never been conquered by a foreign power (not even by Japan during World War II), but the impressive industrial strength of Japan now allows its people to enjoy higher per capita income than that of Australians, and Singapore is not far behind. Moreover, since Australia's economy is based on the export of primary goods (for example, minerals, wheat, beef, and wool) rather than the much more lucrative consumer products manufactured from raw resources, it is likely that Australia will continue to lose ground to the more economically aggressive and heavily populated Asian economies.

This inexorable alteration in socioeconomic status will be a new and difficult experience for Australians, whose standard of living has been the highest in the Pacific Rim for decades. Building on a foundation of sheep (imported in the 1830s and now supplying more than a quarter of the world's supply of wool), mining (gold was discovered in 1851), and agriculture (Australia is nearly self-sufficient in food), the country has developed its manufacturing sector such that Australians are able to enjoy a standard of living equal in most respects to that of North Americans.

But Australians are wary of the growing global tendency to create mammoth regional trading blocs, such as the North American Free Trade Association, consisting of the United States, Canada, Mexico, and others; the European Union (formerly the European Community), eventually including, perhaps, parts of Central/Eastern Europe; the ASEAN nations of Southeast Asia; and an informal "yen bloc" in Asia, headed by Japan. These blocs might ex-

clude Australian products from preferential trade treatment or eliminate them from certain markets altogether. Beginning in 1983, the Labour government of then–prime minister Robert Hawke began to establish collaborative trade agreements with Asian countries, a plan that seemed to have the support of the electorate, even though it meant reorienting Australia's foreign policy away from its traditional posture Westward.

In the early 1990s, under Labour prime minister Paul Keating, the Asianization plan intensified. The Japanese prime minister and the governor of Hong Kong visited Australia, while Australian leaders made calls on the leaders of South Korea, China, Thailand, Vietnam, Malaysia, and Laos. Trade and security agreements were signed with Singapore and Indonesia, and a national-curriculum plan was implemented whereby 60 percent of Australian schoolchildren will be studying Japanese and other Asian languages by the year 2010. The Liberal Party prime minister, John Howard, elected in 1996, has also moderated his views on Asian immigration and now advocates a nondiscriminatory immigration policy rather than the restrictive policy that he championed in the 1980s.

Despite such initiatives (and a few successes: Japan now buys more beef from Australia than from the United States), the economic threat to Australia remains. Even in the islands of the Pacific, an area that Australia and New Zealand generally have considered their own domain for economic investment and foreign aid, investments by Asian countries are beginning to winnow Australia's sphere of influence. U.S. president Bill Clinton, in a 1996 visit, promised Australian leaders that they would not be left out of the emerging economic structures of the region. Indeed, conditions have improved. The 2001 government budget projected a fourth straight year of surpluses, and included a tax cut. Still, years of recession, an unemployment rate sometimes nearing 9 percent, and 2 million people living in poverty leave many Australians concerned about their economic future.

Labor tension erupted in 1998 when dockworkers found themselves locked out of work by employers who claimed they were inefficient workers. Eventually the courts found in favor of the workers, but not until police and workers clashed and national attention was drawn to the protracted sluggish economy.

THE AMERICAN CONNECTION

By any standard, Australia is a democracy solidly embedded in the traditions of the

European exploration of the Australian coastline begins **1600s**	British explorers first land in Australia **1688**	The first shipment of English convicts arrives **1788**	A gold rush lures thousands of immigrants **1851**	Australia becomes independent within the British Commonwealth **1901**	Australia is threatened by Japan during World War II **1940s**
●	●	●	●	●	●

West. Political power is shared back and forth between the Labour Party and the Liberal–National Country Party coalition, and the Constitution is based on both British parliamentary tradition and the U.S.

model. Thus, it has followed that Australia and the United States have built a warm friendship as both political and military allies. A military mutual-assistance agreement, ANZUS (for Australia, New

Zealand, and the United States), was concluded after World War II (New Zealand withdrew in 1986). And just as it had sent troops to fight Germany during World Wars I and II, Australia sent troops to fight in the Korean War in 1950 and the Vietnam War in the 1960s—although anti–Vietnam War sentiment in Australia strained relations with the United States at that time. Australia also joined the United States and other countries in 1954 in establishing the Southeast Asia Treaty Organization, an Asian counterpart to the North Atlantic Treaty Organization designed to contain the spread of communism.

In 1991, when the Philippines refused to renew leases on U.S. military bases there, there was much discussion about transferring U.S. operations to the Cockburn Sound Naval Base in Australia. Singapore was eventually chosen for some of the operations, but the incident reveals the close relationship of the two nations. U.S. military aircraft already land in Australia, and submarines and other naval craft call at Australian ports. The Americans also use Australian territory for surveillance facilities. There is historical precedence for this level of close cooperation: Before the U.S. invasion of the Japanese-controlled Philippines in the 1940s, the United States based its Pacific-theater military headquarters in Australia; moreover, Britain's inability to lead the fight against Japan forced Australia to look to the United States.

A few Australians resent the violation of sovereignty represented by the U.S. bases, but most regard the United States as a solid ally. Indeed, many Australians regard their country as the Southern Hemisphere's version of the United States: Both countries have immense space and vast resources, both were founded as disparate colonies that eventually united and obtained independence from Britain, and both share a common language and a Western cultural heritage.

There is yet another way that some Australians would like to be similar to the United States: They want to be a republic. Polls in advance of a 1999 referendum to decide whether or not Australia should remain a constitutional monarchy, headed by the king or queen of England, showed that more than 60 percent of the population favored severing ties with England.

(San Diego Convention and Visitors Bureau)

Australia has a number of animals that are, in their native form, unique in the world. The koala is found only in the eastern coastal region, where it feeds, very selectively, on the leaves of the eucalyptus tree. It is a marsupial and bears its young every other year. Pictured above is a very rare baby albino koala with its mother.

| Australia proposes the South Pacific Commission 1947 | Australia joins New Zealand and the United States in the ANZUS military security agreement 1951 | Australia joins the South East Asian Treaty Organization 1954 | Relations with the United States are strained over the Vietnam War 1960s | The Australian Labour Party wins for the first time in 23 years; Gough Whitlam is prime minister 1972 | After a constitutional crisis, Whitlam is replaced by opposition leader J. M. Fraser 1975 | Australia begins to strengthen its economic ties with Asian countries 1980s | After 13 years in power, the Labour Party is defeated by Liberal Party leader John Howard 1990s | Australia condemns nuclear testing in the Pacific, recalls the French ambassador |

Aboriginal rights are increasingly part of the public debate in Australia

Sydney hosts the 2000 Summer Olympic Games

However, the actual vote found 55 percent in favor of the status quo—just enough to keep Queen Elizabeth II as head of state. The queen, obviously pleased with the result, visited Australia right after the vote to show her thanks to the residents of working-class and rural areas, where support for the monarchy was strongest.

Unlike New Zealand, which has distanced itself from the United States by refusing to allow nuclear-armed ships to enter its ports and has withdrawn from ANZUS, Australia has joined with the United States in attempting to dissuade South Pacific states from declaring the region a nuclear-free zone. Yet it has also maintained good ties with the small and vulnerable societies of the Pacific through its leadership in such regional associations as the South Pacific Commission, the South Pacific Forum, and the ever-more-influential Asia-Pacific Economic Cooperation group (APEC). It has also condemned nuclear-bomb testing programs in French-controlled territories.

AUSTRALIA AND THE PACIFIC

Australia was not always possessed of good intentions toward the islands around it. For one thing, white Australians thought of themselves as superior to the brown-skinned islanders; and for another, Australia preferred to use the islands' resources for its own economic gain, with little regard for the islanders themselves. At the end of World War I, for example, the phosphate-rich island of Nauru, formerly under German control, was assigned to Australia as a trust territory. Until phosphate mining was turned over to the islanders in 1967, Australian farmers consumed large quantities of the island's phosphates but paid just half the market price. Worse, only a tiny fraction of the proceeds went to the people of Nauru. Similarly, in Papua New Guinea,

Australia controlled the island without taking significant steps toward its domestic development until the 1960s, when, under the guidance of the United Nations, it did an about-face and facilitated changes that advanced the successful achievement of independence in 1975.

In addition to forgoing access to cheap resources, Australia was reluctant to relinquish control of these islands because it saw them as a shield against possible military attack. It learned this lesson well in World War II. In 1941, Japan, taking advantage of the Western powers' preoccupation with Adolf Hitler, moved quickly to expand its imperial designs in Asia and the Pacific. The Japanese first disabled the U.S. Navy by attacking its warships docked in Pearl Harbor, Hawaii. They then moved on to oust the British in Hong Kong and the Gilbert Islands, and the Americans in Guam and Wake Island. Within a few months, the Japanese had taken control of Burma, Malaya, Borneo, the Philippines, Singapore, and hundreds of tiny Pacific islands, which they used to create an immense defensive perimeter around the home islands of Japan. They also had captured part of New Guinea and were keeping a large force there, which greatly concerned the Australians. Yet fighting was kept away from Australia proper when the Japanese were successfully engaged by Australian and American troops in New Guinea. Other Pacific islands were regained from the Japanese at a tremendous cost in lives and military hardware. Japan's defeat came only when islands close enough to Japan to be attacked by U.S. bomber aircraft were finally captured. Japan surrendered in 1945, but the colonial powers had learned that possession of small islands could have strategic importance. This experience is part of the reason for colonial powers' reluctance to grant independence to the vast

array of islands over which they have exercised control. Australia is now faced with the question of whether or not to grant independence to the 4,000 inhabitants of Christmas Island who recently voted to become a self-ruling territory within Australia.

There is no doubt that stressful historical periods and events such as World War II drew the English-speaking countries of the South Pacific closer together and closer to the United States. But recent realignments in the global economic system are creating strains. When the United States insists that Japan take steps to ease the U.S.–Japan trade imbalance, Australia sometimes comes out the loser. For instance, both Australia and the United States are producers of coal. Given the nearly equal distance between those two countries and Japan, it would be logical to expect that Japan would buy coal at about the same price from both countries. In fact, however, Japan pays about $7 a ton more for American coal than for Australian coal, a discrepancy directly attributable to Japan's attempt to reduce the trade imbalance with the United States. Resentment against the United States over such matters is likely to grow, and managing such international tensions will no doubt challenge the skills of the leadership of Australia in the coming years.

DEVELOPMENT

Mining of nickel, iron ore, and other metals continues to supply a substantial part of Australia's gross domestic product. In recent years, Japan has become Australia's primary trading partner rather than Great Britain. Seven out of 10 of Australia's largest export markets are Asian countries.

FREEDOM

Australia is a parliamentary democracy adhering to the ideals incorporated in English common law. Constitutional guarantees of human rights apply to all of Australia's 19.1 million citizens. However, social discrimination continues, and, despite improvements since the 1960s, the Aborigines remain a neglected part of Australian society.

HEALTH/WELFARE

Like New Zealand, Australia has developed a complex and comprehensive system of social welfare. Education is the province of the several states. Public education is compulsory. Australia boasts several world-renowned universities. The world's first voluntary-euthanasia law passed in Northern Territory in 1996, but legal challenges have thus far prevented its use.

ACHIEVEMENTS

The vastness and challenge of Australia's interior lands, called the "outback," have inspired a number of Australian writers to create outstanding poetry and fictional novels. In 1973, Patrick White became the first Australian to win a Nobel Prize in Literature. Jill Ker Conway, Thomas Keneally, and Colleen McCullough are other well-known Australian authors.

Brunei (State of Brunei Darussalam)

GEOGRAPHY
Area in Square Miles (Kilometers):
2,228 (5,770) (about the size
of Delaware)
Capital (Population): Bandar
Seri Begawan (187,000)
Environmental Concerns: water
pollution; seasonals moke/haze
resulting from forest fires in
Indonesia
Geographical Features: flat
coastal plain rises to mountains
in east; hilly lowlands in west
Climate: tropical; hot, humid,
rainy

PEOPLE

Population
Total: 336,400
Annual Growth Rate: 2.17%
Rural/Urban Population (Ratio):
30/70
Major Languages: Malay;
English; Chinese; Iban;
native dialects
Ethnic Makeup: 64% Malay;
20% Chinese; 16% others
Religions: 67% Muslim; 13%
Buddhist; 10% Christian; 10%
indigenous beliefs and others

Health
Life Expectancy at Birth: 71
years (male); 76 years (female)
Infant Mortality Rate (Ratio):
14.8/1,000
Physicians Available (Ratio):
1/1,398

Education
Adult Literacy Rate: 88%
Compulsory (Ages): 5–17; free

COMMUNICATION
Telephones: 70,700 main lines
Daily Newspaper Circulation:
70 per 1,000 people
Televisions: 308 per 1,000 people
Internet Service Provider: 1 (1999)

TRANSPORTATION
Highways in Miles (Kilometers): 676 (1,120)
Railroads in Miles (Kilometers): 8 (13)
Usable Airfields: 2
Motor Vehicles in Use: 165,000

GOVERNMENT
Type: constitutional sultanate (monarchy)
Independence Date: January 1, 1984
(from the United Kingdom)
Head of State/Government: Sultan and
Prime Minister His Majesty Paduka
Seri Baginda Sultan Haji Hassanal
Bolkiah Mu'izzaddin Waddaulah is both
head of state and head of government

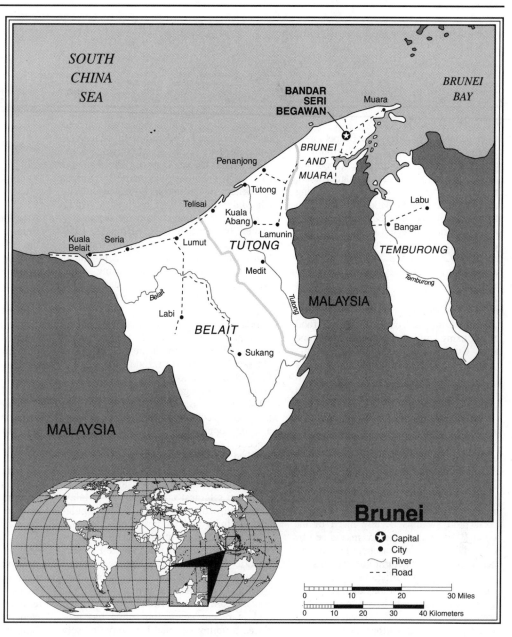

Political Parties: Brunei Solidarity
National Party (the only legal party);
Brunei People's Party (banned); Brunei
National Democratic Party (deregistered)
Suffrage: none

MILITARY
Military Expenditures (% of GDP): 5.1%
Current Disputes: possible dispute over
the Spratley Islands

ECONOMY
Currency ($ U.S. Equivalent): 1.76
Bruneian dollars = $1
Per Capita Income/GDP: $17,400/$5.6 billion
GDP Growth Rate: 2.5%
Inflation Rate: 1%

Unemployment Rate: 4.9%
Labor Force: 144,000 (including military)
Natural Resources: petroleum; natural gas;
timber
Agriculture: rice; cassava (tapioca); ba-
nanas; water buffalo
Industry: petroleum; natural gas; contruction
Exports: $2.04 billion (primary partners Ja-
pan, United Kingdom, United States)
Imports: $1.3 billion (primary partners Sin-
gapore, United Kingdom, Malaysia,
France)

http://www.cia.gov/cia/publications/
factbook/geos/bx.html
http://www.brunet.bn
http://www.brunei.gov.bn/index.html

Brunei is first
visited by
Europeans
A.D. 1521

Brunei is known
as haven for
pirates
1700

Briton James
Brooke is given
Sarawak as
reward for help in
a civil war
1800s

The island of
Labuan is ceded
to Britain
1847

Britain attacks
and ends pirate
activities in Brunei
1849

The remainder of
Brunei becomes
a British
protectorate
1888

Brunei rejects
confederation
with Malaysia
1963

Brunei gains its
independence
1984

Foreign workers
are "imported" to
ease the labor
shortage; Brunei
joins the
International
Monetary Fund
1990s

2000s

The sultan's brother agrees
to return billions of dollars
in stolen state assets

Brunei declares itself
a nuclear-free zone

A WEALTHY COUNTRY

Home to only 336,400 people, Brunei rarely captures the headlines. But perhaps it should, for despite its tiny size (about the size of Delaware or Prince Edward Island), the country boasts one of the highest living standards in the world. Moreover, the sultan of Brunei, with assets of $37 billion, is considered the richest person in the world. The secret? Oil. First exploited in Brunei in the late 1920s, today petroleum and natural gas almost entirely support the sultanate's economy. The government's annual income is nearly twice its expenditures, despite the provision of free education and medical care, subsidized food and housing, and the absence of income taxes. Currently, Brunei is in the middle of a five-year plan designed to diversify its economy, but 98 percent of the nation's revenues continue to derive from the sale of oil and natural gas.

Muslim sultans ruled over the entire island of Borneo and other nearby islands during the sixteenth century. Advantageously located on the northwest coast of the island of Borneo, along the sea lanes of the South China Sea, Brunei was a popular resting spot for traders. During the 1700s, it became known as a haven for pirates. Tropical rain forests and swamps occupy much of the country—conditions that are maintained by heavy monsoon rains for about five months each year. Oil and natural-gas deposits are found both on- and offshore.

In the 1800s, the sultan then in power agreed to the kingdom becoming a protectorate of Britain, in order to safeguard his domain from being further whittled away by aggressors bent on empire-building. The Japanese easily overtook Brunei in 1941, when they launched their Southeast Asian offensive in search of oil and gas for their war machine. Today, the Japanese Mitsubishi Corporation has a one-third interest in the Brunei gas company.

In the 1960s, it was expected that Brunei, which is cut in two and surrounded on three sides by Malaysia, would join the newly proposed Federation of Malaysia; but it refused to do so, preferring to remain under British control. The decision to remain a colony was made by Sultan Sir Omar Ali Saifuddin. Educated in British Malaya, the sultan retained a strong affection for British culture and frequently visited the British Isles. (Brunei's 1959 Constitution, promulgated during Sir Omar's reign, reflected this attachment: It declared Brunei a self-governing state, with its foreign affairs and defense remaining the responsibility of Great Britain.)

In 1967, Sir Omar abdicated in favor of his son, who became the 29th ruler in succession. Sultan (and Prime Minister) Sir Hassanal Bolkiah Mu'izzaddin Waddaulah (a shortened version of his name) oversaw Brunei's gaining of independence, in 1984. Not all Bruneians are pleased with the sultan's control over the political process, but opposition voices have been silenced. There are, in effect, no operative political parties in Brunei, and there have been no elections in the country since 1965, despite a constitutional provision for them.

Brunei's largest ethnic group is Malay, accounting for 64 percent of the population. Indians and Chinese constitute sizable minorities, as do indigenous peoples such as Ibans and Dyaks. Despite Brunei's historic ties with Britain, Europeans make up only a tiny fraction of the population.

Brunei is an Islamic nation with Hindu roots. Islam is the official state religion, and in recent years, the sultan has proposed bringing national laws more closely in line with Islamic ideology. Modern Brunei is officially a constitutional monarchy, headed by the sultan, a chief minister, and a Council; in reality, however, the sultan and his family control all aspects of state decision making. The extent of the sultan's control of the government is revealed by his multiple titles: in addition to sultan, he is Brunei's prime minister, minister of defense, and minister of finance. The Constitution provides the sultan with supreme executive authority in the state. The concentration of near-absolute power in the hands of one family has produced some unsavory results. In 2000, the sultan's brother agreed to return billions of dollars of assets that he had stolen from the state while heading Brunei's overseas investment company. Court documents in the $15 billion lawsuit against him claimed that he had personally consumed $2.7 billion (!) to support his lavish lifestyle.

Brunei, along with several other Southeast Asian nations, claims ownership of the Spratley Islands. In 1999, despite China's refusal to join in, Brunei agreed with the other disputants to moderation in its pursuit of access to the oil-rich isles in the South China Sea. In other international matters, Brunei joined with other ASEAN nations in 1995 to declare its country to be a nuclear-free zone.

In recent years, Brunei has been plagued by a chronic labor shortage. The government and Brunei Shell (a consortium owned jointly by the Brunei government and Shell Oil) are the largest employers in the country. They provide generous fringe benefits and high pay. Non-oil private-sector companies with fewer resources find it difficult to recruit within the country and have, therefore, employed many foreign workers. Indeed, one third of all workers today in Brunei are foreigners. This situation is of considerable concern to the government, which is worried that social tensions between foreigners and residents may flare up at any time.

DEVELOPMENT

Brunei's economy is a mixture of the modern and the ancient: foreign and domestic entrepreneurship, government regulation and welfare statism, and village tradition. Chronic labor shortages are managed by the importation of thousands of foreign workers.

FREEDOM

Although Islam is the official state religion, the government practices religious tolerance. The Constitution provides the sultan with supreme executive authority, which he has used to suppress opposition groups and political parties.

HEALTH/WELFARE

The country's massive oil and natural-gas revenues support wide-ranging benefits to the population, such as subsidized food, fuel, and housing, and free medical care and education. This distribution of wealth is reflected in Brunei's generally favorable quality-of-life indicators.

ACHIEVEMENTS

An important project has been the construction of a modern university accommodating 1,500 to 2,000 students. Since independence, the government has tried to strengthen and improve the economic, social, and cultural life of its people.

Cambodia (Kingdom of Cambodia)

GEOGRAPHY
Area in Square Miles (Kilometers): 69,881 (181,040) (about the size of Oklahoma)
Capital (Population): Phnom Penh (920,000)
Environmental Concerns: habitat loss and declining biodiversity; soil erosion; deforestation; lack of access to potable water
Geographical Features: mostly low, flat plains; mountains in the southwest and north
Climate: tropical; rainy and dry seasons

PEOPLE

Population
Total: 12,212,300
Annual Growth Rate: 2.27%
Rural/Urban Population (Ratio): 79/21
Major Languages: Khmer; French; English
Ethnic Makeup: 90% Khmer (Cambodian); 5% Chinese; 5% others
Religions: 95% Theravada Buddhist; 5% others

Health
Life Expectancy at Birth: 54 years (male); 59 years (female)
Infant Mortality Rate (Ratio): 66.8/1,000
Average Caloric Intake: 85% of FAO minimum
Physicians Available (Ratio): 1/7,900

Education
Adult Literacy Rate: 35%
Compulsory Ages: 6–12

COMMUNICATION
Telephones: 24,300 main lines
Televisions: 8 per 1,000 people
Internet Service Providers: 2

TRANSPORTATION
Highways in Miles (Kilometers): 21,461 (35,769)
Railroads in Miles (Kilometers): 365 (603)
Usable Airfields: 19
Motor Vehicles in Use: 30,000

GOVERNMENT
Type: liberal democracy under a constitutional monarchy
Independence Date: November 9, 1953 (from France)
Head of State/Government: King Norodom Sihanouk; Prime Minister (Premier) Hun Sen

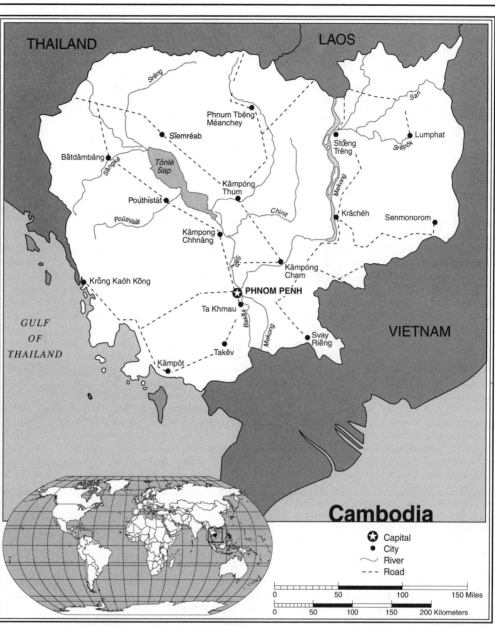

Cambodia

Capital
City
River
Road

0   50   100   150 Miles
0   50   100   150   200 Kilometers

Political Parties: National United Front ... (Funcinpec); Cambodian People's Party; Buddhist Liberal Party; Democratic Kampuchea (Khmer Rouge); Khmer Citizen Party; Cambodian People's Party; Sam Rangsi Party
Suffrage: universal at 18

MILITARY
Military Expenditures (% of GDP): 2.4%
Current Disputes: border disputes with Thailand and Vietnam

ECONOMY
Currency ($ U.S. Equivalent): 3,810 riels = $1
Per Capita Income/GDP: $710/$8.2 billion
GDP Growth Rate: 4%
Inflation Rate: 4.5%
Unemployment Rate: 2.8%
Labor Force: 6,000,000
Natural Resources: timber; gemstones; iron ore; manganese; phosphates; hydropower potential
Agriculture: rice; rubber; corn; vegetables
Industry: rice processing; fishing; wood and wood products; rubber; cement; gem mining; textiles
Exports: $821 million (primary partners United States, Singapore, Japan)
Imports: $1.2 billion (primary partners Singapore, Vietnam, Japan)

http://www.cambodia-web.net
http://www.cia.gov/cia/publications/factbook/geos/cb.html

A LAND OF TRAGEDY

In Khmer (Cambodian), the word *Kampuchea,* which for a time during the 1980s was the official name of Cambodia, means "country where gold lies at the foothill." Always an agricultural economy, Cambodia's monsoon rains contribute to the numerous large rivers that sustain farming on the country's large central plain and produce tropical forests consisting of such useful trees as coconuts, palms, bananas, and rubber. But Cambodia is certainly not a land of gold, nor of food, freedom, or stability. Despite a new Constitution, massive United Nations aid, and a formal cease-fire, the horrific effects of Cambodia's bloody Civil War continue.

Cambodia was not always a place to be pitied. In fact, at times it was the dominant power of Southeast Asia. Around the fourth century A.D., India, with its pacifist Hindu ideology, began to influence in earnest the original Chinese base of Cambodian civilization. The Indian script came to be used, the name of its capital city was an Indian word, its kings acquired Indian titles, and many of its Khmer people believed in the Hindu religion. The mile-square Hindu temple Angkor Wat, built in the twelfth century, still stands as a symbolic reminder of Indian influence, Khmer ingenuity, and the Khmer Empire's glory.

But the Khmer Empire, which at its height included parts of present-day Myanmar (Burma), Thailand, Laos, and Vietnam, was gradually reduced both in size and power until, in the 1800s, it was paying tribute to both Thailand and Vietnam. Continuing threats from these two countries as well as wars and domestic unrest at home led the king of Cambodia to appeal to France for help. France, eager to offset British power in the region, was all too willing to help. A protectorate was established in 1863, and French power grew apace until, in 1887, Cambodia became a part of French Indochina, a conglomerate consisting of the countries of Laos, Vietnam, and Cambodia.

The Japanese temporarily evicted the French in 1945 and, while under Japanese control, Cambodia declared its "independence" from France. Heading the country was the young King Norodom Sihanouk. Controlling rival ideological factions, some of which were pro-West while others were pro-Communist, was difficult for Sihanouk, but he built unity around the idea of permanently expelling the French, who finally left in 1955. King Sihanouk then abdicated his throne in favor of his father so that he could, as premier (prime minister), personally enmesh himself in political governance. He took the title Prince Sihanouk, by which he is known to most

people today, although in 1993 he declared himself, once again, king of Cambodia.

From the beginning, Sihanouk's government was bedeviled by border disputes with Thailand and Vietnam and by the incursion of Communist Vietnamese soldiers into Cambodia. Sihanouk's ideological allegiances were (and remain) confusing at best; but, to his credit, he was able to keep Cambodia officially out of the Vietnam War, which raged for years (1950–1975) on its border. In 1962, Sihanouk announced that his country would remain neutral in the cold war struggle.

Neutrality, however, was not seen as a virtue by the United States, whose people were becoming more and more eager either to win or to quit the war with North Vietnam. A particularly galling point for the U.S. military was the existence of the so-called Ho Chi Minh Trail, a supply route through the tropical mountain forests of Cambodia. For years, North Vietnam had been using the route to supply its military operations in South Vietnam, and Cambodia's neutrality prevented the United States, at least legally, from taking military action against the supply line.

All this changed in 1970, when Sihanouk, out of the country at the time, was evicted from office by his prime minister, General Lon Nol, who was supported by the United States and South Vietnam. Shortly thereafter, the United States, then at its peak of involvement in the Vietnam War, began extensive military action in Cambodia. The years of official neutrality came to a bloody end.

THE KILLING FIELDS

Most of these international political intrigues were lost on the bulk of the Cambodian population, only half of whom could read and write, and almost all of whom survived, as their forebears had before them, by cultivating rice along the Mekong River Valley. The country had almost always been poor, so villagers had long since learned that, even in the face of war, they could survive by hard work and reliance on extended-family networks. Most farmers probably thought that the war next door would not seriously alter their lives. But they were profoundly wrong, for, just as the United States had an interest in having a pro–U.S. government in Cambodia, the North Vietnamese desperately wanted Cambodia to be pro-Communist.

North Vietnam wanted the Cambodian government to be controlled by the Khmer Rouge, a Communist guerrilla army led by Pol Pot, one of a group of former students influenced by the left-wing ideology

taught in Paris universities during the 1950s. Winning control in 1975, the Khmer Rouge launched a hellish 3½-year extermination policy, resulting in the deaths of between 1 million and 3 million fellow Cambodians—that is, between one fifth and one third of the entire Cambodian population. The official goal was to eliminate anyone who had been "polluted" by prerevolutionary thinking, but what actually happened was random violence, torture, and murder.

It is impossible to fully describe the mayhem and despair that engulfed Cambodia during those years. Cities were emptied of people. Teachers and doctors were killed or sent as slaves to work in the rice paddies. Despite the centrality of Buddhism to Cambodian culture (Hinduism having long since been displaced by Buddhist thought), thousands of Buddhist monks were killed or died of starvation as the Khmer Rouge carried out its program of eliminating religion. Some people were killed for no other reason than to terrorize others into submission. Explained Leo Kuper in *International Action Against Genocide*:

> Those who were dissatisfied with the new regime were ... "eradicated," along with their families, by disembowelment, by beating to death with hoes, by hammering nails into the backs of their heads and by other cruel means of economizing on bullets.
>
> Persons associated with the previous regime were special targets for liquidation. In many cases, the executions included wives and children. There were summary executions too of intellectuals, such as doctors, engineers, professors, teachers and students, leaving the country denuded of professional skills.

The Khmer Rouge wanted to alter the society completely. Children were removed from their families, and private ownership of property was eliminated. Money was outlawed. Even the calendar was started over, at year 0. Vietnamese military leader Bui Tin explained just how totalitarian the rulers were:

> [In 1979] there was no small piece of soap or handkerchief anywhere. Any person who had tried to use a toothbrush was considered bourgeois and punished. Any person wearing glasses was considered an intellectual who must be punished.

It is estimated that before the Khmer Rouge came to power in 1975, Cambodia

| France gains control of Cambodia
A.D. 1863 | Japanese invasion; King Norodom Sihanouk is installed
1940s | Sihanouk wins Cambodia's independence of France
1953 | General Lon Nol takes power in a U.S.–supported coup
1970 | The Khmer Rouge, under Pol Pot, overthrows the government and begins a reign of terror
1975 | Vietnam invades Cambodia and installs a puppet government
1978 | Vietnam withdraws troops from Cambodia; a Paris cease-fire agreement is violated by the Khmer Rouge; Pol Pot dies and other Khmer Rouge leaders surrender
1990s |

2000s

Hun Sen is reelected

Prince Ranariddh becomes president of the National Assembly

had 1,200 engineers, 21,000 teachers, and 500 doctors. After the purges, the country was left with only 20 engineers, 3,000 teachers, and 54 doctors.

A kind of bitter relief came in late 1978, when Vietnamese troops (traditionally Cambodia's enemy) invaded Cambodia, drove the Khmer Rouge to the borders of Thailand, and installed a puppet government headed by Hun Sen, a former Khmer Rouge soldier who defected and fled to Vietnam in the 1970s. Although almost everyone was relieved to see the Khmer Rouge pushed out of power, the Vietnamese intervention was almost universally condemned by other nations. This was because the Vietnamese were taking advantage of the chaos in Cambodia to further their aim of creating a federated state of Vietnam, Laos, and Cambodia. Its virtual annexation of Cambodia eliminated Cambodia as a buffer state between Vietnam and Thailand, destabilizing the relations of the region even more.

COALITION GOVERNANCE

The United States and others refused to recognize the Vietnam-installed regime, instead granting recognition to the Coalition Government of Democratic Kampuchea. This entity consisted of three groups: the Communist Khmer Rouge, led by Khieu Samphan and Pol Pot and backed by China; the anti-Communist Khmer People's National Liberation Front, led by former prime minister Son Sann; and the Armee Nationale Sihanoukiste, led by Sihanouk. Although it was doubtful that these former enemies could constitute a workable government for Cambodia, the United Nations granted its Cambodia seat to the coalition and withheld its support from the Hun Sen government.

Vietnam had hoped that its capture of Cambodia would be easy and painless. Instead, the Khmer Rouge and others resisted so much that Vietnam had to send in 200,000 troops, of which 25,000 died. Moreover, other countries, including the United States and Japan, strengthened their resolve to isolate Vietnam in terms of international trade and development financing. After 10 years, the costs to Vietnam of remaining in Cambodia were so great that Vietnam announced it would pull out its troops.

A 1992 diplomatic breakthrough allowed the United Nations to establish a peacekeeping force in the country of some 22,000 troops, including Japanese soldiers—the first Japanese military presence outside Japan since World War II. These troops were to keep the tenacious Khmer Rouge faction under control. The agreement, signed in Paris by 17 nations, called for the release of political prisoners; inspections of prisons; and voter registration for national elections, to be held in 1993. Most important, the warring factions, consisting of some 200,000 troops, including 25,000 Khmer Rouge troops, agreed to disarm under UN supervision.

Unfortunately, the Khmer Rouge, although a signatory to the agreement, refused to abide by its provisions. With revenues gained from illegal trading in lumber and gems with Thailand, it launched new attacks on villages, trains, and even the UN peacekeepers themselves, and it refused to participate in the elections of 1993, although it had been offered a role in the new government if it would cooperate.

Despite a violent campaign, 90 percent of those eligible voted in elections that, after some confusion, resulted in a new Constitution; the reenthronement of Sihanouk as king; and the appointment of Sihanouk's son, Prince Norodom Ranariddh of the Royalist Party, as first prime minister and Hun Sen of the Cambodian People's Party as second prime minister.

The new Parliament outlawed the Khmer Rouge, but relations between the two premiers was rocky at best. Both began negotiating separately with the Khmer Rouge to entice them to lay down arms in return for amnesty. Thousands accepted the offer, fatally weakening the rebel army. But Hun Sen soon claimed that Norodum Ranariddh was recruiting soldiers for a future coup attempt. Hun Sen deposed the prince in a bloody coup of his own. However, pardoned by his father, the king, the prince was eventually allowed to return to Cambodia, where he became head of the National Assembly.

International attempts to bring the aging leadership of the Khmer Rouge to justice have been frustrated by Premier Hun Sen, who apparently has persuaded many of the leaders to lay down arms in return for amnesty. Notorious leaders such as Khieu Samphan and Nuon Chea had surrendered by the late 1990s, and Ta Mok had been captured. The most infamous of all, Pol Pot, had died in 1998. The United Nations and the United States wanted the top 12 leaders to be tried for genocide by an international tibunal, but Hun Sen is still vacillating on the issue. In the midst of these negotiations, it was reported that some 1,000 hours of film footage documenting the genocide had been stolen, jeopardizing any future legal efforts to convict Khmer Rouge leaders.

DEVELOPMENT

In the past, China, the United States, and others built roads and industries in Cambodia, but the country remains an impoverished state whose economy rests on fishing and farming. Continual warfare for 2 decades has prevented industrial development. The economy is sustained primarily by massive foreign aid.

FREEDOM

Few Cambodians can remember political stability, much less political freedom. Every form of human-rights violation has been practiced in Cambodia since even before the arrival of the barbaric Khmer Rouge. Suppression of dissent continues: Journalists have been killed, and opponents of the government— including the king's brother—have been expelled from the country.

HEALTH/WELFARE

Almost all of Cambodia's doctors were killed or died during the Khmer Rouge regime, and warfare disrupted normal agriculture. Thus, disease was rampant, as was malnutrition. The few trained international relief workers remaining in Cambodia today are hard-pressed to make a dent in the country's enormous problems.

ACHIEVEMENTS

Despite violence and intimidation, 90% of the Cambodian people voted in the 1993 election. Successful elections were also held in 1998. These elections restored an elected government, a limited monarchy, and a new Constitution. The government has finally eliminated the Khmer Rouge, and stability is returning to the country.

China (People's Republic of China)

GEOGRAPHY

Area in Square Miles (Kilometers):
3,723,000 (9,596,960)
(slightly smaller than the
United States)

Capital (Population): Beijing
(11,299,000)

Environmental Concerns: air and
water pollution; water shortages; desertification; trade in
endangered species; acid rain;
loss of agricultural land;
deforestation

Geographical Features: mostly
mountains, high plateaus, and
deserts in the west; plains,
deltas, and hills in the east

Climate: extremely diverse;
tropical to subarctic

PEOPLE

Population

Total: 1,262,000,000
Annual Growth Rate: 0.9%
Rural/Urban Population Ratio:
71/29

Major Languages: Standard
Chinese (Putonghua) or
Mandarin; Yue (Cantonese);
Wu (Shanghainese); Minbei
(Fuzhou); Minuan (Hokkien-
Taiwanese); Xiang; Gan;
Hahka

Ethnic Makeup: 92% Han
Chinese; 8% minority groups
(the largest being Chuang,
Hui, Uighur, Yi, and Miao)

Religions: officially atheist; but
Taoism, Buddhism, Islam,
Christianity, ancestor worship,
and animism exist

Health

Life Expectancy at Birth: 70 years (male);
73 years (female)
Infant Mortality Rate (Ratio): 28.9/1,000
Physicians Available (Ratio): 1/628

Education

Adult Literacy Rate: 81.5%
Compulsory (Ages): 7–17

COMMUNICATION

Telephones: 85,020,000 main lines
Daily Newspaper Circulation: 23 per
1,000 people
Televisions: 189 per 1,000 people
Internet Service Providers: 3 (1999)

TRANSPORTATION

Highways in Miles (Kilometers): 726,000
(1,210,000)
Railroads in Miles (Kilometers): 39,390 (65,650)
Usable Airfields: 206
Motor Vehicles in Use: 11,450,000

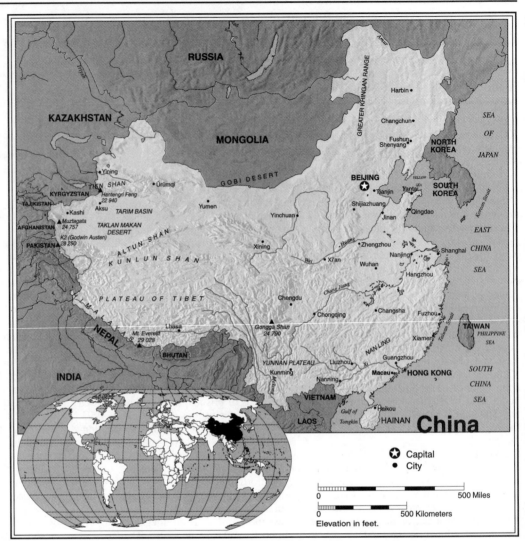

China

● Capital
● City

0 — 500 Miles
0 — 500 Kilometers
Elevation in feet.

GOVERNMENT

Type: one-party Communist state
Independence Date: unification in 221
B.C.; People's Republic established
October 1, 1949
Head of State/Government: President
Jiang Zemin; Premier Zhu Rongji
Political Parties: Chinese Communist
Party; eight registered small parties
controlled by the CCP
Suffrage: universal at 18

MILITARY

Military Expenditures (% of GDP): 1.2%
Current Disputes: boundary disputes with
India, Russia, North Korea, others

ECONOMY

Currency ($ U.S. Equivalent): 8.27 yuan = $1
Per Capita Income/GDP: $3,800/$4.8 trillion
GDP Growth Rate: 7%
Inflation Rate: –1.3%

Unemployment Rate: about 10% in urban
areas
Labor Force: 700,000,000
Natural Resources: coal; petroleum; iron
ore; tin; tungsten; antimony; lead; zinc;
vanadium; magnetite; uranium; hydropower
Agriculture: food grains; cotton; oilseed;
pork; fish; tea; potatoes; peanuts
Industry: iron and steel; coal; machinery;
light industry; textiles and apparel; food
processing; consumer durables and electronics; telecommunications; armaments
Exports: $194.9 billion (primary partners
United States, Hong Kong, Japan)
Imports: $165.8 billion (primary partners
Japan, United States, Taiwan)

 http://www.cia.gov/cia/publications/
factbook/geos/ch.html
http://www.china-embassy.org/
eng/index.html

CHINA

The first important characteristic to note about China is its age. Human civilization appeared in China as early as 20,000 years ago, and the first documented Chinese dynasty, the Shang, began about 1523 B.C. Unproven legends suggest the existence of an even earlier Chinese dynasty (about 2000 B.C.), making China one of the oldest societies with a continuing cultural identity. Over the centuries of years of documented history, the Chinese people have been ruled by a dozen imperial dynasties; have enjoyed hundreds of of stability and amazing cultural progress; and have endured more centuries of chaos, military mayhem, and hunger. Yet China and the Chinese people remain intact—a strong testament to the tenacity of human culture.

A second major characteristic is that the People's Republic of China (P.R.C.) is very big. It is the fourth-largest country in the world, accounting for 6.5 percent of the world's landmass. Much of China—about 40 percent—is mountainous; but large, fertile plains have been created by the country's numerous rivers, most of which flow toward the Pacific Ocean. China is blessed with substantial reserves of oil, minerals, and many other natural resources. Its large size and geopolitical location—it is bordered by Russia, Kazakhstan, Pakistan, India, Nepal, Bhutan, Myanmar, Laos, Vietnam, North Korea, and Mongolia—have caused the Chinese people over the centuries to think of their land as the "Middle Kingdom": that is, the center of world civilization.

However, its unwieldy size has been the undoing of numerous emperors who found it impossible to maintain its borders in the face of outside "barbarians" determined to possess the riches of Chinese civilization. During the Ch'in Dynasty (221–207 B.C.), a 1,500-mile-long, 25-foot-high wall, the so-called Great Wall, was erected along the northern border of China, in the futile hope that invasions from the north could be stopped. Although most of China's national boundaries are now recognized by international law, recent Chinese governments have found it necessary to "pacify" border areas by settling as many Han Chinese there as possible (for example, in Tibet), to prevent secession by China's numerous ethnic minorities.

Another important characteristic of modern China is its huge population. With 1.2 billion people, China is home to about 20 percent of all human beings alive today. About 92 percent of China's people are Han, or ethnic, Chinese; the remaining 8 percent are divided into more than 50 separate minority groups. Many of these ethnic groups speak mutually unintelligible languages, and although they often appear to be Chinese, they derive from entirely different cultural roots; some are Muslims, some are Buddhists, some are animists. As one moves away from the center of Chinese civilization and toward the western provinces, the influence of the minorities increases. The Chinese government has accepted the reality of ethnic influence and has granted a degree of limited autonomy to some provinces with heavy populations of minorities.

One glaring exception to this is Tibet. A land of rugged beauty north of Nepal, India, and Bhutan, Tibet was forcefully annexed by China in 1959. Many Tibetans regard the spiritual leader the Dalai Lama, not the Chinese government, as their true leader, but the Dalai Lama and thousands of other Tibetans live in exile in India, reducing the percentage of Tibetans in Tibet to only 44 percent. China suppresses any dissent against its rule in Tibet (even photos of the Dalai Lama, for example, cannot be displayed in public) and has diluted Tibetan culture by resettling thousands of Han Chinese in the region. Whereas the world community generally would act to protect Taiwan from an aggressive Chinese takeover, it has done very little to protest Chinas's takeover of Tibet.

In the 1950s, Chairman Mao Zedong encouraged couples to have many children, but this policy was reversed in the 1970s, when a formal birth-control program was inaugurated. Urban couples today are permitted to have only one child and are penalized if they have more. Penalties include expulsion from the Chinese Communist Party (CCP), dismissal from work, or a 10 percent reduction in pay for up to 14 years after the birth of the second child. The policy is strictly enforced in the cities, but it has had only a marginal impact on overall population growth because three quarters of China's people live in rural areas, where they are allowed more children in order to help with the farmwork. In the city of Shanghai, with a population of about 13.5 million people, authorities have recently removed second-child privileges for farmers living near the city and for such former exceptional cases as children of revolutionary martyrs and workers in the oil industry. Despite these and other restrictions, it is estimated that 15 million to 17 million people are now born each year in China.

Over the centuries, millions of people have found it necessary or prudent to leave China in search of food, political stability, or economic opportunity. Those who emigrated a thousand or more years ago are now fully assimilated into the cultures of Southeast Asia and elsewhere and identify themselves accordingly. More recent émigrés (in the past 200 years or so), however, constitute visible, often wealthy, minorities in their new host countries, where they have become the backbone of the business community. Ethnic Chinese constitute the majority of the population in Singapore and a sizable minority in Malaysia. Important world figures such as Corazon Aquino, the former president of the Philippines, and Goh Chok Tong, the prime minister of Singapore, are part or full Chinese. The Chinese constituted the first big wave of the 6.5 million Asian Americans to call the United States home. Large numbers of Hong Kong Chinese emigrated to Canada in the mid-1990s. Thus the influence of China continues to spread far beyond its borders, due to the influence of what are called "overseas Chinese."

Another crucial characteristic of China is its history of imperial and totalitarian rule. Except for a few years in the early 1990s, China has been controlled by imperial decree, military order, and patriarchal privilege. Confucius taught that a person must be as loyal to the government as a son should be to his father. Following Confucius by a generation or two was Shang Yang, of a school of governmental philosophy called Legalism, which advocated unbending force and punishment against wayward subjects. Compassion and pity were not considered qualities of good government.

Mao Zedong, building on this heritage as well as that of the Soviet Union's Joseph Stalin and Vladimir Lenin, exercised strict control over both the public and private lives of the Chinese people. Dissidents were summarily executed (generally people were considered guilty once they were arrested), the press was strictly controlled, and recalcitrants were forced to undergo "reeducation" to correct their thinking. Religion of any kind was suppressed, and churches were turned into warehouses. It is estimated that, during the first three years of CCP rule, more than 1 million opponents of Mao's regime were executed. During the Cultural Revolution (1966–1976), Mao, who apparently thought that a new mini-revolution in China might restore his eroding stature in the Chinese Communist Party, encouraged young people to report to the authorities anyone suspected of owning books from the West or having contact with Westerners. Even party functionaries were purged if it were believed that their thinking had been corrupted by Western influences.

(UN photo/John Issac)

In China today, urban couples are permitted to have only one child, and they can be severely penalized if they dare to have a second or if they marry before the legal ages of 22 for men and 20 for women.

THE TEACHINGS OF CONFUCIUS

Confucius (550–478 B.C.) was a Chinese intellectual and minor political figure. He was not a religious leader, nor did he ever claim divinity for himself or divine inspiration for his ideas. As the feudalism of his era began to collapse, he proposed that society could best be governed by paternalistic kings who set good examples. Especially important to a stable society, he taught, were respect and reverence for one's elders. Within the five key relationships of society (ruler and subject, husband and wife, father and son, elder brother and younger brother, and friend and friend), people should always behave with integrity, propriety, and goodness.

The writings of Confucius—or, rather, the works written about him by his followers and entitled the *Analects*—eventually became required knowledge for anyone in China claiming to be an educated person. However, rival ideas such as

Legalism were at times more popular with the elite; at one point, 460 scholars were buried alive for teaching Confucianism. Nevertheless, much of the hierarchical nature of Asian culture today can be traced to Confucian ideas.

ORIGINS OF THE MODERN STATE

Historically, authoritarian rule in China has been occasioned, in part, by China's mammoth size; by its unwieldy population; and by the ideology of some of its intellectuals. The modern Chinese state has arisen from these same pressures as well as some new ones. It is to these that we now turn.

The Chinese had traded with such non-Asian peoples as the Arabs and Persians for hundreds of years before European contact. But in the 1700s and 1800s, the British and others extracted something new from China in exchange for merchandise from the West: the permission for for-

eign citizens to live in parts of China without being subject to Chinese authority. Through this process of granting extraterritoriality to foreign powers, China slowly began to lose control of its sovereignty. The age of European expansion was not, of course, the first time in China's long history that its ability to rule itself was challenged; the armies of Kublai Khan successfully captured the Chinese throne in the 1200s, as did the Manchurians in the 1600s. But these outsiders, especially the Manchurians, were willing to rule China on-site and to imbibe as much Chinese culture as they could. Eventually they became indistinguishable from the Chinese.

The European powers, on the other hand, preferred to rule China (or, rather, parts of it) from afar as a vassal state, with the proceeds of conquest being drained away from China to enrich the coffers of the European monarchs. Aggression against Chinese sovereignty increased in 1843, when the British forced China to cede Hong Kong Island. Britain, France, and the United States all extracted unequal treaties from the Chinese that gave them privileged access to trade and ports along the eastern coast. By the late 1800s, Russia was in control of much of Manchuria, Germany and France had wrested special economic privileges from the ever-weakening Chinese government, and Portugal had long since controlled Macau. Further affecting the Chinese economy was the loss of many of its former tributary states in Southeast Asia. China lost Vietnam to France, Burma (today called Myanmar) to Britain, and Korea to Japan. During the violent Boxer Rebellion of 1900, the Chinese people showed how frustrated they were with the declining fortunes of their country.

Thus weakened internally and embarrassed internationally, the Manchu rulers of China began to initiate reforms that would strengthen their ability to compete with the Western and Japanese powers. A constitutional monarchy was proposed by the Manchu authorities but was preempted by the republican revolutionary movement of Western-trained Sun Yat-sen. Sun and his armies wanted an end to imperial rule; their dreams were realized in 1912, when Sun's Kuomintang (Nationalist Party, or KMT) took control of the new Republic of China.

Sun's Western-style approach to government was received with skepticism by many Chinese who distrusted the Western European model and preferred the thinking of Karl Marx and the philosophy of the Soviet Union. In 1921, Mao Zedong and others organized the Soviet-style Chinese Communist Party (CCP), which grew quickly and began to be seen as an alternative to the Kuomintang. After Sun's death, in 1925, Chiang Kai-shek assumed control of the Kuomintang and waged a campaign to rid the country of Communist influence. Although Mao and Chiang cooperated when necessary—for example, to resist Japanese incursions into Manchuria—they eventually came to be such bitter enemies that they brought a ruinous civil war to all of China.

Mao derived his support from the rural areas of China, while Chiang depended on the cities. In 1949, facing defeat, Chiang Kai-shek's Nationalists retreated to the island of Taiwan, where, under the name Republic of China (R.O.C.), they continued to insist on their right to rule all of China. The Communists, however, controlled the mainland and insisted that Taiwan was just a renegade province of the People's Republic of China. These two an-

tagonists are officially (but not in actuality) still at war. Sometimes tensions between Taiwan and China reach dangerous levels. In the 1940s, the United States had to intervene to prevent an attack from the mainland. In 1996, U.S. warships once again patrolled the 150 miles of ocean named the Taiwan Strait to warn China not to turn its military exercises, including the firing of missiles in the direction of Taiwan, into an actual invasion. China used the blatantly aggressive actions as a warning to the newly elected Taiwanese president not to take any steps toward declaring Taiwan an independent nation.

For many years after World War II, world opinion sided with Taiwan's claim to be the legitimate government of China. Taiwan was granted diplomatic recognition by many nations and given the China seat in the United Nations. In the 1970s, however, many nations, including the United States, came to believe that it was dysfunctional to withhold recognition and standing from such a large and powerful nation as the P.R.C. Because both sides insisted that there could not be two Chinas, nor one China and one Taiwan, the UN proceeded to give the China seat to mainland China, and dozens of countries broke off formal diplomatic relations with Taiwan in order to establish a relationship with China.

PROBLEMS OF GOVERNANCE

The China that Mao came to control was a nation with serious economic and social problems. Decades of civil war had disrupted families and wreaked havoc on the economy. Mao believed that the solution to China's ills was to wholeheartedly embrace socialism. Businesses were nationalized, and state planning replaced private initiative. Slowly, the economy improved. In 1958, however, Mao decided to enforce the tenets of socialism more vigorously so that China would be able to take an economic "Great Leap Forward." Workers were assigned to huge agricultural communes and were denied the right to grow crops privately. All enterprises came under the strict control of the central government. The result was economic chaos and a dramatic drop in both industrial and agricultural output.

Exacerbating these problems was the growing rift between the P.R.C. and the Soviet Union. China insisted that its brand of communism was truer to the principles of Marx and Lenin and criticized the Soviets for selling out to the West. As relations with (and financial support from) the Soviet Union withered, China found itself increasingly isolated from the world community, a circumstance worsened by serious conflicts with India, Indonesia, and

other nations. To gain friends, the P.R.C. provided substantial aid to Communist insurgencies in Vietnam and Laos, thus contributing to the eventual severity of the Vietnam War.

In 1966, Mao found that his power was waning in the face of Communist Party leaders who favored a more moderate approach to internal problems and external relations. To regain lost ground, Mao urged young students called Red Guards to fight against anyone who might have liberal, capitalist, or intellectual leanings. He called it the Great Proletarian Cultural Revolution, but it was an *anti*cultural purge: Books were burned, and educated people were arrested and persecuted. In fact, the entire country remained in a state of domestic chaos for more than a decade.

Soon after Mao died, in 1976, Deng Xiaoping, who had been in and out of Communist Party power several times before, came to occupy the senior position in the CCP. A pragmatist, he was willing to modify or forgo strict socialist ideology if he believed that some other approach would work better. Despite pressure by hard-liners to tighten governmental control, he nevertheless was successful in liberalizing the economy and permitting exchanges of scholars with the West. In 1979, he accepted formalization of relations with the United States— an act interpreted as a signal of China's opening up to the world.

China's opening has been dramatic, not only in terms of its international relations but also internally. During the 1980s, the P.R.C. joined the World Bank, the International Monetary Fund, the Asian Development Bank, and other multilateral organizations. It also began to welcome foreign investment of almost any kind and permitted foreign companies to sell their products within China itself (although many companies have pulled out of China, in frustration over unpredictable business policies or because they were unexpectedly shut down by the authorities). Trade between Taiwan and China—still mostly carried on via Hong Kong, but now also permitted through several small Taiwanese islands adjacent to the mainland—was nearly $6 billion by the early 1990s, and has increased exponentially since then. And while Hong Kong was investing some $25 billion in China, China was investing $11 billion in Hong Kong. More Chinese firms were permitted to export directly and to keep more of the profits. Special Economic Zones (SEZs)—capitalist enclaves adjacent to Hong Kong and along the coast into which were sent the most educated of the Chinese population—were established to catalyze the internal economy. In coastal cities,

(UN/photo by A. Holcombe)

During Mao Zedong's "Great Leap Forward," huge agricultural communes were established, and farmers were denied the right to grow crops privately. The government's strict control of these communes met with chaotic results; there were dramatic drops in agricultural output.

especially in south China, construction of apartment complexes, new manufacturing plants, and roads and highways began in earnest. Indeed, the south China area, along with Hong Kong and Taiwan, seemed to be emerging as a mammoth trading bloc—"Greater China"—which economists began to predict would exceed the economy of Japan by the year 2000 (a prediction that did not come true) and eclipse the U.S. economy by 2012. Stock exchanges opened in Shanghai and Shenzhen. Dramatic changes were implemented even in the inner rural areas. The collectivized farm system imposed by Mao was replaced by a household contract system with hereditary contracts (that is, one step away from actual private land ownership), and free markets replaced most of the system of mandatory agricultural sales to the government. New industries were established in rural villages, and incomes improved such that many families were able to add new rooms onto their homes or to purchase two-story and even three-story homes. Predictions of China's economic dominance have had to be revised downward in the face of the Asian financial crisis, but the country's economy remains a force to be reckoned with.

TIANANMEN SQUARE

Throughout the country in the 1980s, a strong spirit of entrepreneurship took hold. Many people, especially the growing body of educated youth, interpreted economic liberalization as the overture to political democratization. College students, some of whom had studied abroad, pressed the government to crack down on corruption in the Chinese Communist Party and to permit greater freedom of speech and other civil liberties.

In 1989, tens of thousands of college students staged a prodemocracy demonstration in Beijing's Tiananmen Square. The call for democratization received wide international media coverage and soon became an embarrassment to the Chinese leadership, especially when, after several days of continual protest, the students constructed a large statue in the square similar in appearance to the Statue of Liberty in New York Harbor. Some party leaders seemed inclined at least to talk with the students, but hard-liners apparently insisted that the prodemocracy movement be crushed in order that the CCP remain in control of the government. The official policy seemed to be that it would be the Communist Party, and not some prodemocracy movement, that would lead China to capitalism.

The CCP leadership had much to fear; it was, of course, aware of the quickening pace of Communist party power dissolution in the Soviet Union and Central/Eastern Europe, but it was even more concerned about corruption and the breakdown of CCP authority in the rapidly capitalizing rural regions of China, the very areas that had spawned the Communist Party under Mao. Moreover, economic liberalization had spawned inflation, higher prices, and spot shortages, and the general

The Shang Dynasty is the first documented Chinese dynasty 1523–1027 B.C.	The Chou Dynasty and the era of Confucius, Laotze, and Mencius 1027–256 B.C.	The Ch'in Dynasty, from which the word *China* is derived 211–207 B.C.	The Han Dynasty 202 B.C.–A.D. 220	The Three Kingdoms period; the Tsin and Sui Dynasties A.D. 220–618	The T'ang Dynasty, during which Confucianism flourished 618–906	The Five Dynasties and Sung Dynasty periods 906–1279	The Yuan Dynasty is founded by Kublai Khan 1260–1368	The Ming Dynasty 1368–1644	The Manchu or Ch'ing Dynasty 1644–1912	Trading rights and Hong Kong Island are granted to Britain 1834	The Sino-Japanese War 1894–1895

public was disgruntled. Therefore, after several weeks of pained restraint, the authorities moved against the students in what has become known as the Tiananmen Square massacre. Soldiers injured thousands and killed hundreds of students; hundreds more were systematically hunted down and brought to trial for sedition and for spreading counterrevolutionary propaganda.

In the wake of the brutal crackdown, many nations reassessed their relationships with the People's Republic of China. The United States, Japan, and other nations halted or canceled foreign assistance, exchange programs, and special tariff privileges. The people of Hong Kong, anticipating the return of their British colony to P.R.C. control in 1997, staged massive demonstrations against the Chinese government's brutality. Foreign tourism all but ceased, and foreign investment declined abruptly.

The withdrawal of financial support and investment was particularly troublesome to the Chinese leadership, as it realized that China's economy was far behind other nations. Even Taiwan, with a similar heritage and a common history until the 1950s, but having far fewer resources and much less land, had long since eclipsed the mainland in terms of economic prosperity. The Chinese understood that they needed to modernize (although not, they hoped, to Westernize), and they knew that large capital investments from such countries as Japan, Hong Kong, and the United States were crucial to their economic reform program. Moreover, they knew that they could not tolerate a cessation of trade with their new economic partners. By the end of the 1980s, about 13 percent of China's imports came from the United States, 18 percent from Japan, and 25 percent from Hong Kong. Similarly, Japan received 16 percent of China's exports, and Hong Kong received 43 percent.

Fortunately for the Chinese economy, the investment and loan-assistance programs from other countries have been reinstated in most cases as the repercussions of the events of 1989 wane. China was even able to close a $1.2 billion contract with McDonnell Douglas Corporation to build 40 jetliners; and U.S. President Bill Clinton, as a result of a decision to separate China's human-rights issues from

trade issues, repeatedly renewed China's "most favored nation" trade status. Bill Clinton and China's leader, Jiang Zemin, engaged in an unprecedented public debate on Chinese television in June 1998, and Clinton was allowed to engage students and others in direct dialogue in which he urged religious freedom, free speech, and the protection of other human rights. These events and others suggest that China is trying to address some of the concerns voiced against it by the industrialized world—one of which is copyright violations by Chinese companies. Some have estimated that as much as 88 percent of China's exports of CDs consists of illegal copies. A 1995 copyright agreement is having some effect, but still, much of China's trade deficit with the United States, which is now higher than Japan's, comes from illegal products. This has not stopped the United States, Japan, Australia, and other countries from pushing for China to join the World Trade Organization. In fact, most business leaders seem to believe that WTO membership would force China to adhere more faithfully to international rules of fair production and trade.

Improved trade notwithstanding, the Tiananmen Square massacre and the continuing brutality against citizens have convinced many people, both inside and outside China, that the Communist Party has lost, not necessarily its legal, but certainly its moral, authority to govern. Amnesty International's 1996 report claimed that human-rights violations in China occur "on a massive scale" and noted that torture is used on political prisoners held in *laogai,* Chinese gulags similar to those in the former Soviet Union. Although some religions are permitted in China, others are suppressed, the most notable being the Falun Gong, a meditative religion that China banned in 2000 and whose adherents are systematically arrested and jailed.

Despite China's controlled press, reports of other forms of social unrest are occasionally heard. For example, in 1999, some 3,000 farmers in the southern Hunan Province demonstrated against excessive taxation. One protester was killed. In western regions with large Islamic populations, anti-Beijing sentiment sometimes erupts in the form of bombings of govern-

ment buildings and underground antigovernment meetings.

THE SOCIAL ATMOSPHERE
In 1997, the aged Deng Xiaoping died. He was replaced by a decidedly more forward-looking leader, Jiang Zemin. Under his charge, the country was able to avoid many of the financial problems that affected other Asian nations in the late 1990s (although many Chinese banks are dangerously overextended, and real-estate speculation in Shanghai and other major cities has left many high-rise office buildings severely underoccupied). Despite many problems yet to solve, including serious human-rights abuses, it is clear that the Chinese leadership has actively embraced capitalism and has effected a major change in Chinese society. Historically, the loyalty of the masses of the people was placed in their extended families and in feudal warlords, who, at times of weakened imperial rule, were nearly sovereign in their own provinces. Communist policy has been to encourage the masses to give their loyalty instead to the centrally controlled Communist Party. The size of families has been reduced to the extent that "family" as such has come to play a less important role in the lives of ordinary Chinese.

Historical China was a place of great social and economic inequality between the classes. The wealthy feudal lords and their families and those connected with the imperial court and bureaucracy had access to the finest in educational and cultural opportunities, while around them lived illiterate peasants who often could not feed themselves, let alone pay the often heavy taxes imposed on them by feudal and imperial elites. The masses often found life to be bitter, but they found solace in the teachings of the three main religions of China (often adhered to simultaneously): Confucianism, Taoism, and Buddhism. Islam, animism, and Christianity have also been significant to many people in China.

The Chinese Communist Party under Mao, by legal decree and by indoctrination, attempted to suppress people's reliance on religious values and to reverse the ranking of the classes; the values of hard, manual work and rural simplicity were elevated, while the refinement and education of the urban elites were denigrated. Homes of formerly wealthy capitalists

Sun Yat-sen's republican revolution ends centuries of imperial rule; the Republic of China is established
1912

The Chinese Communist Party is organized
1921

Chiang Kai-shek begins a long civil war with the Communists
1926

Mao Zedong's Communist Army defeats Chiang Kai-shek
1949

A disastrous economic reform, the Great Leap Forward, is launched by Mao
1958

The Cultural Revolution; Mao dies
1966–1976

Economic and political liberalization begins under Deng Xiaoping; the P.R.C. and Britain agree to return Hong Kong to the Chinese
1980s

China expands its relationship with Taiwan; the Tiananmen Square massacre provokes international outrage

Crackdowns on dissidents and criminals result in hundreds of arrests and executions
1990s

Deng Xiaoping dies; Jiang Zemin becomes president

2000s

China bans the Falun Gong religion

The pace of China's modernization and political influence accelerates; military spending increases

were taken over by the government and turned into museums, and the opulent life of the capitalists was disparaged. During the Cultural Revolution, high school students who wanted to attend college had first to spend two years doing manual labor in factories and on farms to help them learn to relate to the peasants and the working class. So much did revolutionary ideology and national fervor take precedence over education that schools and colleges were shut down for several years during the 1960s and 1970s and the length of compulsory education was reduced.

One would imagine that after 40 years of communism, the Chinese people would have discarded the values of old China. However, the reverse seems to be true. When the liberalization of the economy began in the late 1970s, many of the former values also came to the fore: the Confucian value of scholarly learning and cultural refinement, the desirability of money, and even Taoist and Buddhist religious values.

Thousands of Chinese are studying abroad with the goal of returning to China to establish or manage profitable businesses. Indeed, some Chinese, especially those with legitimate access to power, such as ranking Communist Party members, have become extremely wealthy. Along with the privatization of state enterprises has come the unemployment of hundreds of thousands of "redundant" workers (2 million workers lost their jobs in one province in a single year in the early 1990s). Many others have had to settle for lower pay or unsafe work conditions as businesses strive to enter the world of competitive production. Demonstrations and more than 300 strikes by angry laborers exploded in early 1994. Even those with good jobs were finding it difficult to keep up with inflation, which in recent years

has been as high as 22 percent. Nevertheless, those with an entrepreneurial spirit were finding ways to make more money than they had ever dreamed possible in an officially communist country.

Some former values may help revitalize Chinese life, while others, once suppressed by the Communists, may not be so desirable. For instance, despite being an unabashed womanizer himself, Mao attempted to eradicate prostitution, eliminate the sale of women as brides, and prevent child marriages. Some of those customs are returning, and gender-based divisions of labor are making their way into the workplace.

Interest in things foreign however, is having a big impact on daily life in China, in ways both large and small. For example, increasingly entranced by Western culture, the Chinese flocked to movie theaters to see *Titanic*. Breaking box-office records across China, the film's theme song reached the top of the charts, and scalpers sold tickets for packed screenings. Other movies are also popular; but, when translated into Chinese characters and then re-translated back into English, their titles are often comical: *The Full Monty* is translated to *Six Naked Pigs;* Oliver Stone's *Nixon* becomes *The Big Liar; The English Patient* becomes *Do Not Ask Me Who I Am, Ever; Fargo* becomes *Mysterious Murder in Snowy Cream;* and *Secrets and Lies* becomes *Dreadful, Difficult People.*

Predicting China's future is difficult, but in recent years, the Chinese government has accelerated the pace of China's modernization and its role as a major power. In 1999, China launched an unmanned space capsule, with the goal of eventually sending Chinese into space. The country is also engaged in the largest public-works project in the world, the controversial

Three Gorges Dam project. It acquired control of the former British colony of Hong Kong in 1997, and of the Portuguese colony of Macau in 1999. China has also been quietly flexing its muscles in the South China Sea, and it bristles at any suggestion of a breach of its sovereignty, the most recent example being the accidental bombing of the Chinese Embassy in Yugoslavia by NATO planes engaged in the Kosovo conflict. Upon news of the bombing, the Chinese government organized anti–U.S. and anti–NATO demonstrations all over China. The U.S. Embassy in Beijing and the consulate in Shanghai were damaged, and the United States, in particular, was vilified in the press.

Regardless of how China develops in the future, every country in the world now recognizes that it will have to find new ways of dealing with Asia's colossus.

DEVELOPMENT

In the early years of Communist control, authorities stressed the value of establishing heavy industry and collectivizing agriculture. More recently, China has attempted to reduce its isolation by establishing trading relationships with the United States, Japan, and others and by constructing free-enterprise zones. The world's largest dam is currently under construction, despite the objections of environmentalists.

FREEDOM

Until the late 1970s, the Chinese people were controlled by Chinese Communist Party cadres who monitored both public and private behavior. Some economic and social liberalization occurred in the 1980s. However, the 1989 Tiananmen Square massacre reminded Chinese and the world that despite some reforms, China is still very much a dictatorship.

HEALTH/WELFARE

The Communist government has overseen dramatic improvements in the provision of social services for the masses. Life expectancy has increased from 45 years in 1949 to 72 years (overall) today. Diverse forms of health care are available at low cost to the patient. The government has attempted to eradicate such diseases as malaria and tuberculosis.

ACHIEVEMENTS

Chinese culture has, for thousands of years, provided the world with classics in literature, art, pottery, ballet, and other arts. Under communism the arts have been marshaled in the service of ideology and have lost some of their dynamism. Since 1949, literacy has increased dramatically and now stands at 73 percent—the highest in Chinese history.

Hong Kong (Hong Kong Special Administrative Region)

GEOGRAPHY
Area in Square Miles (Kilometers): 671 (1,054) (about 6 times the size of Washington, D.C.)
Capital (Population): Victoria (na)
Environmental Concerns: air and water pollution resulting from the pressures of rapid urbanization
Geographical Features: hilly to mountainous, with steep slopes; lowlands in the north
Climate: tropical monsoon

PEOPLE

Population

Total: 7,116,300
Annual Growth Rate: 1.35%
Rural/Urban Population Ratio: 9/91
Major Languages: Chinese (Cantonese); English
Ethnic Makeup: 95% Chinese (mostly Cantonese); 5% others
Religions: 90% a combination of Buddhism and Taoism; 10% Christian

Health
Life Expectancy at Birth: 77 years (male); 82 years (female)
Infant Mortality Rate (Ratio): 5.9/1,000
Physicians Available (Ratio): 1/1,000

Education
Adult Literacy Rate: 92%

COMMUNICATION
Telephones: 3,708,000 main lines
Internet Service Providers: 49 (1999)

TRANSPORTATION
Highways in Miles (Kilometers): 1,135 (1,831)
Railroads in Miles (Kilometers): 22 (34)
Usable Airfields: 3

GOVERNMENT
Type: Special Administrative Region of China
Independence Date: none (Special Administrative Region of China)
Head of State/Government: President Jiang Zemin; Chief Executive Tung Chee-hwa
Political Parties: Democratic Alliance for the Betterment of Hong Kong; Democratic Party; Association for Democracy and People's Livelihood; Hong Kong

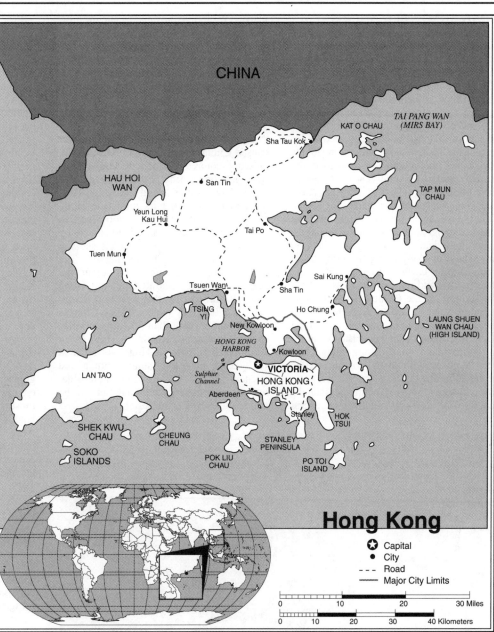

Progressive Alliance; Citizens Party; Frontier Party; Liberal Party
Suffrage: direct elections universal at 18 for residents who have lived in Hong Kong for at least 7 years

MILITARY
Military Expenditures (% of GDP): defense is the responsibility of China
Current Disputes: none

ECONOMY
Currency ($ U.S. Equivalent): 7.79 Hong Kong dollars = $1
Per Capita Income/GDP: $23,100/$158.2 billion
GDP Growth Rate: 1.8%
Inflation Rate: –4%
Unemployment Rate: 6%
Labor Force: 3,360,000
Natural Resources: outstanding deepwater harbor; feldspar
Agriculture: vegetables; poultry
Industry: textiles; clothing; tourism; electronics; plastics; toys; clocks; watches
Exports: $170 billion (primary partners China, United States, Japan)
Imports: $174.4 billion (primary partners China, Japan, United States)

http://www.info.gov.hk/sitemap.htm
http://www.cia.gov/cia/publications/factbook/geos/hk.html

HONG KONG'S BEGINNINGS

Opium started it all for Hong Kong. The addictive drug from which such narcotics as morphine, heroin, and codeine are made, opium had become a major source of income for British merchants in the early 1800s. When the Chinese government declared the opium trade illegal and confiscated more than 20,000 large chests of opium that had been on their way for sale to the increasingly addicted residents of Canton, the merchants persuaded the British military to intervene and restore their trading privileges. The British Navy attacked and occupied part of Canton. Three days later, the British forced the Chinese to agree to their trading demands, including a demand that they be ceded the tiny island of Hong Kong (meaning "Fragrant Harbor"), where they could pursue their trading and military business without the scrutiny of the Chinese authorities.

Initially, the British government was not pleased with the acquisition of Hong Kong; the island, which consisted of nothing more than a small fishing village, had been annexed without the foreknowledge of the authorities in London. Shortly, however, the government found the island's magnificent deepwater harbor a useful place to resupply ships and to anchor military vessels in the event of further hostilities with the Chinese. It turned out to be one of the finest natural harbors along the coast of China. On August 29, 1842, China reluctantly signed the Treaty of Nanking, which ended the first Opium War and gave Britain ownership of Hong Kong Island "in perpetuity."

Twenty years later, a second Opium War caused China to lose more of its territory; Britain acquired permanent lease rights over Kowloon, a tiny part of the mainland facing Hong Kong Island. By 1898, Britain had realized that its miniscule Hong Kong naval base would be too small to defend itself against sustained attack by French or other European navies seeking privileged access to China's markets. The British were also concerned about the scarcity of agricultural land on Hong Kong Island and nearby Kowloon Peninsula. In 1898, they persuaded the Chinese to lease them more than 350 square miles of land adjacent to Kowloon. Thus, Hong Kong consists today of Hong Kong Island (as well as numerous small, uninhabited islands nearby), the Kowloon Peninsula, and the agricultural lands that came to be called the New Territories.

From its inauspicious beginnings, Hong Kong grew into a dynamic, modern society, wealthier than its promoters would have ever dreamed in their wildest imaginations. Hong Kong is now home to 6.5 million people. Most of the New Territories are mountainous or are needed for agriculture, so the bulk of the population is packed into about one tenth of the land space. This gives Hong Kong the dubious honor of being one of the most densely populated human spaces ever created. Millions of people live stacked on top of one another in 30-story-high buildings. Even Hong Kong's excellent harbor has not escaped the population crunch: Approximately 10 square miles of harbor have been filled in and now constitute some of the most expensive real estate on Earth.

Why are there so many people in Hong Kong? One reason is that, after occupation by the British, many Chinese merchants moved their businesses to Hong Kong, under the correct assumption that trade would be given a freer hand there than on the mainland. Eventually, Hong Kong became the home of mammoth trading conglomerates. The laborers in these profitable enterprises came to Hong Kong, for the most part, as political refugees from mainland China in the early 1900s. Another wave of immigrants arrived in the 1930s upon the invasion of Manchuria by the Japanese, and yet another influx came after the Communists took over China in 1949. Thus, like Taiwan, Hong Kong became a place of refuge for those in economic or political trouble on the mainland.

Overcrowding plus a favorable climate for doing business have produced extreme social and economic inequalities. Some of the richest people on Earth live in Hong Kong, alongside some of the most wretchedly poor, notable among whom are recent refugees from China and Southeast Asia (more than 300,000 Vietnamese sought refuge in Hong Kong after the Communists took over South Vietnam, although many have now been repatriated—some forcibly). Some of these refugees have joined the traditionally poor boat peoples living in Aberdeen Harbor. Although surrounded by poverty, many of Hong Kong's economic elites have not found it inappropriate to indulge in ostentatious displays of wealth, such as riding in chauffeured, pink Rolls-Royces or wearing full-length mink coats. Hong Kong resident Li Ka-shing was listed in the year 2000 as one of the world's wealthiest people, with assets of $11.3 billion.

Workers are on the job six days a week, morning and night, yet the average pay for a worker in industry is only about $5,000 per year. With husband, wife, and older

(Photo by Lisa Clyde)

Land is so expensive in Hong Kong that most residences and businesses today are located in skyscrapers. While the buildings are thoroughly modern, construction crews typically erect bamboo scaffolding as the floors mount up, protected by netting. The skyscrapers that appear darker in color in the center of this photo are being built with this technique.

(Photo by Lisa Clyde)

A fishing family lives on this houseboat in Aberdeen Harbor. On the roof, strips of fish are hung up to dry.

children all working, families can survive; some even make it into the ranks of the fabulously wealthy. Indeed, the desire to make money was the primary reason why Hong Kong was settled in the first place. That fact is not lost on anyone who lives there today. Noise and air pollution (which recent studies show have worsened dramatically in the past several years), traffic congestion, and dirty and smelly streets do not deter people from abandoning the countryside in favor of the consumptive lifestyle of the city.

Yet materialism has not wholly effaced the cultural arts and social rituals that are essential to a cohesive society. Indeed, with the vast majority of Hong Kong's residents hailing originally from mainland China, the spiritual beliefs and cultural heritage of China's long history abound. Some residents hang small, eight-sided mirrors outside windows to frighten away malicious spirits, while others burn paper money in the streets each August to pacify the wandering spirits of deceased ancestors. Business owners carefully choose certain Chinese characters for the names of their companies or products, which they hope will bring them luck. Even skyscrapers are designed following ancient Chinese customs so that their entrances are in balance with the elements of nature.

Buddhist and Taoist beliefs remain central to the lives of many residents. In the back rooms of many shops, for example, small religious shrines are erected; joss sticks burning in front of these shrines are thought to bring good fortune to the proprietors. Elaborate festivals, such as those at New Year's, bring the costumes, art, and dance of thousands of years of Chinese history to the crowded streets of Hong Kong. And the British legacy may be found in the cricket matches, ballet troupes, philharmonic orchestras, English-language radio and television broadcasts, and the legal system under which capitalism flourished.

THE END OF AN ERA
Britain was in control of this tiny speck of Asia for nearly 160 years. Except during World War II, when the Japanese occupied Hong Kong for about four years, the territory was governed as a Crown colony of Great Britain, with a governor appointed by the British sovereign. In 1997, China recovered control of Hong Kong from the British. In 1984, British prime minister Margaret Thatcher and Chinese leader Deng Xiaoping concluded two years of acrimonious negotiations over the fate of Hong Kong upon the expiration of the New Territories' lease in 1997. Great Britain claimed the right to control Hong Kong Island and Kowloon forever—a claim disputed by China, which argued that the treaties granting these lands to Britain had been imposed by military force. Hong Kong Island and Kowloon, however,

constituted only about 10 percent of the colony; the other 90 percent was to return automatically to China at the expiration of the lease. The various parts of the colony having become fully integrated, it seemed desirable to all parties to keep the colony together as one administrative unit. Moreover, it was felt that Hong Kong Island and Kowloon could not survive alone.

The British government had hoped that the People's Republic of China would agree to the status quo, or that it would at least permit the British to maintain administrative control over the colony should it be returned to China. Many Hong Kong Chinese felt the same way, since they had, after all, fled to Hong Kong to escape the Communist regime in China. For its part, the P.R.C. insisted that the entire colony be returned to its control by 1997. After difficult negotiations, Britain agreed to return the entire colony to China as long as China would grant important concessions. Foremost among these were that the capitalist economy and lifestyle, including private-property ownership and basic human rights, would not be changed for 50 years. The P.R.C. agreed to govern Hong Kong as a "Special Administrative Region" (SAR) within China and to permit British and local Chinese to serve in the administrative apparatus of the territory. The first direct elections for the 60-member Legislative Council were held in September

| The British begin to occupy and use Hong Kong Island; the first Opium War
A.D. 1839–1842 | The Treaty of Nanking cedes Hong Kong to Britain
1842 | The Chinese cede Kowloon and Stonecutter Island to Britain
1856 | England gains a 99-year lease on the New Territories
1898 | The Boxer Rebellion
1898–1900 | Sun Yat-sen overthrows the emperor of China to establish the Republic of China
1911 | The Japanese attack Pearl Harbor and take Hong Kong
1941 | The Communist victory in China produces massive immigration into Hong Kong
1949 | Great Britain and China agree to the return of Hong Kong to China
1980s | China resumes control of Hong Kong on July 1, 1997; prodemocracy politicians sweep the 1998 elections
1990s |

2000s

Hong Kong's economy continues its recovery from the Asian financial crisis

1991, while the last British governor, Chris Patten, attempted to expand democratic rule in the colony as much as possible before the 1997 Chinese takeover—reforms that the Chinese dismantled to some extent after 1997.

The Joint Declaration of 1984 was drafted by top governmental leaders, with very little input from the people of Hong Kong. This fact plus fears about what P.R.C. control would mean to the freewheeling lifestyle of Hong Kong's ardent capitalists caused thousands of residents, with billions of dollars in assets in tow, to abandon Hong Kong for Canada, Bermuda, Australia, the United States, and Great Britain. Surveys found that as many as one third of the population of Hong Kong wanted to leave the colony before the Chinese takeover. In the year before the change to Chinese rule, so many residents—16,000 at one point—lined up outside the immigration office to apply for British passports that authorities had to open up a nearby sports stadium to accommodate them. About half of Hong Kong residents already held British citizenship, but many of the rest, particularly recent refugees from China, wanted to secure their futures in case life under Chinese rule became repressive. Immigration officials received more than 100,000 applications for British passports in a single month in 1996!

Emigration and unease over the future have unsettled, but by no means ruined, Hong Kong's economy. According to the World Bank, Hong Kong is home to the world's eighth-largest stock market, the fifth-largest banking center and foreign-exchange market (and the second largest in Asia after Japan), and its economy is the sixth richest in the world. Close to 9,000 multinational corporations have offices in Hong Kong, while some of the world's wealthiest people call Hong Kong home. Moreover, over the objections of the Chinese government, the outgoing

British authorities embarked on several ambitious infrastructural projects that will allow Hong Kong to continue to grow economically in the future. Chief among these is the airport on Chek Lap Kok Island. At a cost of $21 billion, the badly needed airport was one of the largest construction projects in the Pacific Rim. Opinion surveys showed that despite fears of angering the incoming Chinese government, most Hong Kong residents supported efforts to improve the economy and to democratize the government by lowering the voting age and allowing direct election rather than appointment by Beijing of more officials. In the 1998 elections, more than 50 percent of registered voters cast ballots—more than voted in Hong Kong's last election under British rule. In 1999, the opposition Democratic Party gained a substantial number of seats in district-level elections. These indicate a desire by the people of Hong Kong for more democracy.

These results as well as anti-Beijing demonstrations before and after Hong Kong's return to Chinese control (some 70,000 people, more than in previous years, demonstrated in Hong Kong in 1999 against the tragedy in Tiananmen Square) might suggest that the people of Hong Kong preferred to remain a British colony. However, while there were large British and American communities in Hong Kong, and although English has been the medium of business and government for many years (China is now proposing the elimination of English as a language of instruction in most schools), many residents over the years had little or no direct emotional involvement with British culture and no loyalty to the British Crown. They asserted that they were, first and foremost, Chinese. This, of course, does not amount to a popular endorsement of Beijing's rule, but it does

imply that some residents of Hong Kong feel that if they have to be governed by others, they would rather it be by the Chinese. Moreover, some believe that the Chinese government may actually help rid Hong Kong of financial corruption and allocate more resources to the poor—although with tourism down since the handover to China and the Hong Kong stock market still suffering the effects of the general Asian financial problems of 1998 and 1999, there may be fewer resources to distribute in the future.

Hong Kong's natural links with China had been expanding steadily for years before the handover. In addition to a shared language and culture, there are in Hong Kong thousands of recent immigrants with strong family ties to the People's Republic. And there are increasingly important commercial ties. Hong Kong has always served as south China's entrepôt to the rest of the world for both commodity and financial exchanges. For instance, for years Taiwan has circumvented its regulations against direct trade with China by transshipping its exports through Hong Kong. Commercial trucks plying the highways between Hong Kong and the P.R.C. form a bumper-to-bumper wall of commerce between the two regions. Already nearly 50 percent of Hong Kong's imports come from China, while 25 percent of its exports go to China. The P.R.C. realizes that Hong Kong needs to remain more or less as it is—therefore, the transition to Chinese rule may be less jarring to residents than was expected. Most people think that Hong Kong will remain a major financial and trading center for Asia.

DEVELOPMENT

Hong Kong is one of the preeminent financial and trading dynamos of the world. Annually, it exports billions of dollars' worth of products. Hong Kong's political future may be uncertain, but its fine harbor as well as its new $21 billion airport, and possibly even a new Disney theme park and information technology "cyberport," are sure to continue to fuel its economy.

FREEDOM

Hong Kong was an appendage to one of the world's foremost democracies for 160 years. Thus, its residents enjoyed the civil liberties guaranteed by British law. Under the new Basic Law of 1997, the Chinese government has agreed to maintain the capitalist way of life and other freedoms for 50 years.

HEALTH/WELFARE

Schooling is free and compulsory in Hong Kong through junior high school. The government has devoted large sums for low-cost housing, aid for refugees, and social services such as adoption. Housing, however, is cramped and inadequate for the population.

ACHIEVEMENTS

Hong Kong has the capacity to hold together a society where the gap between rich and poor is enormous. The so-called boat people have long been subjected to discrimination, but most other groups have found social acceptance and opportunities for economic advancement.

Indonesia (Republic of Indonesia)

GEOGRAPHY
Area in Square Miles (Kilometers): 740,903 (1,919,440) (nearly 3 times the size of Texas)
Capital (Population): Jakarta (9,112,600)
Environmental Concerns: air and water pollution; sewage; deforestation; smoke and haze from forest fires
Geographical Features: the world's largest archipelago; coastal lowlands; larger islands have interior mountains
Climate: tropical; cooler in highlands

PEOPLE

Population
Total: 224,785,000
Annual Growth Rate: 1.63%
Rural/Urban Population Ratio: 64/36
Major Languages: Bahasa Indonesian; English; Dutch; Javanese; many others
Ethnic Makeup: 45% Javanese; 14% Sundanese; 7.5% Madurese; 7.5% coastal Malay; 26% others
Religions: 88% Muslim; 8% Christian; 4% Hindu, Buddhist, and others

Health
Life Expectancy at Birth: 66 years (male); 70 years (female)
Infant Mortality Rate (Ratio): 42.2/1,000
Physicians Available (Ratio): 1/6,570

Education
Adult Literacy Rate: 84%
Compulsory (Ages): 7–16

COMMUNICATION
Telephones: 5,571,600 main lines
Daily Newspaper Circulation: 20 per 1,000 people
Televisions: 145 per 1,000 people
Internet Service Providers: 24

TRANSPORTATION
Highways in Miles (Kilometers): (1999) 212,474 (342,700)
Railroads in Miles (Kilometers): 3,875 (6,450)
Usable Airfields: 446
Motor Vehicles in Use: 4,800,000

GOVERNMENT
Type: republic
Independence Date: December 27, 1949 (legally; from the Netherlands)
Head of State/Government: President Abdurrahman Wahid is both head of state and head of government; Vice-President Megawati Sukarnoputri
Political Parties: Golkar; Indonesia Democracy Party-Struggle; Development Unity Party; Crescent Moon and Star Party; National Awakening Party; others
Suffrage: universal at 17; married persons regardless of age

MILITARY
Military Expenditures (% of GDP): 1.3%
Current Disputes: territorial disputes with Malaysia; internal strife

ECONOMY
Currency ($ U.S. Equivalent): 8,050 Indonesian rupiahs = $1
Per Capita Income/GDP: $2,800/$610 billion
GDP Growth Rate: 0%
Inflation Rate: 2%
Unemployment Rate: 15%–20%
Labor Force: 67,000,000
Natural Resources: petroleum; tin; natural gas; nickel; timber; bauxite; copper; fertile soils; coal; gold; silver
Agriculture: rice; cassava; peanuts; rubber; cocoa; coffee; copra; other tropical; livestock products; poultry; beef; pork; eggs
Industry: petroleum; natural gas; textiles; mining; cement; chemical fertilizers; food; rubber; wood
Exports: $48 billion (primary partners Japan, European Union, United States)
Imports: $24 billion (primary partners Japan, United States, Singapore)

http://www.uni-stuttgart.de/indonesia
http://www.indonesia.elga.net.id

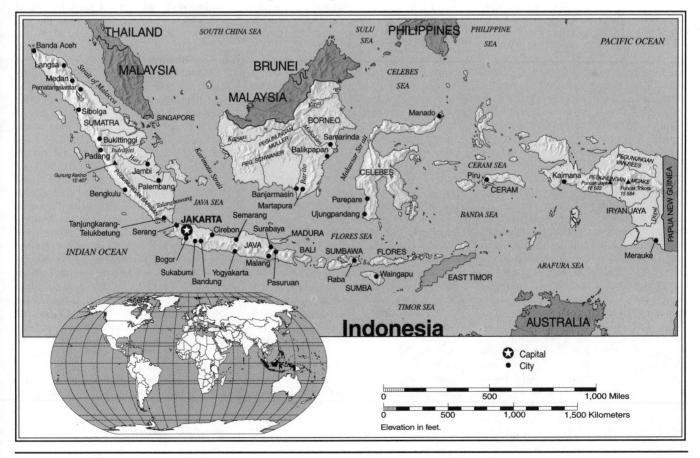

A KALEIDOSCOPIC CULTURE

Present-day Indonesia is a kaleidoscope of some 300 languages and more than 100 ethnic groups. Beginning about 5000 B.C., people of Mongoloid stock settled the islands that today constitute Indonesia, in successive waves of migration from China, Thailand, and Vietnam. Animism—the nature-worship religion of these peoples—was altered substantially (but never completely lost) about A.D. 200, when Hindus from India began to settle in the area and wield the dominant cultural influence. Several hundred years later, Buddhist missionaries and settlers began converting Indonesians in a proselytizing effort that produced strong political and religious antagonisms. In the thirteenth century, Muslim traders began the Islamization of the Indonesian people; today, 87 percent of the population claim the Muslim faith—meaning that there are more Muslims in Indonesia than in any other country of the world, including the states of the Middle East. Commingling with all these influences are cultural inputs from the islands of Polynesia.

The real roots of the Indonesian people undoubtedly go back much further than any of these historic cultures. In 1891, the fossilized bones of a hominid who used stone tools, camped around a fire, and probably had a well-developed language were found on the island of Java. Named *Pithecanthropus erectus* ("erect ape-man"), these important early human fossils, popularly called Java Man, have been dated at about 750,000 years of age. Fossils similar to Java Man have been found in Europe, Africa, and Asia.

Modern Indonesia was sculpted by the influence of many outside cultures. Portuguese Catholics, eager for Indonesian spices, made contact with Indonesia in the 1500s and left 20,000 converts to Catholicism, as well as many mixed Portuguese–Indonesian communities and dozens of Portuguese "loan words" in the Indonesian-style Malay language. In the following century, Dutch Protestants established the Dutch East India Company to exploit Indonesia's riches. Eventually the Netherlands was able to gain complete political control; it reluctantly gave it up in the face of insistent Indonesian nationalism only as recently as 1950. Before that, however, the British briefly controlled one of the islands, and the Japanese ruled the country for three years during the 1940s.

Indonesians, including then-president Sukarno, initially welcomed the Japanese as helpers in their fight for independence from the Dutch. Everyone believed that the Japanese would leave soon. Instead, the Japanese military forced farmers to give food to the Japanese soldiers, made everyone worship the Japanese emperor, neglected local industrial development in favor of military projects, and took 270,000 young men away from Indonesia to work elsewhere as forced laborers (fewer than 70,000 survived to return home). Military leaders who attempted to revolt against Japanese rule were executed. Finally, in August 1945, the Japanese abandoned their control of Indonesia, according to the terms of surrender with the Allied powers.

Consider what all these influences mean for the culture of modern Indonesia. Some of the most powerful ideologies ever espoused by humankind—supernaturalism, Islam, Hinduism, Buddhism, Christianity, mercantilism, colonialism, and nationalism—have had an impact on Indonesia. Take music, for example. Unlike Western music, which most people just listen to, Indonesian music, played on drums and gongs, is intended as a somewhat sacred ritual in which all members of a community are expected to participate. The instruments themselves are considered sacred. Dances are often the main element in a religious service whose goal might be a good rice harvest, spirit possession, or exorcism. Familiar musical styles can be heard here and there around the country. In the eastern part of Indonesia, the Nga'dha peoples, who were converted to Christianity in the early 1900s, sing Christian hymns to the accompaniment of bronze pot gongs and drums. On the island of Sumatra, Minang Kabau peoples, who were converted to Islam in the 1500s, use local instruments to accompany Islamic poetry singing. Communal feasts in Hindu Bali, circumcision ceremonies in Muslim Java, and Christian baptisms among the Bataks of Sumatra all represent borrowed cultural traditions. Thus, out of many has come one rich culture.

But the faithful of different religions are not always able to work together in harmony. For example, in the 1960s, when average Indonesians were trying to distance themselves from radical Communists, many decided to join Christian faiths. Threatened by this tilt toward the West and by the secular approach of the government, many fundamentalist Muslims resorted to violence. They burned Christian churches, threatened Catholic and Baptist missionaries, and opposed such projects as the construction of a hospital by Baptists. Indonesia is one of the most predominately Muslim countries in the world, and the hundreds of Islamic socioreligious and political organizations intend to keep it that way.

A LARGE LAND, LARGE DEBTS

Unfortunately, Indonesia's economy is not as rich as its culture. Three quarters of the population live in rural areas; more than half of the people engage in fishing and small-plot rice and vegetable farming. The average income per person is only $2,800 a year, based on gross domestic product. A 1993 law increased the minimum wage in Jakarta to $2.00 *per day*.

Also worrisome is the level of government debt. Indonesia is blessed with large oil reserves (Pertamina is the state-owned oil company) and minerals and timber of every sort (also state-owned), but to extract these natural resources has required massive infusions of capital, most of it borrowed. In fact, Indonesia has borrowed more money than any other country in Asia. The country must allocate 40 percent of its national budget just to pay the interest on loans. Low oil prices in the 1980s made it difficult for the country to keep up with its debt burden. Extreme political unrest and an economy that contracted nearly 14 percent (!) in 1998 have seriously exacerbated Indonesia's economic headaches in recent years.

To cope with these problems, Indonesia has relaxed government control over foreign investment and banking, and it seems to be on a path toward privatization of other parts of the economy. Still, the gap between the modernized cities and the traditional countryside continues to plague the government.

Indonesia's financial troubles seem puzzling, because in land, natural resources, and population, the country appears quite well-off. Indonesia is the second-largest country in Asia (after China). Were it superimposed on a map of the United States, its 13,677 tropical islands would stretch from California, past New York, and out to Bermuda in the Atlantic Ocean. Oil and hardwoods are plentiful, and the population is large enough to constitute a viable internal consumer market. But transportation and communication are problematic and costly in archipelagic states. Before the financial crisis of 1997–1998 hit, Indonesia's national airline, Garuda Indonesia, had hoped to launch a $3.6 billion development program that would have brought into operation 50 new aircraft stopping at 13 new airports. New seaports are also under construction. But the cost of linking together the 6,000 inhabited islands is a major drain on the economy. Moreover, exploitation of Indonesia's amazing panoply of resources is drawing the ire of more and more people around the world who fear the destruction of the world's ecosystem.

Indonesia's population of nearly 225 million is one of the largest in the world, but 16 percent of adults cannot read or write. Only about 600 people per 100,000 attend college, as compared to 3,580 in nearby Philippines. Moreover, since almost 70 percent of the population reside on or near the island of Java, on which the capital city, Jakarta, is located, educational and development efforts have concentrated there, at the expense of the communities on outlying islands. Many children in the out-islands never complete the required six years of elementary school. Some ethnic groups, on the islands of Irian Jaya (New Guinea) and Kalimantan (Borneo), for example, continue to live isolated in small tribes, much as they did thousands of years ago. By contrast, the modern city of Jakarta, with its classical European-style buildings, is home to millions of people. Over the past 20 years, poverty has been reduced from 60 to 15 percent, but Indonesia was seriously damaged by the Asian financial crisis of 1997–1998, and experts expect that the economy will not return to normal in the near future.

With 2.3 million new Indonesians entering the labor force every year and half the population under age 20, serious efforts must be made to increase employment opportunities. For the 1990s, the government earmarked millions of dollars to promote tourism. Nevertheless, the most pressing problem was to finish the many projects for which World Bank and Asian Development Bank loans had already been received. With considerable misgivings, the World Bank, the Asian Development Bank, and the government of Japan agreed in 2000 to provide more than $4 billion in additional aid to Indonesia to alleviate poverty and help decentralize government authority.

MODERN POLITICS

Establishing the current political and geographic boundaries of the Republic of Indonesia has been a bloody and protracted task. So fractured is the culture that many people doubt whether there really is a single country that one can call Indonesia. During the first 15 years of independence (1950–1965), there were revolts by Muslims and pro-Dutch groups, indecisive elections, several military coups, battles against U.S.–supported rebels, and serious territorial disputes with Malaysia and the Netherlands. In 1966, nationalistic President Sukarno, who had been a founder of Indonesian independence, lost power to Army General Suharto. (Many Southeast Asians had no family names until influenced by Westerners; Sukarno and Suharto have each used only one name.) Anti-

(UN Photo 155517/John Issac)

Great Buddhist temple at Borobudur, Indonesia, was built in the ninth century in the midst of Java's volcanic peaks.

Communist feeling grew during the 1960s, and thousands of suspected members of the Indonesian Communist Party (PKI) and other Communists were killed before the PKI was banned in 1966.

In 1975, ignoring the disapproval of the United Nations, President Suharto invaded and annexed East Timor, a Portuguese colony. Although the military presence in East Timor was subsequently reduced, separatists were beaten and killed by the Indonesian Army as recently as 1991; and in 1993, a separatist leader was sentenced to 20 years in prison. In late 1995, Amnesty International accused the Indonesian military of raping and executing human-rights activists in East Timor, while the 20th anniversary of the Indonesian takeover was marked by Timorese storming foreign embassies and demanding asylum and redress for the kidnapping and killing of protesters. In 1996, antigovernment rioting in Jakarta resulted in the arrest of more than 200 opposition leaders and the disappearance of many others. The rioters were supporters of the Indonesian Democracy Party and its leader, Megawati Sukarnoputri, daughter of Sukarno, and the person who would be, in effect, ruling the country by 2000.

In August 1999, the 800,000 residents of East Timor voted overwhelmingly (78.5 percent) for independence of Indonesia, in a peaceful referendum. The struggle of the Timorese, however, is far from over as they try to establish order and prosperity in Asia's newest country.

Suharto's so-called New Order government ruled with an iron hand, suppressing student and Muslim dissent and controlling the press and the economy. With the economy in serious trouble in 1998, and with the Indonesian people tired of government corruption and angry at the control of Suharto and his six children over much of the economy, rioting broke out all over the country. Some 15,000 people took to the streets, occupied government offices, burned cars, and fought with police. The International Monetary Fund suspended vital aid because it appeared that Suharto would not conform to the belt-tightening required of IMF recipients. With unemployed migrant workers streaming back to Indonesia from Malaysia and surrounding countries, with the government unable to control forest fires burning thousands of acres and producing a haze all over Southeast Asia, with even his own lifetime political colleagues calling for him to step down, Suharto at last resigned, ending a 32-year dictatorship. The new leader, President Bacharuddin Jusuf Habibie, pledged to honor IMF commitments and restore dialogue on the East Timor dispute. But protests dogged Habibie, because he was seen as too closely allied with the Suharto leadership. In the first democratic elections in years, a respected Muslim cleric, Abdurrahman Wahid, was elected president, with Megawati Sukarnoputri, daughter of Indonesia's founding father, as vice-president. In mid-2000, in the face of mounting criticism of his seeming inability to restore order and jump-start the economy, Wahid, who was nearly blind due to a serious eye disease, announced that he was turning over day-to-day administration of the country to his vice-president. Under the Wahid administration, charges of corruption against Suharto finally resulted in his house arrest, although his lawyers claimed he was too old and frail to stand trial.

Among other urgent measures, the new government will need to encourage foreign investment, particularly from such important trading partners as Japan. (Indonesia sends 27 percent of its exports to Japan and buys 23 percent of its imports from Japan.) And Japanese investment money, now much scarcer than before, will need to continue to flow to Indonesia, as elsewhere in the Pacific Rim. In the

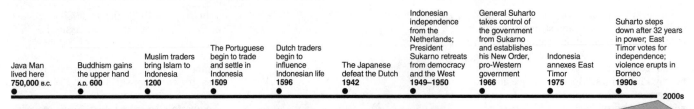

Java Man lived here 750,000 B.C. ●	Buddhism gains the upper hand A.D. 600 ●	Muslim traders bring Islam to Indonesia 1200 ●	The Portuguese begin to trade and settle in Indonesia 1509 ●	Dutch traders begin to influence Indonesian life 1596 ●	The Japanese defeat the Dutch 1942 ●	Indonesian independence from the Netherlands; President Sukarno retreats from democracy and the West 1949–1950 ●	General Suharto takes control of the government from Sukarno and establishes his New Order, pro-Western government 1966 ●	Indonesia annexes East Timor 1975 ●	Suharto steps down after 32 years in power; East Timor votes for independence; violence erupts in Borneo 1990s ●	2000s

The economy remains stalled

President Wahid turns over day-to-day adminstration to Vice-President Sukarnoputri

past, companies like Toyota invested millions in Indonesia's ASTRA automobile company, while Japanese banks supported the expansion of Indonesia's tourist industry. If the new government can prove to the world that it will govern wisely, it is possible that foreign investment will help lead Indonesia out of its current quagmire.

CRISIS IN BORNEO

In early 1999, violence broke out on the Indonesian island of Borneo. By the time it ended months later, 50,000 people had fled the island, and hundreds of men, women, and children had been killed, their decapitated heads displayed in towns and villages. The attackers even swaggered through towns victoriously holding up the dismembered body parts of their victims.

What caused this violence? It started with overpopulation and poverty. Some 50 years ago, the Indonesian government decided that it had to do something about overpopulation on the soil-poor island of Madura. Located near the island of Java, Madura, with its white-sand beaches and its scores of related islets, was the home of the Madurese, Muslims who had long since found it difficult to survive by farming the rocky ground. The government thus decided to move some of the Madurese to the island of Borneo, where land was more plentiful and fertile. It seemed like a logical solution, but the government failed to consider the ethnic context—that is, the island of Borneo was inhabited by the Dayaks, a people who considered the island their tribal homeland. They had little or no interest in the government in Jakarta, regarding themselves as Dayaks first and Indonesians a distant second, if at all.

Almost immediately the Dayaks began harassing the newcomers, whom they judged to be "hot-headed" and crude. They resented both the loss of their land and, later, the loss of jobs in the villages. Over the years, hundreds of people were killed in sporadic attacks, but the violence

in 1999 was worse than ever, in part because the Dayaks were, for the first time, joined by the Malays in "cleansing" the island of the hated Madurese. At first, the Indonesian government seemed to do little to stop the carnage. Eventually, when the Madurese had been chased from their homes, President Wahid promised aid money and assistance with relocation. But he wanted the Madurese to return to Borneo. The Dayaks, with little regard for the Indonesian government, responded that they would kill any who returned. As of this writing, thousands of Madurese remain as refugees on the island of Java.

This incident illustrates at once the problems so common to the Pacific Rim: overpopulation, poverty, ethnic conflict, tribal versus national identity, and distrust of legally constituted government.

EAST TIMOR DECLARES INDEPENDENCE

In a peaceful referendum in August 1999, the 800,000 residents of East Timor voted overwhelmingly (78.5 percent) for independence from Indonesia. Almost immediately, anti-independence militias in the region, together with Indonesian troops, began destroying East Timor, and especially Dili, the capital city. Residents were driven from their homes, beaten, killed, and their homes and businesses burned. The violence was so severe, and Indonesia's response so weak, that outside nations, particularly nearby Australia, felt compelled to send in some 8,000 troops to prevent a wholesale bloodbath. What caused this hemorrhage of brutality? And what is this place called East Timor?

The island of Timor, Indonesia's easternmost island, is about 280 miles long and rests just above Australia. Covered in places by stands of bamboo, teak, sandalwood, and other trees, the island provides residents a modest living, mostly through farming. In the early 1500s, the Portuguese established settlements there and began converting the population to Chris-

tianity. The Dutch did likewise, and at first Portugal and the Netherlands competed for control of the island. Portugal established an official colony in the eastern half of the island in 1859 and, with continued missionary efforts, produced an island of Christianity (95 percent Catholic) in an otherwise very Muslim country. In 1974, after a coup in Portugal that ousted the right-wing dictatorship there, Portugal offered independence to all of its colonies, including East Timor. In 1975, civil war broke out between the Christians and the Muslims, providing a justification for Indonesia to invade and officially annex East Timor, during the process of which Indonesian soldiers killed an estimated 100,000 people. The East Timorese would not tolerate this state of affairs and many, like freedom fighter Jose Alexandre (Xanana) Gusmao, began a life of underground resistance to Indonesian occupation. When Indonesian strongman Suharto was chased from office in 1998, calls for East Timor's separation from Indonesia became more strident, and the United Nations offered to oversee a vote on the question—partly to find a way to stop the violence between Christians and Muslims. Indonesia reluctantly agreed to the vote, but the military decided that should the vote go the "wrong" way, it would destroy the country before it had to pull out. And that is what it did.

In October 1999, the Indonesian Legislature approved the results of the East Timor vote, paving the way for United Nations-led elections to choose leaders and a constitution for Asia's newest emerging country.

DEVELOPMENT

Indonesia continues to be hamstrung by its heavy reliance on foreign loans, a burden inherited from the Sukarno years. Current Indonesian leaders speak of "stabilization" and "economic dynamism."

FREEDOM

Demands for Western-style human rights are frequently heard, but only the army has the power to impose order on the numerous and competing political groups, many imbued with religious fervor.

HEALTH/WELFARE

Indonesia has one of the highest birth rates in the Pacific Rim. Many children will grow up in poverty, never learning even to read or write their national language, Bahasa Indonesian.

ACHIEVEMENTS

Balinese dancers' glittering gold costumes and unique choreography epitomize the "Asian-ness" of Indonesia.

Laos (Lao People's Democratic Republic)

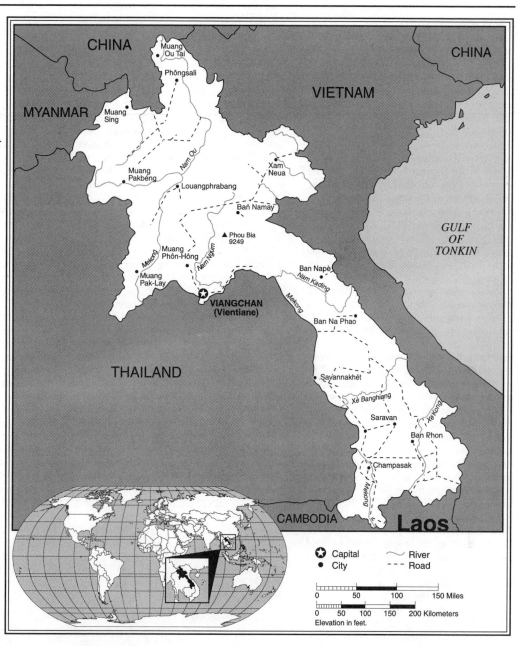

GEOGRAPHY

Area in Square Miles (Kilometers):
91,400 (236,800) (about the
size of Utah)
Capital (Population): Viangchan
(Vientiane) (531,800)
Environmental Concerns:
unexploded ordnance;
deforestation; soil erosion;
lack of access to potable water
Geographical Features: mostly
rugged mountains; some
plains and plateaus
Climate: tropical monsoon

PEOPLE

Population
Total: 5,498,000
Annual Growth Rate: 2.5%
Rural/Urban Population (Ratio):
79/21
Major Languages: Lao; French;
English; ethnic languages
Ethnic Makeup: 50% Lao; 20%
tribal Thai; 15% Phoutheung
(Kha); 15% Meo, Hmong,
Yao, and others
Religions: 60% Buddhist; 40%
indigenous beliefs and others

Health
Life Expectancy at Birth: 51
years (male); 55 years (female)
Infant Mortality Rate (Ratio):
94.8/1,000
Physicians Available (Ratio):
1/3,555

Education
Adult Literacy Rate: 57%
Compulsory (Ages): for 5 years
between ages 6–15

COMMUNICATION
Telephones: 28,500 main lines
Televisions: 17 per 1,000 people
Internet Service Providers: na

TRANSPORTATION
Highways in Miles (Kilometers): 13,029
(21,716)
Railroads in Miles (Kilometers): none
Usable Airfields: 52
Motor Vehicles in Use: 18,000

GOVERNMENT
Type: Communist state
Independence Date: July 19, 1949 (from
France)
Head of State/Government: President
Khamtai Siphandon; Prime Minister
Sisavat Keobounphan
Political Parties: Lao People's Revolution-
ary Party; other parties proscribed
Suffrage: universal at 18

MILITARY
Military Expenditures (% of GDP): 4.2%
Current Disputes: internal strife; indefinite
border with Thailand

ECONOMY
Currency ($ U.S. Equivalent):
7,600 kip = $1
Per Capita Income/GDP: $1,307/$7 billion
GDP Growth Rate: 5.2%
Inflation Rate: 140%
Unemployment Rate: 5.7%
Labor Force: 1,000,000–1,500,000
Natural Resources: timber; hydropower;
gypsum; tin; gold; gemstones

Agriculture: rice; sweet potatoes;
vegetables; coffee; tobacco; sugarcane;
cotton; livestock
Industry: mining; timber; garments;
electric power; agricultural processing;
construction
Exports: $271 million (primary partners
Vietnam, Thailand, Germany)
Imports: $497 million (primary partners
Thailand, Japan, Vietnam)

http://www.global.lao.net/laoVL.html
http://www.stockton.edu/~gilmorew/
consorti/1reasia.htm

A NATION DIVIDED

Laos seems a sleepy place. Almost everyone lives in small villages where the only distraction might be the Buddhist temple gong announcing the day. Water buffalo plow quietly through centuries-old rice paddies, while young Buddhist monks in saffron-colored robes make their silent rounds for rice donations. Villagers build their houses on stilts for safety from annual river flooding, and top them with thatch or tin. Barefoot children play under the palm trees or wander to the village Buddhist temple for school in the outdoor courtyard. Mothers work at home, weaving brightly colored cloth for the family and preparing meals—on charcoal or wood stoves—of rice, bamboo shoots, pork, duck, and snakes seasoned with hot peppers and ginger.

Below this serene surface, however, Laos is a nation divided. The name Laos is taken from the dominant ethnic group, but there are actually about 70 ethnic groups in the country. Over the centuries, they have battled one another for supremacy, for land, and for tribute money. The constant feuding has weakened the nation and served as an invitation for neighboring countries to annex portions of Laos forcibly, or to align themselves with one or another of the Laotian royal families or generals for material gain. China, Burma (today called Myanmar), Vietnam, and especially Thailand—with which Laotian people share many cultural and ethnic similarities—have all been involved militarily in Laos.

Historically, palace jealousies often led one member of the royal family to fight against a relative for dominance creating unrest. More recently, Laos has been seen as a pawn in the battle of the Western powers for access to the rich natural resources of Southeast Asia or as a "domino" that some did and others did not want to fall to communism. Former members of the royal family continue to find themselves on the opposite sides of many issues.

The results of these struggles have been devastating. Laos is now one of the poorest countries in the world. There are few industries in the country, so most people survive by subsistence farming and fishing, raising or catching just what they need to eat rather than growing food to sell. In fact, some "hill peoples" (about two thirds of the Laotian people live in the mountains) in the long mountain range that separates Laos from Vietnam continue to use the most ancient farming technique known, slash-and-burn farming, an unstable method of land use that allows only a few years of good crops before the soil is depleted and the farmers must move to new

ground. Today, soil erosion and deforestation pose significant threats to economic growth.

Even if all Laotian farmers used the most modern techniques and geared their production to cash crops, it would still be difficult to export food (or, for that matter, anything else) because of Laos's woefully inadequate transportation network. There are no railroads, and muddy, unpaved roads make many mountain villages completely inaccessible by car or truck. Only one bridge in Laos, the Thai-Lao Friendship Bridge near the capital city of Viangchan (Vientiane), spans the famous Mekong River. Moreover, Laos is landlocked. In a region of the world where wealth flows toward those countries with the best ports, having no direct access to the sea is a serious impediment to economic growth. In addition, for years the economy has been strictly controlled by the government. Foreign investment and trade have not been welcomed; tourists were not allowed into the country until 1989. But the economy began to open up in the late 1980s. The government's "New Economic Mechanism" (NEP) in 1986 called for foreign investment in all sectors and anticipated gross domestic product growth of 8 percent per year. The year 1999 was declared "Visit Laos Year," and the government set a goal of drawing 1 million tourists. With its technological infrastructure abysmally underdeveloped, tourism seemed like the easiest way for the country to gain some foreign currency and

create jobs. Some likely tourist sites, such as Luan Prabang, an ancient royal city with a 112-pound gold statue of Buddha, have changed little in a thousand years, so tourists wishing to see "Old Asia" may indeed find Laos a draw.

Some progress has been made in the past decade. Laos is now self-sufficient in its staple crop, rice; and surplus electricity generated from dams along the Mekong River is sold to Thailand to earn foreign exchange. Laos imports various commodity items from Thailand, Vietnam, Singapore, Japan, and other countries, and it has received foreign aid from the Asian Development Bank and other organizations. Exports to Thailand, China, and the United States include teakwood, tin, and various minerals. In 1999, a state-owned bank in Vietnam agreed to set up a joint venture with the Laotian Bank for Foreign Trade. The new bank will mainly deal with imports and exports, particularly between Vietnam and Laos. In 2000, leaders of the world's wealthiest nations, the G-7 group, agreed to help Laos by offering various types of debt relief, but it will be many years before the country can claim that its economy is solid.

Despite the 1995 "certification" by the United States that Laos is a cooperating country in the world antidrug effort, Laos is also the source of many controlled substances, such as opium, cannabis, and heroin, much of which finds its way to Europe and the United States. The Laotian government is now trying to prevent hill

(UPI/Bettmann)

Laos is one of the poorest countries in the world. With few industries, most people survive by subsistence farming and fishing. These fishermen spend their days catching tiny fish, measuring two to five inches, that must suffice to feed their families.

The first Laotian nation is established
A.D. **1300s**

Laos is under French control
1890s

The Japanese conquer Southeast Asia
1940s

France grants independence to Laos
1949

South Vietnamese troops, with U.S. support, invade Laos
1971

Pathet Lao Communists gain control of the government
1975

Laos signs military and economic agreements with Vietnam
1977

The government begins to liberalize some aspects of the economy
1980s

The Pathet Lao government maintains firm control; efforts to maintain high GDP growth are threatened by deforestation and soil erosion
1990s

2000s

Laos works for closer economic ties to Vietnam

The G-7 nations promise debt relief for Laos

peoples from cutting down valuable forests for opium-poppy cultivation.

HISTORY AND POLITICS

The Laotian people, originally migrating from south China, settled Laos in the thirteenth century A.D., when the area was controlled by the Khmer (Cambodian) Empire. Early Laotian leaders expanded the borders of Laos through warfare with Cambodia, Thailand, Burma, and Vietnam. Internal warfare, however, led to a loss of autonomy in 1833, when Thailand forcibly annexed the country (against the wishes of Vietnam, which also had designs on Laos). In the 1890s, France, determined to have a part of the lucrative Asian trade and to hold its own against growing British strength in Southeast Asia, forced Thailand to give up its hold on Laos. Laos, Vietnam, and Cambodia were combined into a new political entity, which the French termed *Indochina*. Between these French possessions and the British possessions of Burma and Malaysia lay Thailand; thus, France, Britain, and Thailand effectively controlled mainland Southeast Asia for several decades.

There were several small uprisings against French power, but these were easily suppressed until the Japanese conquest of Indochina in the 1940s. The Japanese, with their "Asia for Asians" philosophy, convinced the Laotians that European domination was not a given. In the Geneva Agreement of 1949, Laos was granted independence, although full French withdrawal did not take place until 1954.

Prior to independence, Prince Souphanouvong (who died in 1995 at the age of 82) had organized a Communist guerrilla army, with help from the revolutionary Ho Chi Minh of the Vietnamese Communist group Viet Minh. This army called itself the Pathet Lao (meaning "Lao Country").

In 1954, it challenged the authority of the government in the Laotian capital. Civil war ensued, and by 1961, when a cease-fire was arranged, the Pathet Lao had captured about half of Laos. The Soviet Union supported the Pathet Lao, whose strength was in the northern half of Laos, while the United States supported a succession of pro-Western but fragile governments in the south. A coalition government consisting of Pathet Lao, pro-Western, and neutralist leaders was installed in 1962, but it collapsed in 1965, when warfare once again broke out.

During the Vietnam War, U.S. and South Vietnamese forces bombed and invaded Laos in an attempt to disrupt the North Vietnamese supply line known as the Ho Chi Minh Trail. Americans flew nearly 600,000 bombing missions over Laos (many of the small cluster bombs released during those missions remain unexploded in fields and villages and present a continuing danger). Communist battlefield victories in Vietnam encouraged and aided the Pathet Lao Army, which became the dominant voice in a new coalition government established in 1974. The Pathet Lao controlled the government exclusively by 1975. In the same year, the government proclaimed a new "Lao People's Democratic Republic." It abolished the 622-year-old monarchy and sent the king and the royal family to a detention center to learn Marxist ideology.

Vietnamese Army support, and flight by many of those opposed to the Communist regime, have permitted the Pathet Lao to maintain control of the government. The ruling dictatorship is determined to prevent the democratization of Laos: In 1993, several cabinet ministers were jailed for 14 years for trying to establish a multiparty democracy.

The Pathet Lao government was sustained militarily and economically by the Soviet Union and other East bloc nations for more than 15 years. However, with the end of the Cold War and the collapse of the Soviet Union, Laos has had to look elsewhere, including non-Communist countries, for support. In 1992, Laos signed a friendship treaty with Thailand to facilitate trade between the two historic enemy countries. In 1994, the Australian government, continuing its plan to integrate itself more fully into the strong Asian economy, promised to provide Laos with more than $33 million in aid. In 1995, Laos joined with ASEAN nations to declare the region a nuclear-free zone.

Trying to teach communism to a devoutly Buddhist country has not been easy. Popular resistance has caused the government to retract many of the regulations it has tried to impose on the Buddhist Church (technically, the Sangara, or order of the monks—the Buddhist equivalent of a clerical hierarchy). As long as the Buddhist hierarchy limits its activities to helping the poor, it seems to be able to avoid running afoul of the Communist leadership.

Intellectuals, especially those known to have been functionaries of the French administration, have fled Laos, leaving a leadership vacuum. As many as 300,000 people are thought to have left Laos for refugee camps in Thailand and elsewhere. Many have taken up permanent residence in foreign countries. The exodus has exacted a significant drain on Laos's intellectual resources.

DEVELOPMENT

Communist rule after 1975 isolated Laos from world trade and foreign investment. The planned economy has not been able to gain momentum on its own. In 1986, the government loosened restrictions so that government companies could keep a portion of their profits. A goal is to integrate Laos economically with Vietnam and Cambodia.

FREEDOM

Laos is ruled by the political arm of the Pathet Lao Army. Opposition parties and groups as well as opposition newspapers and other media are outlawed. Lack of civil liberties as well as poverty have caused many thousands of people to flee the country.

HEALTH/WELFARE

Laos is typical of the least developed countries in the world. The birth rate is high, but so is infant mortality. Most citizens eat less than an adequate diet. Life expectancy is low, and many Laotians die from illnesses for which medicines are available in other countries. Many doctors fled the country when the Communists came to power.

ACHIEVEMENTS

The original inhabitants of Laos, the Kha, have been looked down upon by the Lao, Thai, and other peoples for centuries. But under the Communist regime, the status of the Kha has been upgraded and discrimination formally proscribed.

Macau (Macau Special Administrative Region)

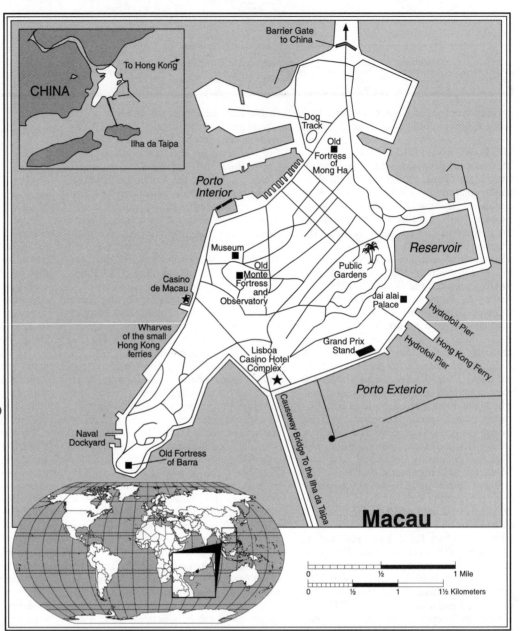

GEOGRAPHY

Area in Square Miles (Kilometers): 8 (21) (about ¹/₁₀ the size of Washington, D.C.)

Capital (Population): Macau (445,600)

Environmental Concerns: air and water pollution

Geographical Features: generally flat

Climate: subtropical; marine with cool winters, warm summers

PEOPLE

Population

Total: 445,600

Annual Growth Rate: 1.83%

Rural/Urban Population Ratio: 0/100

Major Languages: Portuguese; Cantonese

Ethnic Makeup: 95% Chinese; 5% Macanese, Portuguese, and others

Religions: 50% Buddhist; 15% Roman Catholic; 35% unaffiliated or other

Health

Life Expectancy at Birth: 79 years (male); 85 years (female)

Infant Mortality Rate (Ratio): 4.5/1,000

Physicians Available (Ratio): 1/2,470

Education

Adult Literacy Rate: 90%

TRANSPORTATION

Highways in Miles (Kilometers): 31 (50)

Railroads in Miles (Kilometers): none

Usable Airfields: 1

GOVERNMENT

Type: Special Administrative Region of China

Independence Date: none; Special Administrative Region of China

Head of State/Government: President (of China) Jiang Zemin; Chief Executive Edmund Ho Hau-Wah

Political Parties: there are no formal political parties; civic associations are used instead

Suffrage: direct election universal at 18 for permanent residents living in Macau for 7 years; indirect election limited to organizations registered as "corporate voters" and a 300-member Election Committee

MILITARY

Military Expenditures (% of GDP): defense is the responsibility of China

Current Disputes: none

ECONOMY

Currency ($ U.S. Equivalent): 8.03 pataca = $1 (tied to Hong Kong dollar)

Per Capita Income/GDP: $17,500/$7.65 billion

GDP Growth Rate: –4%

Inflation Rate: –3%

Unemployment Rate: 7%

Labor Force: 281,200

Natural Resources: fish

Agriculture: rice; vegetables

Industry: clothing; textiles; toys; tourism; electronics; footwear; gambling

Exports: $1.7 billion (primary partners United States, Europe, Hong Kong)

Imports: $1.5 billion (primary partners China, Hong Kong, Europe)

 http://www.cia.gov/cia/publications/factbook/geos/mc.html

| A Portuguese trading colony is established at Macau A.D. **1557** | Portugal declares sovereignty over Macau **1849** | China signs a treaty recognizing Portuguese sovereignty over Macau **1887** | Immigrants from China flood into the colony **1940s** | Pro-Communist riots in Macau **1967** | Portugal begins to loosen direct administrative control over Macau **1970s** | Macau becomes a Chinese territory but is still administered by Portugal **1976** | China and Portugal sign an agreement scheduling the return of Macau to Chinese control **1987** | Portugal sends troops to help tamp down gambling-related crime; Macau reverts to Chinese control on December 20, 1999 **1990s** |

2000s

The reversion to Chinese control goes smoothly   Macau fights to recover from the Asian financial crisis

MACAU'S HISTORY

Just 17 miles across the Pearl River estuary from Hong Kong is the world's most densely populated territory: the former Portuguese colony of Macau (sometimes spelled Macao). Consisting of only six square miles of land, the peninsula and two tiny islands are home to nearly half a million people, 95 percent of whom are Chinese. Until December 20, 1999, when it reverted to China as a "Special Administrative Region" (SAR), it had been the oldest outpost of European culture in the Far East, with a 442-year history of Portuguese administration.

Macau's population has varied over the years, depending on conditions in China. During the Japanese occupation of China during the 1940s, for instance, Macau's Chinese population is believed to have doubled, and more refugees streamed in when the Communists took over China in 1949.

Macau was frequented by Portuguese traders as early as 1516, but it was not until 1557 that the Chinese agreed to Portuguese settlement of the land; it did not, however, acknowledge Portuguese sovereignty. Indeed, the Chinese government did not recognize the Portuguese right of "perpetual occupation" until 1887.

In 1987, Chinese and Portuguese officials signed an agreement, effective December 20, 1999, to end European control of the first—and last—colonial outpost in China. Actually, Portugal had offered to return Macau on two earlier occasions (in 1967, during the Chinese Cultural Revolution; and in 1974, after the political coup in Portugal that ended the dictatorship there), but China refused. The 1999 transition went smoothly, although Portugal's president announced that he would not attend the hand-over ceremony if China sent in troops before the official transition date. China relented, but when the troops arrived on December 20, they were generally cheered by crowds who hoped that they would bring some order to the gang-infested society.

The transition agreement is similar to that signed by Great Britain and China over the fate of Hong Kong. China agreed to allow Macau to maintain its capitalist way of life for 50 years, to permit local elections, and to allow its residents to travel freely without Chinese intervention. Unlike Hong Kong residents, who staged massive demonstrations against future Chinese rule or emigrated from Hong Kong before its return to China, Macau residents—some of whom have been openly pro-Communist—have not seemed bothered by the new arrangements. In fact, a new airport has been opened to bolster the economy, and plans for the reversion to China have gone smoothly. Indeed, businesses in Macau and Hong Kong have contributed to a de facto merging with the mainland by investing more than $20 billion in China since the mid-1990s.

Since it was established in the sixteenth century as a trading colony with interests in oranges, tea, tobacco, and lacquer, Macau has been heavily influenced by Roman Catholic priests of the Dominican and Jesuit orders. Christian churches, interspersed with Buddhist temples, abound. The name of Macau itself reflects its deep and enduring religious roots; the city's official name is "City of the Name of God in China, Macau, There Is None More Loyal." Macau has perhaps the highest density of churches and temples per square mile in the world. Buddhist immigrants from China have reduced the proportion of Christians in the population.

A HEALTHY ECONOMY

Macau's modern economy is a vigorous blend of light industry, fishing, tourism, and gambling. Revenues from the latter two sources are impressive, accounting for 25 percent of gross domestic product. There are five major casinos and many other gambling opportunities in Macau, which, along with the considerable charms of the city itself, attract more than 5 million foreign visitors a year, more than 80 percent of them Hong Kong Chinese with plenty of money to spend (gambling is illegal on the mainland). Macau's gambling industry is run by a syndicate of Chinese businesspeople operating under the name Macau Travel & Amusement Company, which won monopoly rights on all licensed gambling in Macau in 1962. But not everyone is content with that arrangement. Rival gangs, fighting to control parts of the gambling business, produced such a crime spree in 1998 that military troops had to be sent from Portugal to restore order.

Export earnings derived from light-industry products such as textiles, fireworks, plastics, and electronics are also critical to the colony. Macau's leading export markets are the United States, China, Germany, France, and Hong Kong; ironically, Portugal consumes only about 3 percent of Macau's exports.

As might be expected, the general success of the economy has a downside. In Macau's case, the hallmarks of modernization—crowded apartment blocks and bustling traffic—are threatening to eclipse the remnants of the old, serene, Portuguese-style seaside town.

DEVELOPMENT

The development of industries related to gambling and tourism (tourists are primarily from Hong Kong) has been very successful. Most of Macau's foods, energy, and fresh water are imported from China; Japan and Hong Kong are the main suppliers of raw materials.

FREEDOM

Under the new "Basic Law," China agreed to maintain Macau's separate legal, political, and economic system. The Legislature is partly elected and partly appointed.

HEALTH/WELFARE

Macau has very impressive quality-of-life statistics. It has a low infant mortality rate and very high life expectancy for both males and females. Literacy is 90 percent.

ACHIEVEMENTS

Considering its unfavorable geographical characteristics, such as negligible natural resources and a port so shallow and heavily silted that oceangoing ships must anchor offshore, Macau has had stunning economic success.

Malaysia

GEOGRAPHY

Area in Square Miles (Kilometers):
121,348 (329,750) (slightly larger than New Mexico)

Capital (Population): Kuala Lumpur (1,236,000)

Environmental Concerns: air and water pollution; deforestation; smoke/haze from Indonesian forest fires

Geographical Features: coastal plains rising to hills and mountains

Climate: tropical; annual monsoons

PEOPLE

Population

Total: 21,794,000

Annual Growth Rate: 2.01%

Rural/Urban Population Ratio: 46/54

Major Languages: Peninsular Malaysia: Bahasa Malaysia, English, Chinese dialects, Tamil; Sabah: English, Malay, numerous tribal dialects, Mandarin and Hakka dialects; Sarawak: English, Malay, Mandarin, numerous tribal dialects, Arabic, others

Ethnic Makeup: 58% Malay and other indigenous; 26% Chinese; 7% Indian; 9% others

Religions: Peninsular Malaysia: Malays nearly all Muslim, Chinese mainly Buddhist, Indians mainly Hindu; Sabah: 33% Muslim, 17% Christian, 45% others; Sarawak: 35% traditional indigenous, 24% Buddhist and Confucian, 20% Muslim, 16% Christian, 5% others

Health

Life Expectancy at Birth: 68 years (male); 74 years (female)

Infant Mortality Rate (Ratio): 23.2/1,000

Physicians Available (Ratio): 1/2,153

Education

Adult Literacy Rate: 84%

Compulsory (Ages): 6–16; free

COMMUNICATION

Telephones: 4,384,000 main lines

Daily Newspaper Circulation: 139 per 1,000 people

Televisions: 454 per 1,000 people

Internet Service Providers: 8 (1999)

TRANSPORTATION

Highways in Miles (Kilometers): 58,590 (94,500)

Railroads in Miles (Kilometers): 1,116 (1,800)

Usable Airfields: 115

Motor Vehicles in Use: 3,948,000

GOVERNMENT

Type: constitutional monarchy

Independence Date: August 31, 1957 (from the United Kingdom)

Head of State/Government: Paramount Ruler (King) Jaafar bin Abdul Rahman; Prime Minister Datuk Mahathir bin Mohamad

Political Parties: Peninsular Malaysia: National Front and others; Sabah: National Front and others; Sarawak: National Front and others

Suffrage: universal at 21

MILITARY

Military Expenditures (% of GDP): 1.6%

Current Disputes: dispute over the Spratly Islands; Sabah is claimed by the Philippines; other territorial disputes

ECONOMY

Currency ($ U.S. Equivalent): 3.80 ringgits = $1

Per Capita Income/GDP: $10,700/$229.1 billion

GDP Growth Rate: 6% predicted

Inflation Rate: 2.8%

Unemployment Rate: 3%

Labor Force: 9,300,000

Natural Resources: tin; petroleum; timber; natural gas; bauxite; iron ore; copper; fish

Agriculture: rubber; palm oil; rice; coconut oil; pepper; timber

Industry: rubber and palm oil manufacturing and processing; light manufacturing; electronics; tin mining and smelting; logging and timber processing; petroleum; food processing

Exports: $83.5 billion (primary partners United States, Singapore, Japan)

Imports: $61.5 billion (primary partners Japan, United States, Singapore)

http://st-www.cs.uiuc.edu/users/chai/malaysia.html

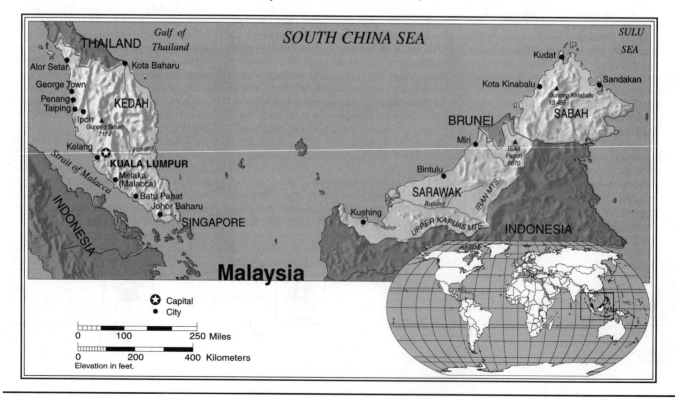

A FRACTURED NATION

About the size of Japan and famous for its production of natural rubber and tin, Malaysia sounds like a true political, economic, and social entity. Although it has all the trappings of a modern nation-state, Malaysia is one of the most fragmented nations on Earth.

Consider its land. West Malaysia, wherein reside 86 percent of the population, is located on the Malay Peninsula between Singapore and Thailand; but East Malaysia, with 60 percent of the land, is located on the island of North Borneo, some 400 miles of ocean away.

Similarly, Malaysia's people are divided along racial, religious, and linguistic lines. Fifty-eight percent are Malays and other indigenous peoples, many of whom adhere to the Islamic faith or animist beliefs; 26 percent are Chinese, most of whom are Buddhist, Confucian, or Taoist; 7 percent are Indians and 9 percent are Pakistanis and others, some of whom follow the Hindu faith. Bahasa Malaysia is the official language, but English, Arabic, two forms of Chinese, Tamil, and other languages are also spoken. Thus, although the country is called Malaysia (a name adopted only 35 years ago), many people living in Kuala Lumpur, the capital, or in the many villages in the countryside do not think of themselves first and foremost as Malaysians.

Malaysian culture is further fragmented because each ethnic group tends to replicate the architecture, social rituals, and norms of etiquette peculiar to itself. The Chinese, whose ancestors were imported in the 1800s from south China by the British to work the rubber plantations and tin mines, have become so economically powerful that their cultural influence extends far beyond their actual numbers.

Malaysian history is equally fragmented. Originally controlled by numerous sultans who gave allegiance to no one or only reluctantly to various more powerful states in surrounding regions, Malaysia first came to Western attention in 1511, when the prosperous city of Malacca, founded on the west coast of the Malay Peninsula about A.D. 1400, was conquered by the Portuguese. The Dutch took Malacca away from the Portuguese in 1641. The British seized it from the Dutch in 1824 (the British had already acquired an island off the coast and had established the port of Singapore). By 1888, the British were in control of most of the area that is now Malaysia.

However, British hegemony did not mean total control, for each of the many sultanates—the origin of the 13 states that constitute Malaysia today—continued to act more or less independently of the British, engaging in wars with one another and maintaining an administrative apparatus apart from the British. And some groups, such as the Dayaks, an indigenous people living in the jungles of Borneo, remained more or less aloof from the various intrigues of modern state-making and developed little or no identity of themselves as citizens of any modern nation.

It is hardly surprising, then, that Malaysia has had a difficult time emerging as a nation. Indeed, it is not likely that there would have been an independent Malaysia had it not been for the Japanese, who defeated the British in Southeast Asia during World War II and promulgated their alluring doctrine of "Asia for Asians."

After the war, Malaysian demands for independence from European domination grew more persuasive; Great Britain attempted in 1946 to meet these demands by proposing a partly autonomous Malay Union. However, ethnic rivalries and power-sensitive sultans created such enormous tension that the plan was scrapped. In an uncharacteristic display of cooperation, some 41 different Malay groups organized the United Malay National Organization (UMNO) to oppose the British plan. In 1948, a new Federation of Malaya was attempted. It granted considerable freedom within a framework of British supervision, allowed sultans to retain power over their own regions, and placed certain restrictions on the power of the Chinese living in the country.

Opposing any agreement short of full independence, a group of Chinese Communists, with Indonesian support, began a guerrilla war against the government and against capitalist ideology. Known as "The Emergency," the war lasted more than a decade and involved some 250,000 government troops. Eventually, the insurgents withdrew.

The three main ethnic groups—Malayans, represented by UMNO; Chinese, represented by the Malayan Chinese Association, or MCA; and Indians, represented by the Malayan Indian Congress, or MIC—were able to cooperate long enough in 1953 to form a single political party under the leadership of Abdul Rahman. This party demanded and received complete independence for the Federation in 1957, although some areas, such as Brunei, refused to join. Upon independence, the Federation of Malaya (not yet called Malaysia), excluding Singapore and the territories on the island of Borneo, became a member of the British Commonwealth of Nations and was admitted to the United Nations. In 1963, a new Federation was proposed that included Singapore and the lands on Borneo. Again, Brunei refused to join. Singapore joined but withdrew in 1965. Thus, what is known as Malaysia acquired its current form in 1966. It is regarded today as a rapidly developing nation that, despite an increasingly despotic prime minister, seems to try to govern itself according to democratic principles.

Political troubles stemming from the deep ethnic divisions in the country, however, remain a constant feature of Malaysian life. With nine of the 13 states controlled by independent sultans, every election is a test of the ability of the National Front, a multiethnic coalition of 11 different parties that has a two-thirds majority in Parliament, to maintain political stability. Particularly troublesome has been the state of Sabah (an area claimed by the Philippines), many of whose residents have wanted independence or, at least, greater autonomy from the federal government. In 1994, however, the National Front was able to gain a slight majority in Sabah elections, indicating the growing confidence that people have in the federal government's economic development policies.

ECONOMIC DEVELOPMENT

For years, Malaysia's "miracle" economy kept social and political instability in check. Although it had to endure normal fluctuations in market demand for its products, the economy grew at 5 to 8 percent per year from the 1970s to the late 1990s, making it one of the world's top 20 exporters/importers. The manufacturing sector developed to such an extent that it accounted for 70 percent of exports. Then, in 1998, a financial crisis hit. Malaysia was forced to devalue its currency, the ringgit, making it more difficult for consumers to buy foreign products, and dramatically slowing the economy. The government found it necessary to deport thousands of illegal Indonesian and other workers (dozens of whom fled to foreign embassies to avoid deportation) in order to find jobs for Malaysians. In the 1980s and early 1990s, up to 20 percent of the Malaysian workforce had been foreign workers, but the downturn produced "Operation Get Out," in which at least 850,000 "guest workers" were to be deported to their home countries.

Malaysia continues to be rich in raw materials. Therefore, it is not likely that the crisis of the late 1990s will permanently cripple its economy. Moreover, the Malaysian government has a good record of active planning and support of business ventures—directly modeled after Japan's export-oriented strategy. Malaysia launched a "New Economic Policy" (NEP) in the 1970s that welcomed foreign direct investment and sought to diversify the economic base. Japan, Taiwan, and the United States invested heavily in Malaysia. So successful was this

| The city of Malacca is established; it becomes a center of trade and Islamic conversion A.D. 1403 | The Portuguese capture Malacca 1511 | The Dutch capture Malacca 1641 | The British obtain Malacca from the Dutch 1824 | Japan captures the Malay Peninsula 1941 | The British establish the Federation of Malaya; a Communist guerrilla war begins, lasting for a decade 1948 | The Federation of Malaya achieves independence under Prime Minister Tengku Abdul Rahman 1957 | The Federation of Malaysia, including Singapore but not Brunei, is formed 1963 | Singapore leaves the Federation of Malaysia 1965 | Malaysia attempts to build an industrial base 1980s | The NEP is replaced with Vision 2020; economic crisis 1990s |

2000s

The economy rebounds; the environment suffers

Former deputy prime minister Anwar Ibrahim is arrested and convicted under questionable circumstances

strategy that economic growth targets set for the mid-1990s were actually achieved several years early. In 1991, the government replaced NEP with a new plan, "Vision 2020." Its goal was to bring Malaysia into full "developed nation" status by the year 2020. Sectors targeted for growth included the aerospace industry, biotechnology, microelectronics, and information and energy technology. The government expanded universities and encouraged the creation of some 170 industrial and research parks, including "Free Zones," where export-oriented businesses were allowed duty-free imports of raw materials. Some of Malaysia's most ambitious projects, including a $6 billion hydroelectric dam (strongly opposed by environmentalists), have been shelved, at least until the full effects of the Asian financial crisis are overcome. That may not be long, for while the economy nosedived in 1998, planners predicted a growth rate of more than 6 percent by the year 2001.

Despite Malaysia's substantial economic successes, serious social problems remain. These problems stem not from insufficient revenues but from inequitable distribution. The Malay portion of the population in particular continues to feel economically deprived as compared to the wealthier Chinese and Indian segments. Furthermore, most Malays are farmers, and rural areas have not benefited from Malaysia's economic boom as much as urban areas have.

In the 1960s and 1970s, riots involving thousands of college students were headlined in the Western press as having their basis in ethnicity. This was true to some degree, but the core issue was economic inequality. Included in the economic master plan of the 1970s were plans (similar to affirmative action in the United States) to change the structural barriers that prevented many Malays from fully enjoying the benefits of the economic boom. Under the leadership of Prime Minister Datuk

Mahathir bin Mohamad, plans were developed that would assist Malays until they held a 30 percent interest in Malaysian businesses. In 1990, the government announced that the figure had already reached an impressive 20 percent. Unfortunately, many Malays have insufficient capital to maintain ownership in businesses, so the government has been called upon to acquire many Malay businesses in order to prevent their being purchased by non-Malays. In addition, the system of preferential treatment for Malays has created a Malay elite, detached from the Malay poor, who now compete with the Chinese and Indian elites; interethnic and interracial goodwill is still difficult to achieve. Nonetheless, social goals have been attained to a greater extent than most observers have thought possible. Educational opportunities for the poor have been increased, farmland development has proceeded on schedule, and the poverty rate has dropped below 10 percent.

THE LEADERSHIP

In a polity so fractured as Malaysia's, one would expect rapid turnover among political elites. But Prime Minister Mahathir, a Malay, has had the support of the electorate for more than a decade. In 2000, despite some defections and public protests, his United Malay National Organization reelected him as party head. His primary challenger has been the Chinese Democratic Action Party (DAP), which has sometimes reduced Mahathir's majority in Parliament but has not been able to top his political strength. The policies that have sustained Mahathir's reputation as a credible leader include the NEP and his nationalist—but moderate—foreign policy. Malaysia has been an active member of ASEAN and has courted Japan and other Pacific Rim nations for foreign investment and export markets (Mahathir's "Look East" policy).

Malaysia's success has not been achieved without some questionable practices. The government seems unwilling to regulate economic growth, even though strong voices have been raised against industrialization's deleterious effects on the old-growth teak forests and other parts of the environment. Moreover, the blue-collar workers who are the muscle behind Malaysia's economic success are prohibited from forming labor unions, and outspoken critics have been silenced. The most outspoken critic was Anwar Ibrahim. He had been the deputy prime minister and heir-apparent to Mahathir; but when he challenged Mahathir's policies, he was fired, arrested, beaten, and eventually sent to prison for 14 years on various charges. Protesters frequently took to the streets in his defense, but they were beaten by police and sprayed with tear gas and water cannons. The largest opposition newspaper came to Anwar's defense (for which the editor was charged with sedition), and Anwar's wife started a new political party to challenge the government. The event has severely tarnished Mahathir's reputation at home (his overseas reputation has been at a low ebb for years), and charges of government corruption are now heard everywhere. The environmentalists' case was substantially strengthened in 1998 when forest and peat-bog fires in Malaysia and Indonesia engulfed Kuala Lumpur in a thick haze for weeks. The government, unable to snuff out the fires, ordered sprinklers installed atop the city's skyscrapers to settle the dust and lower temperatures.

DEVELOPMENT

Efforts to move the economy away from farming and toward industrial production have been very successful. Manufacturing now accounts for 30% of GDP, and Malaysia is the third-largest producer of semiconductors in the world. With Thailand, Malaysia will build a $1.3 billion, 530-mile natural-gas pipeline.

FREEDOM

Malaysia is attempting to govern according to democratic principles. Ethnic rivalries, however, severely hamper the smooth conduct of government and limit such individual liberties as the right to form labor unions. Protests have increased against Mahathir's strong-arm tactics to silence opponents.

HEALTH/WELFARE

City dwellers have ready access to educational, medical, and social opportunities, but the quality of life declines dramatically in the countryside. Malaysia has one of the highest illiteracy rates in the Pacific Rim. It spends only a small percentage of its GDP on education.

ACHIEVEMENTS

Malaysia has made impressive economic advancements. The government's New Economic Policy has achieved a measure of wealth redistribution to the poor. Since the 1970s, the economy has grown at an impressive rate. Malaysia has also made impressive social and political gains.

Myanmar (Union of Myanmar; commonly known as Burma)

GEOGRAPHY

Area in Square Miles (Kilometers): 261,901 (678,500) (slightly smaller than Texas)

Capital (Population): Yangon (Rangoon) (3,873,000)

Environmental Concerns: deforestation; air, soil, and water pollution; inadequate sanitation and water treatment

Geographical Features: central lowlands ringed by steep, rugged highlands

Climate: tropical monsoon

PEOPLE

Population

Total: 41,735,000

Annual Growth Rate: 0.64%

Rural/Urban Population Ratio: 74/26

Major Languages: Burmese; various minority languages

Ethnic Makeup: 68% Burman; 9% Shan; 7% Karen; 4% Rakhine; 3% Chinese; 9% Mon, Indian, and others

Religions: 89% Buddhist; 4% Muslim; 4% Christian; 3% others

Health

Life Expectancy at Birth: 54 years (male); 56 years (female)

Infant Mortality Rate (Ratio): 78.5/1,000

Physicians Available (Ratio): 1/3,554

Education

Adult Literacy Rate: 83%

Compulsory (Ages): 5–10; free

COMMUNICATION

Telephones: 213,500 main lines

Daily Newspaper Circulation: 23 per 1,000 people

Televisions: 22 per 1,000 people

Internet Service Providers: none (1999)

TRANSPORTATION

Highways in Miles (Kilometers): 17,484 (28,200)

Railroads in Miles (Kilometers): 2,474 (3,991)

Usable Airfields: 80

Motor Vehicles in Use: 69,000

GOVERNMENT

Type: military regime

Independence Date: January 4, 1948 (from the United Kingdom)

Head of State/Government: Prime Minister Chairman of the State Peace and Development Council (General) Than

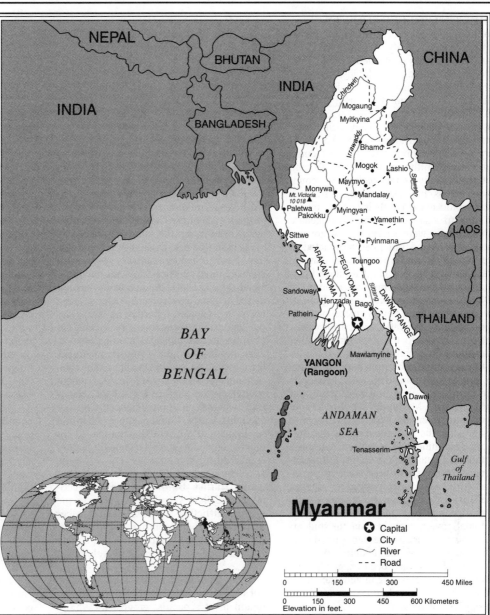

Myanmar

Shwe is both head of state and head of government

Political Parties: Union Solidarity and Development Association; National League for Democracy; National Unity Party; others

Suffrage: universal at 18

MILITARY

Military Expenditures (% of GDP): 2.1%

Current Disputes: internal strife; sporadic border conflict with Thailand

ECONOMY

Currency ($ U.S. Equivalent): 6.57 kyat = $1

Per Capita Income/GDP: $1,200/$59.4 billion

GDP Growth Rate: 4.6%

Inflation Rate: 38%

Unemployment Rate: 7.1%

Labor Force: 19,700,000

Natural Resources: petroleum; tin; timber; antimony; zinc; copper; tungsten; lead; coal; marble; limestone; precious stones; natural gas

Agriculture: rice; corn; oilseed; sugarcane; pulses; hardwood

Industry: agricultural processing; textiles; footwear; wood and wood products; petroleum refining; mining; construction materials; pharmaceuticals; fertilizer

Exports: $1.2 billion (primary partners India, China, Singapore)

Imports: $2.5 billion (primary partners Singapore, Japan, Thailand)

 http://www.myanmar.com

THE CONTROLLED SOCIETY

For more than three decades, Myanmar (as Burma was officially renamed in 1989) has been a tightly controlled society. Telephones, radio stations, railroads, and many large companies have been under the direct control of a military junta that has brutalized its opposition and forced many to flee the country. For many years, tourists were allowed to stay only two weeks (for a while the limit was 24 hours), could stay only at military-approved hotels, and could visit only certain parts of the country. Citizens too were highly restricted: They could not leave their country by car or train to visit nearby countries because all the roads were sealed off by government decree and rail lines terminated at the border. Even Western-style dancing was declared illegal. Until a minor liberalization of the economy was achieved in 1989, all foreign exports—every grain of rice, every peanut, every piece of lumber—though generally owned privately, had to be sold to the government rather than directly to consumers.

Observers attribute this state of affairs to military commanders who overthrew the legitimate government in 1962, but the roots of Myanmar's political and economic dilemma actually go back to 1885, when the British overthrew the Burmese government and declared Burma a colony of Britain. In the 1930s, European-educated Burmese college students organized strikes and demonstrations against the British. Seeing that the Japanese Army had successfully toppled other European colonial governments in Asia, the students determined to assist the Japanese during their invasion of the country in 1941. Once the British had been expelled, however, the students organized the Anti-Fascist People's Freedom League (AFPFL) to oppose Japanese rule.

When the British tried to resume control of Burma after World War II, they found that the Burmese people had given their allegiance to U Aung San, one of the original student leaders. He and the AFPFL insisted that the British grant full independence to Burma, which they reluctantly did in 1948. So determined were the Burmese to remain free of foreign domination that, unlike most former British colonies, they refused to join the British Commonwealth of Nations, an economic association of former British colonies. This was the first of many decisions that would have the effect of isolating Burma from the global economy.

Unlike Japan, with its nearly homogeneous population and single national language, Myanmar is a multiethnic state; in fact, only about 60 percent of the people

speak Burmese. The Burman people are genetically related to the Tibetans and the Chinese; the Chin are related to peoples of nearby India; the Shan are related to Thais; and the Mon migrated to Burma from Cambodia. In general, these ethnic groups live in separate political states within Myanmar—the Kachin State, the Shan State, the Karen State, and so on; and for hundreds of years, they have warred against one another for dominance. Upon the withdrawal of the British in 1948, some ethnic groups, particularly the Kachins, the Karens, and the Shans, embraced the Communist ideology of change through violent revolution. Their rebellion against the government in the capital city of Yangon (the new name of Rangoon) had the effect of removing from government control large portions of the country. Headed by U Nu (U Aung San and several of the original government leaders having been assassinated shortly before independence), the government considered its position precarious and determined that to align itself with the Communist forces then ascendant in the People's Republic of China and other parts of Asia would strengthen the hand of the ethnic separatists, whereas to form alliances with the capitalist world would invite a repetition of decades of Western domination. U Nu thus attempted to steer a decidedly neutral course during the cold war era and to be as tolerant as possible of separatist groups within Burma. Burma refused U.S. economic aid, had very little to do with the warfare afflicting Vietnam and the other Southeast Asian countries, and was not eager to join the Southeast Asian Treaty Organization or the Asian Development Bank.

Some factions of Burmese society were not pleased with U Nu's relatively benign treatment of separatist groups. In 1958, a political impasse allowed Ne Win, a military general, to assume temporary control of the country. National elections were held in 1962, and a democratically elected government was installed in power. Shortly thereafter, however, Ne Win staged a military coup. The military has controlled Burma/Myanmar ever since. Under Ne Win, competing political parties were banned, the economy was nationalized, and the country's international isolation became even more pronounced.

Years of ethnic conflict, inflexible socialism, and self-imposed isolation have severely damaged economic growth in Myanmar. In 1987, despite Burma's abundance of valuable teak and rubber trees in its forests, sizable supplies of minerals in the mountains to the north, onshore oil, rich farmland in the Irrawaddy Delta, and

a reasonably well-educated population, the United Nations declared Burma one of the least developed countries in the world (it had once been the richest country in Southeast Asia). Debt incurred in the 1970s exacerbated the country's problems, as did the government's fear of foreign investment. Thus, by 2000, Myanmar's per capita income was only $1,200 a year.

Myanmar's industrial base is still very small; about two thirds of the population of nearly 42 million make their living by farming (rice is a major export) and by fishing. The tropical climate yields abundant forest cover, where some 250 species of valuable trees abound. Good natural harbors and substantial mineral deposits of coal, natural gas, and others also bless the land. Only about 10 percent of gross domestic product comes from the manufacturing sector (as compared to, for example, approximately 45 percent in wealthy Taiwan). In the absence of a strong economy, black marketeering has increased, as have other forms of illegal economic transactions. It is estimated that 80 percent of the heroin smuggled into New York City comes from the jungles of Myanmar and northern Thailand.

Over the years, the Burmese have been advised by economists to open up their country to foreign investment and to develop the private sector of the economy. They have resisted the former idea because of their deep-seated fear of foreign domination; they have similar suspicions of the private sector because it was previously controlled almost completely by ethnic minorities (the Chinese and Indians). The government has relied on the public sector to counterbalance the power of the ethnic minorities.

Beginning in 1987, however, the government began to admit publicly that the economy was in serious trouble. To counter massive unrest in the country, the military authorities agreed to permit foreign investment from countries such as Malaysia, South Korea, Singapore, and Thailand and to allow trade with China and Thailand. In 1989, the government signed oil-exploration agreements with South Korea, the United States, the Netherlands, Australia, and Japan. Both the United States and the former West Germany withdrew foreign aid in 1988, but Japan did not; in 1991, Japan supplied $61 million—more than any other country—in aid to Myanmar.

POLITICAL STALEMATE

Despite these reforms, Myanmar has remained in a state of turmoil. In 1988, thousands of students participated in six months of demonstrations to protest the lack of democracy in the country and to demand multiparty elections. General Saw Maung brutally suppressed the demonstrators, imprisoning many students—and killing more than 3,000 of them. He then took control of the government and reluctantly agreed to multiparty elections. About 170 political parties registered for the elections, which were held in 1990—the first elections in 30 years. Among these were the National Unity Party (a new name for the Burma Socialist Program Party, the only legal party since 1974) and the National League for Democracy, a new party headed by Aung San Suu Kyi, daughter of slain national hero U Aung San.

The campaign was characterized by the same level of military control that had existed in all other aspects of life since the 1960s. Martial law, imposed in 1988, remained in effect; all schools and universities were closed; opposition-party workers were intimidated; and, most significantly, the three most popular opposition leaders were placed under house arrest and barred from campaigning. The United Nations began an investigation of civil-rights abuses during the election and, once again, students demonstrated against the military government. Several students even hijacked a Burmese airliner to demand the release of Aung San Suu Kyi, who had been placed under house arrest.

As the votes were tallied, it became apparent that the Burmese people were eager to end military rule; the National League for Democracy won 80 percent of the seats in the National Assembly. But the military junta refused to step down and remains in control of the government. Under General Than Shwe, who replaced General Saw Maung in 1992, the military has organized various operations against Karen rebels and has so oppressed Muslims that some 40,000 to 60,000 of them have fled to Bangladesh. Hundreds of students who fled the cities during the 1988 crackdown on student demonstrations have now joined rural guerrilla organizations, such as the Burma Communist Party and the Karen National Union, to continue the fight against the military dictatorship. Among those most vigorously opposed to military rule are Buddhist monks. Five months after the elections, monks in the capital city of Yangon boycotted the government by refusing to conduct religious rituals for soldiers. Tens of thousands of people joined in the boycott. The government responded by threatening to shut down monasteries in Yangon and Mandalay.

The military government calls itself the State Law and Order Restoration Council (SLORC) and appears determined to stay in power. SLORC has kept Aung San Suu

Kyi under house arrest off and on for years, and watches her every move. For several years, even her husband and children were forbidden to visit her. In 1991, she was awarded the Nobel Peace Prize; in 1993, several other Nobelists gathered in nearby Thailand to call for her release from house arrest—a plea ignored by SLORC. The United Nations has shown its displeasure with the military junta by substantially cutting development funds, as has the United States, which, on the basis of Myanmar's heavy illegal-drug activities, has disqualified the country from receiving most forms of economic aid.

But perhaps the greatest pressure on the dictatorship is from within the country itself. Despite brutal suppression, the military seems to be losing control of the people. Both the Kachin and Karen ethnic groups have organized guerrilla move-

ments against the regime; in some cases, they have coerced foreign lumber companies to pay them protection money, which they have used to buy arms to fight against the junta. Opponents of SLORC control one third of Myanmar, especially along its eastern borders with Thailand and China and in the north alongside India. With the economy in shambles, the military appears to be involved with the heroin trade as a way of acquiring needed funds; it reportedly engages in bitter battles with drug lords periodically for control of the trade. To ease economic pressure, the military rulers have ended their monopoly of some businesses and have legalized the black market, making products from China, India, and Thailand available on the street.

Still, for ordinary people, especially those in the countryside, life is anything but pleasant. A 1994 human-rights study

In 1990, Myanmar's first elections in 30 years were held; a new opposition party, the National League for Democracy, headed by Aung San Suu Kyi, pictured above, won 80 percent of the seats in the National Assembly, but she was never permitted to take office. Instead, she was placed under house arrest for several years. In 1991, Aung San Suu Kyi was awarded the Nobel Peace Prize.

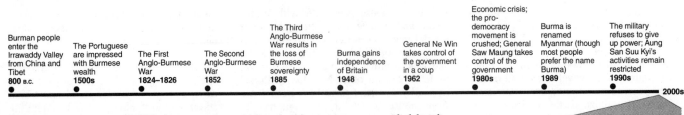

| Burman people enter the Irrawaddy Valley from China and Tibet **800 B.C.** | The Portuguese are impressed with Burmese wealth **1500s** | The First Anglo-Burmese War **1824–1826** | The Second Anglo-Burmese War **1852** | The Third Anglo-Burmese War results in the loss of Burmese sovereignty **1885** | Burma gains independence of Britain **1948** | General Ne Win takes control of the government in a coup **1962** | Economic crisis; the pro-democracy movement is crushed; General Saw Maung takes control of the government **1980s** | Burma is renamed Myanmar (though most people prefer the name Burma) **1989** | The military refuses to give up power; Aung San Suu Kyi's activities remain restricted **1990s** |

2000s

The world community increasingly registers its disapproval of the Myanmar junta

found that as many as 20,000 women and girls living in Myanmar near the Thai border had been abducted to work as prostitutes in Thailand. For several years, SLORC has carried out an "ethnic-cleansing" policy against villagers who have opposed their rule; thousands of people have been carried off to relocation camps, forced to work as slaves or prostitutes for the soldiers, or simply killed. Some 400,000 members of ethnic groups have fled the country, including 300,000 Arakans who escaped to Bangladesh and 5,000 Karenni, 12,000 Mon, and 50,000 Karens who fled to Thailand. Food shortages plague certain regions of the country, and many young children are forced to serve in the various competing armies rather than acquire an education or otherwise enjoy a normal childhood. Indeed, warfare and violence are the only reality many youth know. In 1999, a group of "Burmese Student Warriors" seized the Myanmar Embassy in Thailand, taking hostages and demanding talks between the military junta and Aung San Suu Kyi. The next year, a youth group calling itself God's Army and led by 12-year-old twin brothers, seized a hospital in Thailand and held 800 patients and staff hostage. The boys were from the Karen people, a Christian subculture long persecuted in Burma. The boys attacked Thai residents to protest their villages having been shelled by the Thai military, which is increasingly uneasy with the large number of Burmese refugees inside and along its border. Despite the lifting of martial law and some minor liberalization of the economy, it will be a long time before democracy will take hold in Myanmar.

THE CULTURE OF BUDDHA

For a brief period in the 1960s, Buddhism was the official state religion of Burma.

Although this status was repealed by the government in order to weaken the power of the Buddhist leadership, or *Sangha,* vis-à-vis the polity, Buddhism, representing the belief system of 89 percent of the population, remains the single most important cultural force in the country. Even the Burmese alphabet is based, in part, on Pali, the sacred language of Buddhism. Buddhist monks joined with college students after World War II to pressure the British government to withdraw from Burma, and they have brought continual pressure to bear on the current military junta.

Historically, so powerful has been the Buddhist Sangha in Burma that four major dynasties have fallen because of it. This has not been the result of ideological antagonism between church and state (indeed, Burmese rulers have usually been quite supportive of Buddhism) but, rather, because Buddhism soaks up resources that might otherwise go to the government or to economic development. Believers are willing to give money, land, and other resources to the religion, because they believe that such donations will bring them spiritual merit; the more merit one acquires, the better one's next life will be. Thus, all over Myanmar, but especially in older cities such as Pagan, one can find large, elaborate Buddhist temples, monuments, or monasteries, some of them built by kings and other royals on huge, untaxed parcels of land. These monuments drained resources from the government but brought to the donor unusual amounts of spiritual merit. As Burmese scholar Michael Aung-Thwin explained it: "One built the largest temple because one was spiritually superior, and one was spiritually superior because one built the largest temple."

Today, the Buddhist Sangha is at the forefront of the opposition to military rule. This is a rather unusual position for Bud-

dhists, who generally prefer a more passive attitude toward "worldly" issues. Monks have joined college students in peaceful-turned-violent demonstrations against the junta. Other monks have staged spiritual boycotts against the soldiers by refusing to accept merit-bringing alms from them or to perform weddings and funerals. The junta has retaliated by banning some Buddhist groups altogether and purging many others of rebellious leaders. The military regime now seems to be relaxing its intimidation of the Buddhists, has reopened universities, and has invited some foreign investment. Efforts by the junta to enhance tourism have been opposed by the outlawed National League for Democracy, on the grounds that any improvement in the economy would strengthen the hand of the military rulers. All 15 European Union members as well as the national U.S. Chamber of Commerce supported a 2000 Massachusetts state law that would have penalized companies for doing business with Myanmar; the law was thrown out by the U.S. Supreme Court, but the sentiment of broad opposition to the Myanmar dictatorship remained. Although the Japanese have invested in Myanmar throughout the military dictatorship, some potential investors from other countries refuse to invest in the brutal regime.

DEVELOPMENT

Primarily an agricultural nation, Myanmar has a poorly developed industrial sector. Until recently, the government forbade foreign investment and severely restricted tourism. In 1989, recognizing that the economy was on the brink of collapse, the government permitted foreign investment and signed contracts with Japan and others for oil exploration.

FREEDOM

Myanmar is a military dictatorship. Until 1989, only the Burma Socialist Program Party was permitted. Other parties, while now legal, are intimidated by the military junta. The democratically elected National League for Democracy has not been permitted to assume office. The government has also restricted the activities of Buddhist monks and has carried out "ethnic cleansing" against minorities.

HEALTH/WELFARE

The Myanmar government provides free health care and pensions to citizens, but the quality and availability of these services are erratic, to say the least. Malnourishment and preventable diseases are common, and infant mortality is high. Overpopulation is not a problem; Myanmar is one of the most sparsely populated nations in Asia.

ACHIEVEMENTS

Myanmar is known for the beauty of its Buddhist architecture. Pagodas and other Buddhist monuments and temples dot many of the cities, especially Pagan, one of Burma's earliest cities. Politically, it is notable that the country was able to remain free of the warfare that engulfed much of Indochina during the 1960s and 1970s.

New Zealand (Dominion of New Zealand)

GEOGRAPHY
Area in Square Miles (Kilometers):
98,874 (268,680) (about the size of Colorado)
Capital (Population): Wellington (335,500)
Environmental Concerns: deforestation; soil erosion; damage to native flora and fauna from outside species
Geographical Features: mainly mountainous with some large coastal plains
Climate: temperate; sharp regional contrasts

PEOPLE

Population
Total: 3,820,000
Annual Growth Rate: 1.17%
Rural/Urban Population (Ratio): 14/86
Major Languages: English; Maori
Ethnic Makeup: 88% New Zealand and other European; 9% Maori; 3% Pacific islander
Religions: 81% Christian; 18% unaffiliated; 1% others

Health
Life Expectancy at Birth: 75 years (male); 81 years (female)
Infant Mortality Rate (Ratio): 6.4/1,000
Physicians Available (Ratio): 1/318

Education
Adult Literacy Rate: 100%
Compulsory (Ages): 6–16; free

COMMUNICATION
Telephones: 1,870,000 main lines
Daily Newspaper Circulation: 239 per 1,000 people
Televisions: 514 per 1,000 people
Internet Service Providers: 56 (1999)

TRANSPORTATION
Highways in Miles (Kilometers): 57,164 (92,200)
Railroads in Miles (Kilometers): 2,383 (3,813)
Usable Airfields: 111
Motor Vehicles in Use: 2,053,000

GOVERNMENT
Type: parliamentary democracy
Independence Date: September 26, 1907 (from the United Kingdom)
Head of State/Government: Queen Elizabeth II; Prime Minister Helen Clark
Political Parties: ACT; Alliance (a coalition); National Party; New Zealand Labour Party; New Zealand First Party; Democratic Party; New Zealand Liberal Party; Green Party; Mana Motuhake; others
Suffrage: universal at 18

MILITARY
Military Expenditures (% of GDP): 1.1%
Current Disputes: disputed territorial claim in Antarctica

ECONOMY
Currency ($ U.S. Equivalent): 2.44 New Zealand dollars = $1
Per Capita Income/GDP: $17,400/$63.8 billion
GDP Growth Rate: 3.1%
Inflation Rate: 1.3%
Unemployment Rate: 7%
Labor Force: 1,860,000

Natural Resources: natural gas; iron ore; sand; coal; timber; hydropower; gold; limestone
Agriculture: wool; beef; dairy products; wheat; barley; potatoes; pulses; fruits; vegetables; fishing
Industry: food processing; wood and paper products; textiles; machinery; transportation equipment; banking; insurance; tourism; mining
Exports: $12.2 billion (primary partners Australia, Japan, United States)
Imports: $11.2 billion (primary partners Australia, United States, Japan)

http://www.cia.gov/cia/publications/factbook/geos/nz.html
http://dir.yahoo.com/regional/countries/New_Zealand/

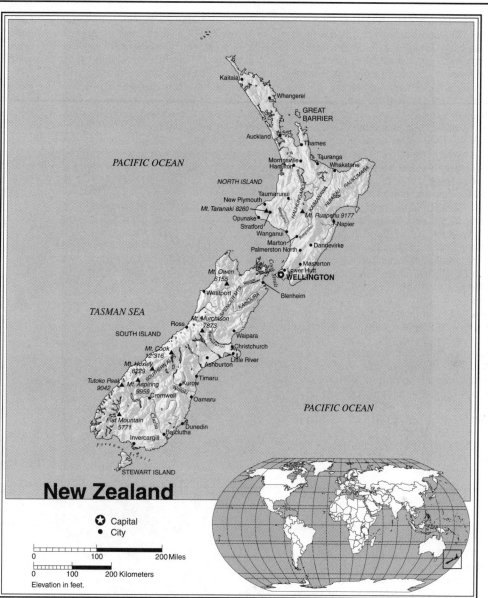

New Zealand

⭐ Capital
● City

0   100   200 Miles
0   100   200 Kilometers
Elevation in feet.

ITS PLACE IN THE WORLD

New Zealand, like Australia, is decidedly an anomaly among Pacific Rim countries. Eighty-eight percent of the population are of European descent (primarily British), English is the official language, and most people, even many of the original Maori inhabitants, are Christians. Britain claimed the beautiful, mountainous islands officially in 1840, after agreeing to respect the property rights of Maoris, most of whom lived on the North Island. Blessed with a temperate climate and excellent soils for crop and dairy farming, New Zealand—divided into two main islands, North Island and South Island—became an important part of the British Empire.

New Zealand, although largely self-governing since 1907 and fully independent as of 1947, has always maintained very close ties with the United Kingdom and is a member of the Commonwealth of Nations. It has, in fact, attempted to re-create British culture—customs, architecture, even vegetation—in the Pacific. So close were the links with Great Britain in the 1940s, for example, that England purchased fully 88 percent of New Zealand's exports (mostly agricultural and dairy products), while 60 percent of New Zealand's imports came from Britain. And believing itself to be very much a part of the British Empire, New Zealand always sided with the Western nations on matters of military defense.

Efforts to maintain a close cultural link with Great Britain do not stem entirely from the common ethnicity of the two nations; they also arise from New Zealand's extreme geographical isolation from the centers of European and North American activity. Even Australia is more than 1,200 miles away. Therefore, New Zealand's policy—until the 1940s—was to encourage the British presence in Asia and the Pacific, by acquiring more lands or building up naval bases, to make it more likely that Britain would be willing and able to defend New Zealand in a time of crisis. New Zealand had involved itself somewhat in the affairs of some nearby islands in the late 1800s and early 1900s, but its purpose was not to provide development assistance or defense. Rather, its aim was to extend the power of the British Empire and put New Zealand in the middle of a mini-empire of its own. To that end, New Zealand annexed the Cook Islands in 1901 and took over German Samoa in 1914. In 1925, it assumed formal control over the atoll group known as the Tokelau Islands.

REGIONAL RELIANCE

During World War II (or, as the Japanese call it, the Pacific War), Japan's rapid conquest of the Malay Archipelago, its seizure of many Pacific islands, and its plans to attack Australia demonstrated to New Zealanders the futility of relying on the British to guarantee their security. After the war, and for the first time in its history, New Zealand began to pay serious attention to the real needs and ambitions of the peoples nearby rather than to focus on Great Britain. In 1944 and again in 1947, New Zealand joined with Australia and other colonial nations to create regional associations on behalf of the Pacific islands. One of the organizations, the South Pacific Commission, has itself spawned many regional subassociations dealing with trade, education, migration, and cultural and economic development. Al-

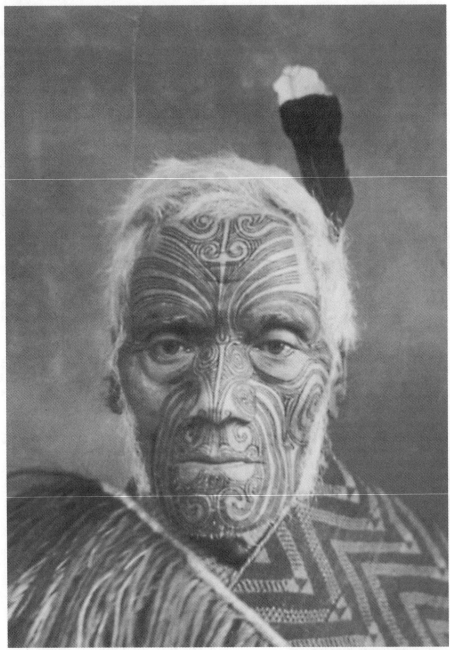

(The Peabody Museum of Salem photo)

Maoris occupied New Zealand long before the European settlers moved there. The Maoris quickly realized that the newcomers were intent on depriving them of their land, but it was not until the 1920s that the government finally regulated unscrupulous land-grabbing practices. Today, the Maoris pursue a lifestyle that preserves key parts of their traditional culture while incorporating skills necessary for survival in the modern world.

though it had neglected the islands that it controlled during its imperial phase, in the early 1900s, New Zealand cooperated fully with the United Nations in the islands' decolonization during the 1960s (although Tokelau, by choice, and the Ross dependency remain under New Zealand's control), while at the same time increasing development assistance. New Zealand's first alliance with Asian nations came in 1954, when it joined the Southeast Asian Treaty Organization.

New Zealand's new international focus certainly did not mean the end of cooperation with its traditional allies, however. In fact, the common threat of the Japanese during World War II strengthened cooperation with the United States to the extent that, in 1951, New Zealand joined a three-way, regional security agreement known as ANZUS (for Australia, New Zealand, and the United States). Moreover, because the United States was, at war's end, a Pacific/Asian power, any agreement with the United States was likely to bring New Zealand into more, rather than less, contact with Asia and the Pacific. Indeed, New Zealand sent troops to assist in all of the United States' military involvements in Asia: the occupation of Japan in 1945, the Korean War in 1950, and the Vietnam War in the 1960s. And, as a member of the British Commonwealth, New Zealand sent troops in the 1950s and 1960s to fight Malaysian Communists and Indonesian insurgents.

A NEW INTERNATIONALISM

Beginning in the 1970s, especially when the Labour Party of Prime Minister Norman Kirk was in power, New Zealand's orientation shifted even more markedly toward its own region. Under the Labour Party, New Zealand defined its sphere of interest and responsibility as the Pacific, where it hoped to be seen as a protector and benefactor of smaller states. Of immediate concern to many island nations was the issue of nuclear testing in the Pacific. Both the United States and France had undertaken tests by exploding nuclear devices on tiny Pacific atolls. In the 1960s, the United States ceased these tests, but France continued. On behalf of the smaller islands, New Zealand argued before the United Nations against testing, but France still did not stop. Eventually, the desire to end testing congealed into

the more comprehensive position that the entire Pacific should be declared a nuclear-free zone. Not only testing but also the transport of nuclear weapons through the area would be prohibited under the plan.

New Zealand's Labour government issued a ban on the docking of ships with nuclear weapons in New Zealand, despite the fact that such ships were a part of the ANZUS agreement. When the National Party regained control of the government in the late 1970s, the nuclear ban was revoked, and the foreign policy of New Zealand tipped again toward its traditional allies. The National Party government argued that, as a signatory to ANZUS, New Zealand was obligated to open its docks to U.S. nuclear ships. However, under the subsequent Labour government of Prime Minister David Lange, New Zealand once again began to flex its muscles over the nuclear issue. Lange, like his Labour Party predecessors, was determined to create a foreign policy based on moral rather than legal rationales. In 1985, a U.S. destroyer was denied permission to call at a New Zealand port, even though its presence there was due to joint ANZUS military exercises. Because the United States refused to say whether or not its ship carried nuclear weapons, New Zealand insisted that the ship could not dock. Diplomatic efforts to resolve the standoff were unsuccessful; and in 1986, New Zealand, claiming that it was not fearful of foreign attack, formally withdrew from ANZUS.

The issue of use of the Pacific for nuclear-weapons testing by superpowers is still of major concern to the New Zealand government. The nuclear test ban treaty signed by the United States in 1963 has limited U.S. involvement in that regard, but France has continued to test atmospheric weapons, and at times both the United States and Japan have proposed using uninhabited Pacific atolls to dispose of nuclear waste. In 1995, when France ignored the condemnation of world leaders and detonated a nuclear device in French Polynesia, New Zealand recalled its ambassador to France out of protest.

In the early 1990s, a new issue came to the fore: nerve-gas disposal. With the end of the Cold War, the U.S. military proposed disposing of most of its European stockpile of nerve gas on an atoll in the Pacific. The atoll is located within the

trust territory granted to the United States at the conclusion of World War II. The plan is to burn the gas away from areas of human habitation, but those islanders living closest (albeit hundreds of miles away) worry that residues from the process could contaminate the air and damage humans, plants, and animals. The religious leaders of Melanesia, Micronesia, and Polynesia have condemned the plan, not only on environmental grounds but also on grounds that outside powers should not be permitted to use the Pacific region without the consent of the inhabitants there—a position with which the Labour government of New Zealand strongly concurs.

ECONOMIC CHALLENGES

The New Zealand government's new foreign-policy orientation has caught the attention of observers around the world, but more urgent to New Zealanders themselves is the state of their own economy. Until the 1970s, New Zealand had been able to count on a nearly guaranteed export market in Britain for its dairy and agricultural products. Moreover, cheap local energy supplies as well as inexpensive oil from the Middle East had produced several decades of steady improvement in the standard of living. Whenever the economy showed signs of being sluggish, the government would artificially protect certain industries to ensure full employment.

All of this came to a halt in 1973, when Britain joined the European Union (then called the European Economic Community) and when the Organization of Petroleum Exporting Countries sent the world into its first oil shock. New Zealand actually has the potential of near self-sufficiency in oil, but the easy availability of Middle East oil over the years has prevented the full development of local oil and natural-gas reserves. As for exports, New Zealand had to find new outlets for its agricultural products, which it did by contracting with various countries throughout the Pacific Rim. Currently, about one third of New Zealand's trade occurs within the Pacific Rim. In the transition to these new markets, farmers complained that the manufacturing sector—intentionally protected by the government as a way of diversifying New Zealand's reliance on agriculture—was getting unfair favorable treatment. Subsequent changes in government policy toward industry resulted in a

Socialized
medicine is
implemented
1941

New Zealand
becomes fully
independent
within the British
Commonwealth
of Nations
1947

New Zealand
backs creation of
the South Pacific
Commission

Restructuring of
export markets
1950s

The National Party
takes power; New
Zealand forges
foreign policy more
independent of
traditional allies
1970s

The Labour Party
regains power;
New Zealand
withdraws from
ANZUS
1980s

New Zealanders
consider withdrawing
from the Commonwealth;
Maoris and white New
Zealanders face
economic challenges
from other Pacific
Rimmers
1990s

2000s

New Zealand
is led by its
second
woman prime
minister,
Helen Clark

New Zealand
champions
cultural and
environmental
goals

Parliament
debates the
merits of a
proposed
parent-leave
law

new phenomenon for New Zealand: high unemployment. Moreover, New Zealand had constructed a rather elaborate social-welfare system since World War II, so, regardless of whether economic growth was high or low, social-welfare checks still had to be sent. This untenable position has made for a difficult political situation, for, when the National Party cut some welfare benefits and social services, it lost the support of many voters. The welfare issue, along with a change to a mixed member proportional voting system that enhanced the influence of smaller parties, threatened the National Party's political power. Thus, in order to remain politically dominant, in 1996 the National Party was forced to form a coalition with the United Party—the first such coalition government in more than 60 years.

In the 1970s, for the first time, New Zealanders began to notice a decline in their standard of living. Two decades later, the economy is not greatly improved. New Zealand's economic growth rate is estimated at only 3.1 percent per year; the unemployment rate is a worrisome 7 percent; and its per capita income is lower than in Hong Kong, Australia, and Japan. In 2000, the Labour Party prime minister, Helen Clark, had to cancel a large purchase of F-16 fighter jets from the United States. She said that New Zealand could not afford the purchase. But a few good economic signs have appeared: Inflation in 1999 was at its lowest levels since the 1930s (1.3 percent), interest rates had been cut, and the government was able to realize a budget surplus—prompting calls for a national tax cut.

New Zealanders are well aware of Japan's economic strength and its potential for benefiting their own economy through joint ventures, loans, and trade. Yet they also worry that Japanese wealth may constitute a symbol of New Zealand's declining strength as a culture. For instance, in the 1980s, as Japanese tourists began traveling en masse to New Zealand, complaints were raised about the quality of New Zealand's hotels. Unable to find the funds for a massive upgrading of the hotel industry, New Zealand agreed to allow Japan to build its own hotels; it reasoned that the local construction industry could use an economic boost and that the better hotels would encourage well-heeled Japanese to spend even more tourist dollars in the country. However, they also worried that, with the Japanese owning the hotels, New Zealanders might be relegated to low-level jobs.

Concern about their status vis-à-vis non-whites had never been much of an issue to many Anglo-Saxon New Zealanders; they always simply assumed that non-whites were inferior. Many settlers of the 1800s believed in the Social Darwinistic philosophy that the Maori and other brown- and black-skinned peoples would gradually succumb to their European "betters." It did not take long for the Maoris to realize that, land guarantees notwithstanding, the whites intended to deprive them of their land and culture. Violent resistance to these intentions occurred in the 1800s, but Maori landholdings continued to be gobbled up, usually deceptively, by white farmers and sheep herders. Government control of these unscrupulous practices was lax until the 1920s. Since that time, many Maoris (whose population has increased to about 260,000) have intentionally sought to create a lifestyle that preserves key parts of traditional culture while incorporating the skills necessary for survival in a white world. In 1999, Maoris on the South Island, using land-loss funds provided by the government, made such a large land purchase (nearly

300,000 acres) that they became the largest land owner on the island.

More than Australians, New Zealanders are attempting to rectify the historic discrimination against the country's indigenous peoples. The current cabinet, for example, includes four Maoris and a member of Pacific Island descent (as well as 11 women). The new Labour/Green Party coalition government has also attempted to eliminate some of the colonial cobwebs from their society by abolishing knighthoods bestowed by the British Crown. Local honors are awarded instead. So, New Zealand has been reaching out to change the social landscape.

Now, though, Maoris and whites alike feel the social leveling that is the consequence of years of economic stagnation. Moreover, both worry that the superior financial strength of the Japanese and newly industrializing Asian and Southeast Asian peoples may diminish in some way the standing of their own cultures. The Maoris, complaining recently about Japanese net fishing and its damage to their own fishing industry, have a history of accommodation and adjustment to those who would rule over them; but for the whites, submissiveness, even if it is imposed from afar and is largely financial in nature, will be a new and challenging experience.

DEVELOPMENT

Government protection of manufacturing has allowed this sector to grow at the expense of agriculture. Nevertheless, New Zealand continues to export large quantities of dairy products, wool, meat, fruits, and wheat. Full development of the country's oil and natural-gas deposits could alleviate New Zealand's dependence on foreign oil.

FREEDOM

New Zealand partakes of the democratic heritage of English common law and subscribes to all the human-rights protections that other Western nations have endorsed. Maoris, originally deprived of much of their land, are now guaranteed the same legal rights as whites. Social discrimination against Maoris is much milder than with many other colonized peoples.

HEALTH/WELFARE

New Zealand established pensions for the elderly as early as 1898. Child-welfare programs were started in 1907, followed by the Social Security Act of 1938, which augmented the earlier benefits and added a minimum-wage requirement and a 40-hour work week. A national health program was begun in 1941. The government began dispensing free birth-control pills to all women in 1996 in an attempt to reduce the number of abortions.

ACHIEVEMENTS

New Zealand is notable for its efforts on behalf of the smaller islands of the Pacific. In addition to advocating a nuclear-free Pacific, New Zealand has promoted interisland trade and has established free-trade agreements with Western Samoa, the Cook Islands, and Niue. It provides educational and employment opportunities to Pacific islanders who reside within its borders.

North Korea (Democratic People's Republic of Korea)*

GEOGRAPHY

Area in Square Miles (Kilometers): 44,358 (120,540) (about the size of Mississippi)

Capital (Population): P'yongyang (2,750,000)

Environmental Concerns: air and water pollution; insufficient potable water

Geographical Features: mostly hills and mountains separated by deep, narrow valleys; coastal plains

Climate: temperate

PEOPLE

Population

Total: 21,688,000

Annual Growth Rate: 1.35%

Rural/Urban Population (Ratio): 38/62

Major Language: Korean

Ethnic Makeup: homogeneous Korean

Religions: mainly Buddhist and Confucianist (autonomous religious activities now almost nonexistent)

Health

Life Expectancy at Birth: 68 years (male); 74 years (female)

Infant Mortality Rate (Ratio): 24.2/1,000

Physicians Available (Ratio): 1/370

Education

Adult Literacy Rate: 99%

Compulsory (Ages): 6–17; free

COMMUNICATION

Telephones: 1,100,000 main lines

Daily Newspaper Circulation: 213 per 1,000 people

Televisions: 85 per 1,000 people

Internet Service Providers: na

TRANSPORTATION

Highways in Miles (Kilometers): 19,345 (31,200)

Railroads in Miles (Kilometers): 3,000 (4,800)

Usable Airfields: 49

GOVERNMENT

Type: authoritarian socialist; one-man dictatorship

Independence Date: September 9, 1948

Head of State/Government: President Kim Jong-Il; Premier Hong Song-nam

Political Parties: Korean Workers' Party; Korean Social Democratic Party; Chondoist Chongu Party

Suffrage: universal at 17

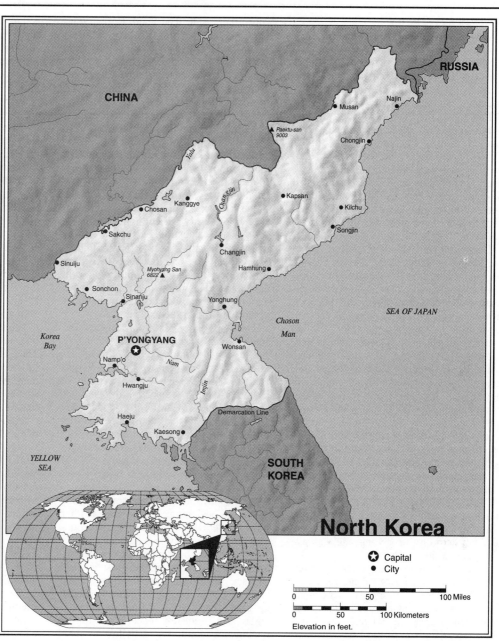

North Korea

❂ Capital
● City

Elevation in feet.

MILITARY

Military Expenditures (% of GDP): 25%–33%

Current Disputes: Demarcation Line with South Korea; unclear border with China

ECONOMY

Currency ($ U.S. Equivalent): 2.2 won = $1

Per Capita Income/GDP: $1,000/$22.6 billion

GDP Growth Rate: 1%

Labor Force: 9,615,000

Natural Resources: hydropower; iron ore; copper; lead; salt; zinc; coal; magnesite; gold; tungsten; graphite; pyrites; fluorspar

Agriculture: rice; corn; potatoes; soybeans; pulses; livestock

Industry: machinery; military products; electric power; chemicals; mining; metallurgy; textiles; food processing; tourism

Exports: $680 million (primary partners Japan, South Korea, China)

Imports: $954 million (primary partners China, Japan, Russia)

Note: Statistics for North Korea are generally estimated due to unreliable official information.

http://www.skas.org
http://memory.loc.gov/frd/cs/kptoc.html

A COUNTRY APART
The area that we now call North Korea has, at different times in Korea's long history, been separated from the South. In the fifth century A.D., the Koguryo Kingdom in the North was distinct from the Shilla, Paekche, and Kaya Kingdoms of the South. Later, the Parhae Kingdom in the North remained separate from the expanded Shilla Kingdom in the South. Thus, the division of Korea in 1945 into two unequal parts was not without precedent. Yet this time, the very different paths of development that the North and South chose rendered the division more poignant and, to those separated with little hope of reunion, more emotionally painful.

Beginning in 1945, Kim Il-Song, with the strong backing of the Soviet Union, pursued a hard-line Communist policy for both the political and economic development of North Korea. The Soviet Union's involvement on the Korean Peninsula arose from its opportunistic entry into the war against Japan, just eight days before Japan's surrender to the Allies. Thus, when Japan withdrew from its long colonial rule over Korea, the Soviets, an allied power, were in a position to be one of the occupying armies. Reluctantly, the United States allowed the Soviet Union to move troops into position above the 38th Parallel, a temporary dividing line for the respective occupying forces. It was the Soviet Union's intention to establish a Communist buffer state between itself and the capitalist West. Therefore, it moved quickly to establish the area north of the 38th Parallel as a separate political entity. The northern city of P'yongyang was established as the capital.

When United Nations representatives arrived in 1948 to oversee elections and ease the transition from military occupation and years of Japanese rule to an independent Korea, the Soviets would not cooperate. Kim Il-Song took over the reins of power in the North. Separate elections were held in the South, and the beginning of separate political systems got underway. The 38th Parallel came to represent not only the division of the Korean Peninsula but also the boundary between the worlds of capitalism and communism.

THE KOREAN WAR (1950–1953)
Although not pleased with the idea of division, the South, without a strong army, resigned itself to the reality of the moment. In the North, a well-trained military, with Soviet and Chinese help, began preparations for a full-scale invasion of the South. The North attacked in June 1950, a year after U.S. troops had vacated the South, and quickly overran most of the

Korean Peninsula. The South Korean government requested help from the United Nations, which dispatched personnel from 19 nations, under the command of U.S. general Douglas MacArthur. (A U.S. intervention was ordered on June 27 by President Harry Truman.)

MacArthur's troops advanced against the North's armies and by October were in control of most of the peninsula. However, with massive Chinese help, the North once again moved south. In response, UN troops decided to inflict heavy destruction on the North through the use of jet fighter/bomber planes. Whereas South Korea was primarily agricultural, North Korea was the industrialized sector of the peninsula. Bombing of the North's industrial targets severely damaged the economy, forcing several million North Koreans to flee south to escape both the war and the Communist dictatorship under which they found themselves.

Eventually, the UN troops recaptured South Korea's capital, Seoul. Realizing that further fighting would lead to an expanded Asian war, the two sides agreed to cease-fire talks. They signed a truce in 1953 that established a 2.5-mile-wide demilitarized zone (DMZ) for 155 miles across the peninsula and more or less along the former 38th Parallel division. The Korean War took the lives of more than 54,000 American soldiers, 58,000 South Koreans, and 500,000 North Koreans—but when it was over, both sides occupied about the same territory as they had at the beginning. Yet, because neither side has ever declared peace, the two countries remain officially in a state of war.

For years, the border between North and South has been one of the most volatile in Asia. The North staged military exercises along the border in 1996; breaking the cease-fire of 1953, it fired shots in the DMZ. The South responded by raising its intelligence-monitoring activities to their highest level in years, and by requesting U.S. AWACS surveillance planes to monitor military movements in the North.

Scholars are still debating whether the Korean War should be called the United States' first losing war and whether or not the bloodshed was really necessary. To understand the Korean War, one must remember that in the eyes of the world, it was more than a civil war among different kinds of Koreans. The United Nations, and particularly the United States, saw North Korea's aggression against the South as the first step in the eventual communization of the whole of Asia. Just a few months before North Korea attacked, China had fallen to the Communist forces of Mao Zedong, and Communist guerrilla

activity was being reported throughout Southeast Asia. The "Red Scare" frightened many Americans, and witchhunting for suspected Communist sympathizers—a college professor who might have taught about Karl Marx in class or a news reporter who might have praised the educational reforms of a Communist country—became the everyday preoccupation of such groups as the John Birch Society and the supporters of U.S. senator Joseph McCarthy.

In this highly charged atmosphere, it was relatively easy for the U.S. military to promote a war whose aim it was to contain communism. Containment rather than defeat of the enemy was the policy of choice, because the West was weary after battling Germany and Japan in World War II. The containment policy also underlay the United States' approach to Vietnam. Practical though it may have been, this policy denied Americans the opportunity of feeling satisfied in victory, since there was to be no victory, just a stalemate. Thus, the roots of the United States' dissatisfaction with the conduct of the Vietnam War actually began in the policies shaping the response to North Korea's offensive in 1950. North Korea was indeed contained, but the communizing impulse of the North remained.

COLLECTIVE CULTURE
With Soviet backing, North Korean leaders moved quickly to repair war damage and establish a Communist culture. The school curriculum was rewritten to emphasize nationalism and equality of the social classes. Traditional Korean culture, based on Confucianism, had stressed strict class divisions, but the Communist authorities refused to allow any one class to claim privileges over another (although eventually the families of party leaders came to constitute a new elite). Higher education at the more than 600 colleges and training schools was redirected to technical rather than analytical subjects. Industries were nationalized; farms were collectivized into some 3,000 communes; and the communes were invested with much of the judicial and executive powers that other countries grant to cities, counties, and states. To overcome labor shortages, nearly all women were brought into the workforce, and the economy slowly returned to prewar levels.

Today, many young people bypass formal higher education in favor of service in the military. Although North Korea has not published economic statistics for nearly 30 years, it is estimated that military expenses consume one quarter to one third of the entire national budget—this despite near-starvation conditions in many parts of the country.

With China and the former Communist-bloc nations constituting natural markets for North Korean products, and with substantial financial aid from both China and the former Soviet Union in the early years, North Korea was able to regain much of its former economic, and especially industrial, strength. Today, North Korea successfully mines iron and other minerals and exports such products as cement and cereals. China has remained North Korea's only reliable ally; trade between the two countries is substantial. In one Chinese province, more than two thirds of the people are ethnic Koreans, most of whom take the side of the North in any dispute with the South.

Tensions with the South have remained high since the war. Sporadic violence along the border has left patrolling soldiers dead, and the assassination of former South Korean president Park Chung Hee and attempts on the lives of other members of the South Korean government have been attributed to North Korea, as was the bombing of a Korean Airlines flight in 1987. Both sides have periodically accused each other of attempted sabotage. In 1996, North Korea tried to send spies to the South via a small submarine; the attempt failed, and most of the spies were killed.

The North, has long criticized the South for its suppression of dissidents. Although the North's argument is bitterly ironic, given its own brutal suppression of human rights, it is nonetheless accurate in its view that the government in the South has been blatantly dictatorial. To suppress opponents, the South Korean government has, among other things, abducted its own students from Europe, abducted opposition leader Kim Dae Jung from Japan, tortured dissidents, and violently silenced demonstrators. All of this is said to be necessary because of the need for unity in the face of the threat from the North; as pointed out by scholar Gavan McCormack, the South seems to use the North's threat as an excuse for maintaining a rigid dictatorial system.

Under these circumstances, it is not surprising that the formal reunification talks, begun in 1971 with much fanfare, have just recently started to bear fruit. Visits of residents separated by the war were approved in 1985—the first time in 40 years that an opening of the border had even been considered. In 1990, in what many saw as an overture to the United States, North Korea returned the remains of five American soldiers killed during the Korean War. But real progress came in late 1991, when North Korean premier Yon Hyong Muk and South Korean premier Chung Won Shik signed a nonaggression and reconciliation pact, whose goal was the eventual declaration of a formal peace treaty between the two governments. In 1992, the governments established air, sea, and land links and set up mechanisms for scientific and environmental cooperation. North Korea also signed the nuclear non-proliferation agreement with the International Atomic Energy Agency. This move placated growing concerns about North Korea's rumored development of nuclear weapons and opened the way for investment by such countries as Japan, which had refused to invest until they received assurances on the nuclear question.

THE NUCLEAR ISSUE FLARES UP

The goodwill deteriorated quickly in 1993 and 1994, when North Korea refused to allow inspectors from the International Atomic Energy Agency (IAEA) to inspect its nuclear facilities, raising fears in the United States, Japan, and South Korea that the North was developing a nuclear bomb. When pressured to allow inspections, the North responded by threatening to withdraw from the IAEA and expel the inspectors. Tensions mounted, with all parties engaging in military threats and posturing and the United States, South Korea, and Japan (whose shores could be reached in minutes by the North's new ballistic missiles) threatening economic sanctions. Troops in both Koreas were put on high alert. Former U.S. president Jimmy Carter helped to defuse the issue by making a private goodwill visit to Kim Il-Song in P'yongyang, the unexpected result of which was a promise by the North to hold a first-ever summit meeting with the South. Then, in a near-theatrical turn of events, Kim Il-Song, at five decades the longest national office-holder in the world, died, apparently of natural causes. The summit was canceled and international diplomacy was frozen while the North Korean government mourned the loss of its "Great Leader" and informally selected a new one, "Dear Leader" Kim Jong-Il, Kim Il-Song's son. Eventually, the North agreed to resume talks, a move interpreted as evidence that, for all its bravado, the North wanted to establish closer ties with the West. In 1994, North Korea agreed to a freeze on nuclear power-plant development as long as the United States would supply fuel oil; and in 1996, it agreed to open its airspace to all airlines. Unfortunately, tensions increased when North Korea launched a missile over Japan in 1998 and promised to keep doing so.

In the midst of these developments, a most amazing breakthrough occurred: the North agreed to a summit. In June 2000, President Kim Dae Jung of the South flew to P'yongyang in the North for talks with President Kim Jung-Il. Unlike anything the South had expected, the South Korean president was greeted with cheers from the crowds lining the streets, and was feted at a state banquet. An agreement was reached that seemed to pave the way for peace and eventual reunification. To reward the North for this dramatic improvement in relations, the United States immediately lifted trade sanctions that had been in place for 50 years. Both the North

(UPI/Bettmann photo by Norman Williams)

Pictured above are U.S. Marines with North Koreans captured during the Korean War.

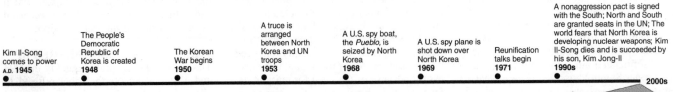

Kim Il-Song comes to power A.D. **1945**

The People's Democratic Republic of Korea is created **1948**

The Korean War begins **1950**

A truce is arranged between North Korea and UN troops **1953**

A U.S. spy boat, the *Pueblo*, is seized by North Korea **1968**

A U.S. spy plane is shot down over North Korea **1969**

Reunification talks begin **1971**

A nonaggression pact is signed with the South; North and South are granted seats in the UN; The world fears that North Korea is developing nuclear weapons; Kim Il-Song dies and is succeeded by his son, Kim Jong-Il **1990s**

2000s

Famine causes thousands to flee the country

North and South Korea meet in a dramatic summit in 2000; North and South Koreans march together in the 2000 Summer Olympics

U.S secretary of state Madeleine Albright visits North Korea; tensions increase with the Bush administration

and the South suspended propaganda broadcasts and began plans for reuniting families long separated by the fortified DMZ. The North even went so far as to establish diplomatic ties with Australia and Italy.

THE CHANGING INTERNATIONAL LANDSCAPE

North Korea has good reason to promote better relations with the West, because the world of the 1990s is not the world of the 1950s. In 1989, for instance, several former Soviet-bloc countries cut into the North's economic monopoly by welcoming trade initiatives from South Korea; some even established diplomatic relations. At the same time, the disintegration of the Soviet Union meant that North Korea lost its primary political and military ally. Perhaps most alarming to the North is its declining economy; it has suffered a negative growth for several years. Severe flooding in 1995 destroyed much of the rice harvest and forced the North to do the unthinkable: accept rice donations from the South. More than 100 North Koreans have defected to the South in the past several years, all of them complaining of near-famine conditions. Even more, perhaps as many as 2,000 per month, have fled to China. With the South's economy consistently booming and the example of the failed economies of Central/Eastern Europe as a danger signal, the North appears to understand that it must break out of its decades of isolation or lose its ability to govern. Nevertheless, it is not likely that North Koreans will quickly retreat from the Communist model of development that they have espoused for so long.

Kim Il-Song, who controlled North Korea for nearly 50 years, promoted the development of heavy industries, the collectivization of agriculture, and strong linkages with the then–Communist bloc. Governing with an iron hand, Kim denied basic civil rights to his people and forbade

any tendency of the people to dress or behave like the "decadent" West. He kept tensions high by asserting his intention of communizing the South. His son, Kim Jong-Il, who had headed the North Korean military but was barely known outside his country, was eventually named successor to his father—the first dynastic power transfer in the Communist world. How the younger Kim will influence the direction of North Korea is still unclear, but the somewhat more liberal authorities at his side know that the recent diplomatic initiatives of the South require a response. The North Korean government hopes that recent actions will bring it some badly needed international goodwill. But more than good public relations will be needed if North Korea is to prosper in the new, post–Cold War climate in which it can no longer rely on the generosity or moral support of the Soviet bloc. When communism was introduced in North Korea in 1945, the government nationalized major companies and steered economic development toward heavy industry. In contrast, the South concentrated on heavy industry to balance its agricultural sector until the late 1970s but then geared the economy toward meeting consumer demand. Thus, the standard of living in the North for the average resident remains far behind that of the South. Indeed, Red Cross, United Nations, and other observers have documented widespread malnutrition and starvation in North Korea, conditions that are likely to continue well into the twenty-first century unless the North dramatically alters its current economic policies. Conditions are so bad in some areas that there is no electricity nor chlorine to run water-treatment plants, resulting in contaminated water supplies for about 60 percent of the population.

Even before the June 2000 summit, North Korea exhibited signs of liberalization. The government agreed to allow foreign companies to establish joint ventures inside the country, tourism was being promoted as a way of earning foreign cur-

rency, and two small Christian churches allowed to be established. Nevertheless, years of a totally controlled economy in the North and shifting international alliances indicate many difficult years ahead for North Korea.

RECENT TRENDS

Although political reunification still seems to be years away, social changes are becoming evident everywhere as a new generation, unfamiliar with war, comes to adulthood, and as North Koreans are being exposed to outside sources of news and ideas. Many North Koreans now own radios that receive signals from other countries. South Korean stations are now heard in the North, as are news programs from the Voice of America. Modern North Korean history, however, is one of repression and control, first by the Japanese and then by the Kim government, who used the same police surveillance apparatus as did the Japanese during their occupation of the Korean Peninsula. It is not likely, therefore, that a massive push for democracy will be forthcoming soon from a people long accustomed to dictatorship.

DEVELOPMENT

Already more industrialized than South Korea at the time of the Korean War, North Korea built on this foundation with massive assistance from China and the Soviet Union. Heavy industry was emphasized to the detriment of consumer goods. Economic isolation presages more negative growth ahead.

FREEDOM

The mainline Communist approach has meant that the human rights commonplace in the West have never been enjoyed by North Koreans. Through suppression of dissidents, a controlled press, and restrictions on travel, the regime has kept North Koreans isolated from the world.

HEALTH/WELFARE

Under the Kim Il-Song government, illiteracy was greatly reduced. Government housing is available at low cost, but shoppers are often confronted with empty shelves and low-quality goods. Malnutrition is widespread, and mass starvation has been reported in some regions.

ACHIEVEMENTS

North Korea has developed its resources of aluminum, cement, and iron into solid industries for the production of tools and machinery while developing military superiority over South Korea, despite a population numbering less than half that of South Korea.

Papua New Guinea (Independent State of Papua New Guinea)

GEOGRAPHY

Area in Square Miles (Kilometers): 178,612 (461,690) (about the size of California)

Capital (Population): Port Moresby (192,000)

Environmental Concerns: deforestation; pollution from mining projects; drought

Geographical Features: mostly mountains; coastal lowlands and rolling foothills

Climate: tropical monsoon

PEOPLE

Population

Total: 4,927,000

Annual Growth Rate: 2.47%

Rural/Urban Population Ratio: 84/16

Major Languages: English; New Guinea Pidgin; Motu; 715 indigenous languages

Ethnic Makeup: predominantly Melanesian and Papuan; some Negrito, Micronesian, and Polynesian

Religions: 66% Christian; 34% indigenous beliefs

Health

Life Expectancy at Birth: 61 years (male); 65 years (female)

Infant Mortality Rate (Ratio): 59.8/1,000

Physicians Available (Ratio): 1/5,584

Education

Adult Literacy Rate: 72%

COMMUNICATION

Telephones: 47,000 main lines

Daily Newspaper Circulation: 15 per 1,000 people

Televisions: 23 per 1,000 people

Internet Service Providers: 2 (1999)

TRANSPORTATION

Highways in Miles (Kilometers): 11,904 (19,200)

Railroads in Miles (Kilometers): none

Usable Airfields: 492

Motor Vehicles in Use: 99,000

GOVERNMENT

Type: parliamentary democracy

Independence Date: September 16, 1975 from the Australian-administered UN trusteeship)

Head of State/Government: Queen Elizabeth II; Prime Minister Sir Mekere Moruata

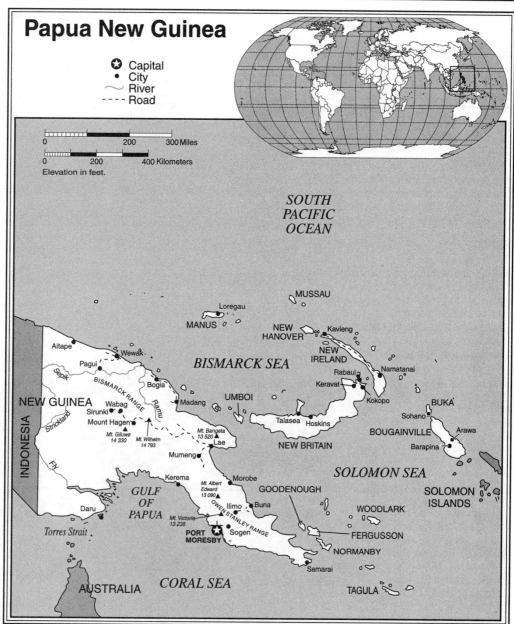

Political Parties: Black Party; People's Democratic Movement; People's Action Party; People's Progress Party; United Party; Papua Party; National Party; Melanesian Alliance; others

Suffrage: universal at 18

MILITARY

Military Expenditures (% of GDP): 1%

Current Disputes: none

ECONOMY

Currency ($ U.S. Equivalent): 3.33 kina = $1

Per Capita Income/GDP: $2,500/$11.6 billion

GDP Growth Rate: 3.6%

Inflation Rate: 16.5%

Labor Force: 1,941,000

Natural Resources: gold; copper; silver; natural gas; timber; petroleum; fisheries

Agriculture: coffee; cocoa; coconuts; palm kernels; tea; rubber; sweet potatoes; fruit; vegetables; poultry; pork

Industry: copra crushing; palm oil processing; wood processing and production; mining; construction; tourism

Exports: $1.9 billion (primary partners Australia, Japan, Germany)

Imports: $1 billion (primary partners Australia, Singapore, Japan)

http://www.cia.gov/cia/publications/factbook/geos/pp.html

TERRA INCOGNITA

Papua New Guinea is an independent nation and a member of the British Commonwealth. Occupying the eastern half of New Guinea (the second-largest island in the world) and many outlying islands, Papua New Guinea is probably the most overlooked of all the nations in the Pacific Rim.

It was not always overlooked, however. Spain claimed the vast land in the mid-sixteenth century, followed by Britain's East India Company in 1793. The Netherlands laid claim to part of the island in the 1800s and eventually came to control the western half (now known as Irian Jaya, a province of the Republic of Indonesia). In the 1880s, German settlers occupied the northeastern part of the island; and in 1884, Britain signed a treaty with Germany, which gave the British about half of what is now Papua New Guinea. In 1906, Britain gave its part of the island to Australia. Australia invaded and quickly captured the German area in 1914. Eventually, the League of Nations and, later, the United Nations gave the captured area to Australia to administer as a trust territory.

During World War II, the northern part of New Guinea was the scene of bitter fighting between a large Japanese force and Australian and U.S. troops. The Japanese had apparently intended to use New Guinea as a base for the military conquest of Australia. Australia resumed control of the eastern half of the island after Japan's defeat, and it continued to administer Papua New Guinea's affairs until 1975, when it granted independence. The capital is Port Moresby, where, in addition to English, the Motu language and a hybrid language known as New Guinea Pidgin are spoken.

STONE-AGE PEOPLES MEET THE TWENTIETH CENTURY

Early Western explorers found the island's resources difficult to reach. The coastline and some of the interior are swampy and mosquito- and tick-infested, while the high, snow-capped mountainous regions are densely forested and hard to traverse. But perhaps most daunting to early would-be settlers and traders were the local inhabitants. Consisting of hundreds of sometimes warring tribes with totally different languages and customs, the New Guinea populace was determined to prevent outsiders from settling the island. Many adventurers were killed, their heads displayed in villages as victory trophies. The origins of the Papuan people are unknown, but some tribes share common practices with Melanesian islanders. Others appear to be Negritos, and some may be related to the Australian Aborigines. More than 700 languages are spoken in Papua New Guinea.

Australians and other Europeans found it beneficial to engage in trade with coastal tribes who supplied them with unique tropical lumbers, such as sandal-

(Photo credit UN/Witlin)

The interior of Papua New Guinea is very difficult to reach. Achieving easier access to the country's valuable minerals and exotic timber have caused a push for the development of transportation services. The island has nearly 500 airstrips, some in very isolated areas, along with an increasing development of a road network. The impact of this development on the environment is of great concern.

The main island is sighted by Portuguese explorers A.D. 1511	The Dutch annex the west half of the island 1828	A British protectorate over part of the eastern half of the island; the Germans control the northeast 1884	Gold is discovered in Papua New Guinea 1890	Australia assumes control of the island 1906–1914	Australia is given the former German areas as a UN trust territory 1920	Japan captures the northern part of the island; Australia resumes control in 1945 1940s	Australia grants independence to Papua New Guinea 1975	A revolt against the government begins on the island of Bougainville 1988	An economic blockade of Bougainville is lifted, but violence continues, claiming 3,000 lives 1990s

2000s

Bougainville independence referendum is set

The new government moves to privatize industries

wood and bamboo, and foodstuffs such as sugarcane, coconut, and nutmeg. Rubber and tobacco were also traded. Tea, which grows well in the highland regions, is an important cash crop.

But the resource that was most important for the economic development of Papua New Guinea was gold. It was discovered there in 1890; two major gold rushes occurred, in 1896 and 1926. Prospectors came mostly from Australia and were hated by the local tribes; some prospectors were killed and cannibalized. A large number of airstrips in the otherwise undeveloped interior eventually were built by miners who needed a safe and efficient way to receive supplies. Today, copper is more important than gold—copper is, in fact, the largest single earner of export income for Papua New Guinea.

Meanwhile, pollution from mining is increasingly of concern to environmentalists, as is deforestation of Papua New Guinea's spectacular rain forests. A diplomatic flap occurred in 1992, when Australian environmentalists protested that a copper and gold mine in Papua New Guinea was causing enormous environmental damage. They called for the mine to be shut down. The Papuan government strongly resented the verbal intrusion into its sovereignty and reminded the protesters and the Australian government that it alone would establish environmental standards for companies operating inside its borders. The Papuan government holds a 20 percent interest in the mining company.

The tropical climate that predominates in all areas except the highest mountain peaks produces an impressive variety of plant and animal life. Botanists and other naturalists have been attracted to the island for scientific study for many years. Despite extensive contacts with these and other outsiders over the past century, and despite the establishment of schools and a university by the Australian government,

some inland mountain tribes continue to live much as they probably did in the Stone Age. Thus, the country lures not only miners and naturalists but also anthropologists and archaelogists looking for clues to humankind's early lifestyles. One of the most famous of these was Bronislaw Malinowski, the Polish-born founder of the field of social anthropology. In the early 1900s, he spent several years studying the cultural practices of the tribes of Papua New Guinea, particularly those of the Trobriand Islands.

Most of the 5 million Papuans live by subsistence farming. Agriculture for commercial trade is limited by the absence of a good transportation network: Most roads are unpaved, and there is no railway system. Travel on tiny aircraft and helicopters is common, however; New Guinea boasts nearly 500 airstrips, most of them unpaved and dangerously situated in mountain valleys. The harsh conditions of New Guinea life have produced some unique ironies. For instance, Papuans who have never ridden in a car or truck may have flown in a plane dozens of times.

In 1998, 23-foot-high tidal waves caused by offshore, undersea earthquakes inundated dozens of villages along the coast and drowned 6,000 people. A social earthquake, of sorts, had also occurred in the late 1980s, when secessionist rebels on Bougainville island began to fight against the government in an effort to establish independence. A cease-fire was arranged in 1998; and in 2000, the government agreed to an independence referendum, prior to which a provincial government would be established, followed by elections.

The government has had to handle other touchy issues in recent years. For instance, former prime minister Bill Skate had established diplomatic ties with Taiwan—which, as usual, raised a political storm from the mainland Chinese. The contro-

versy eventually caused Skate's resignation as prime minister and the defection of several members of his party to the opposition. The new prime minister, Sir Mekere Moruata, immediately reversed the decision over Taiwan, but barely had he been in office when he fired three of his cabinet ministers, claiming that they were plotting against him. Nevertheless, he moved quickly to privatize almost all state-owned enterprises, including the airlines, telecommunications, and others.

Given the differences in socialization of the Papuan peoples and the difficult conditions of life on their island, it will likely be many decades before Papua New Guinea, which joined the Asia-Pacific Economic Cooperation group in 1993, is able to participate fully in the Pacific Rim community.

DEVELOPMENT

Agriculture (especially coffee and copra) is the mainstay of Papua New Guinea's economy. Copper, gold, and silver mining are also important, but large-scale development of other industries is inhibited by rough terrain, illiteracy, and a huge array of spoken languages—more than 700. There are substantial reserves of untapped oil.

FREEDOM

Papua New Guinea is a member of the British Commonwealth and officially follows the English heritage of law. However, in the country's numerous, isolated small villages, effective control is wielded by village elites with personal charisma; tribal customs take precedence over national law—of which many inhabitants are virtually unaware.

HEALTH/WELFARE

Three quarters of Papua New Guinea's population have no formal education. Daily nutritional intake falls far short of recommended minimums, and tuberculosis and malaria are common diseases.

ACHIEVEMENTS

Papua New Guinea, lying just below the equator, is world-famous for its astoundingly varied and beautiful flora and fauna, including orchids, birds of paradise, butterflies, and parrots. Dense forests cover 70 percent of the country. Some regions receive as much as 350 inches of rain a year.

Philippines (Republic of the Philippines)

GEOGRAPHY

Area in Square Miles (Kilometers):
110,400 (300,000) (about the size of Arizona)
Capital (Population): Manila (1,655,000)
Environmental Concerns: deforestation; air and water pollution; soil erosion; pollution of mangrove swamps
Geographical Features: mostly mountainous; coastal lowlands
Climate: tropical marine; monsoonal

PEOPLE

Population
Total: 81,160,000
Annual Growth Rate: 2.07%
Rural/Urban Population Ratio: 45/55
Major Languages: Pilipino (based on Tagalog); English
Ethnic Makeup: 95% Malay; 5% Chinese and others
Religions: 83% Roman Catholic; 9% Protestant; 5% Muslim; 3% Buddhist and others

Health
Life Expectancy at Birth: 65 years (male); 70 years (female)
Infant Mortality Rate (Ratio): 29.5/1,000
Physicians Available (Ratio): 1/849

Education
Adult Literacy Rate: 95%
Compulsory (Ages): 7–12; free

COMMUNICATION
Telephones: 2,078,000 main lines
Daily Newspaper Circulation: 65 per 1,000 people
Televisions: 125 per 1,000 people
Internet Service Providers: 93 (1999)

TRANSPORTATION
Highways in Miles (Kilometers): 124,000 (200,000)
Railroads in Miles (Kilometers): 499 (800)
Usable Airfields: 266
Motor Vehicles in Use: 2,050,000

GOVERNMENT
Type: republic
Independence Date: July 4, 1946 (from the United States)
Head of State/Government: President Gloria Macapagal-Arroyo is both head of state and head of government

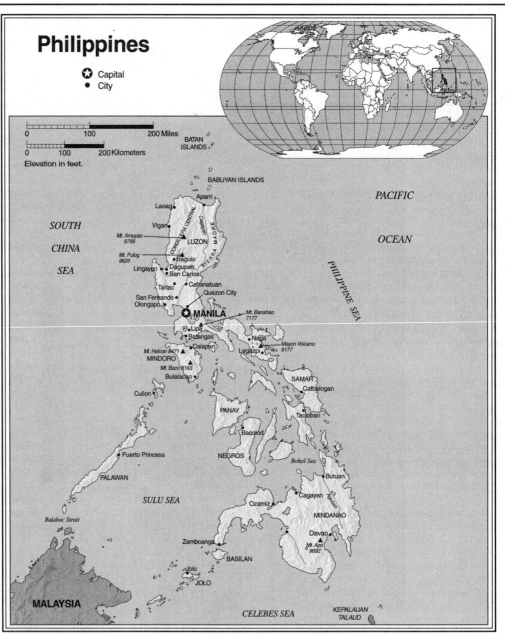

Philippines

★ Capital
● City

Elevation in feet.

Political Parties: Democratic Action; Struggle of the Filipino Masses; Liberal Party; Lakas; People's Reform Party
Suffrage: universal at 18

MILITARY
Military Expenditures (% of GDP): 1.5%
Current Disputes: territorial disputes with China, Malaysia, Taiwan, Vietnam, and possibly Brunei

ECONOMY
Currency ($ U.S. Equivalent): 48.55 Philippine pesos = $1
Per Capita Income/GDP: $3,600/$282 billion
GDP Growth Rate: 2.9%
Inflation Rate: 6.8%

Unemployment Rate: 9.6%
Labor Force: 32,000,000
Natural Resources: timber; petroleum; nickel; cobalt; silver; gold; salt; copper
Agriculture: rice; coconuts; corn; sugarcane; fruit; animal products; fish
Industry: food processing; chemicals; textiles; pharmaceuticals; wood products; electronics assembly; petroleum refining; fishing
Exports: $34.8 billion (primary partners United States, Europe, Japan, and Singapore)
Imports: $30.7 billion (primary partners United States, Europe, Japan)

THIS IS ASIA?

The Philippines is a land with close historic ties to the West. Eighty-three percent of Filipinos, as the people of the Philippines are known, are Roman Catholics, and most speak at least some English. Many use English daily in business and government. In fact, English is the language of instruction at school. Moreover, when they discuss their history as a nation, Filipinos will mention Spain, Mexico, the Spanish-American War, the United States, and cooperative Filipino–American attempts to defeat the Japanese in World War II. The country was even named after a European, King Philip II of Spain. (Currently, 4.4 percent of the United States' foreign-born population, second only to Mexico, are from the Philippines.) If this does not sound like a typical Asian nation, it is because Philippine nationhood essentially began with the arrival of Westerners. That influence continues to dominate the political and cultural life of the country.

Yet the history of the region certainly did not begin with European contact; indeed, there is evidence of human habitation in the area as early as 25,000 B.C. Beginning about 2,000 B.C., Austronesians, Negritos, Malays, and other tribal peoples settled many of the 7,107 islands that constitute the present-day Philippines. Although engaged to varying degrees in trade with China and Southeast Asia, each of these ethnic groups (nearly 60 distinct groups still exist) lived in relative isolation from one another, speaking different languages, adhering to different religions,

and, for good or ill, knowing nothing of the concept of national identity.

Although 5 million ethnic peoples remain marginated from the mainstream today, for most islanders the world changed forever in the mid-1500s, when soldiers and Roman Catholic priests from Spain began conquering and converting the population. Eventually, the disparate ethnic groups came to see themselves as one entity, Filipinos, a people whose lives were controlled indirectly by Spain from Mexico—a fact that, unique among Asian countries, linked the Philippines with the Americas. Thus, the process of national-identity formation for Filipinos actually began in Europe and North America.

Some ethnic groups assimilated rather quickly, marrying Spanish soldiers and administrators and acquiring the language and cultural outlook of the West. The descendants of these mestizos ("mixed" peoples, including local/Chinese mixes) have become the cultural, economic, and political elite of the country. Others, particularly among the Islamic communities on the Philippine island of Mindanao, resisted assimilation right from the start and continue to challenge the authority of Manila. Indeed, the Communist insurgency, reported so often in the news, does not seem to go away. Several groups, among them the Moro Islamic Liberation Front, the Moro National Liberation Front, and the Islamic Abu Sayyat Group, continue to take tourists and others as hostages, set off bombs, and inflict violence on villagers. The violence is, in part, an attempt by

marginated ethnics and others to regain the cultural independence that their peoples lost some 400 years ago. The current government, led by President Gloria Macapagal-Arroyo, is hoping to revive stalled peace talks with these groups.

As in other Asian countries, the Chinese community has played an important but controversial role in Philippine life. Dominating trade for centuries, the Philippine Chinese have acquired clout (and enemies) that far exceeds their numbers (fewer than 1 million). Former president Corazon Aquino was of part-Chinese ancestry, and some of the resistance to her presidency stemmed from her ethnic lineage. The Chinese-Philippine community, in particular, has been the target of ethnic violence—kidnappings and abductions—because their wealth, relative to other Filipino groups, makes them easy prey.

FOREIGN INTERESTS

Filipinos occupy a resource-rich, beautiful land. Monsoon clouds dump as much as 200 inches of rain on the fertile, volcanic soil. Rice and corn grow well, as do hemp, coconut, sugarcane, and tobacco. Tuna, sponges, shrimp, and hundreds of other kinds of marine life flourish in the surrounding ocean waters. Part of the country is covered with dense tropical forests yielding bamboo and lumber and serving as habitat to thousands of species of plant and animal life. The northern part of Luzon Island is famous for its terraced rice paddies.

Given this abundance, it is not surprising that several foreign powers have taken

(United Nations photo by J. M. Micaud)

The Philippines has suffered from the misuse of funds entrusted to the government over the past several decades. The result has been a polarity of wealth, with many citizens living in severe poverty. Slums, such as Tondo in Manila, pictured above, are a common sight in many of the urban areas of the Philippines.

Negritos and others begin settling the islands **25,000 B.C.**	Malays arrive in the islands **2,000 B.C.**	Chinese, Arabs, and Indians control parts of the economy and land **A.D. 400–1400**	The islands are named for the Spanish king Philip II **1542**	Local resistance to Spanish rule **1890s**	A treaty ends the Spanish-American War **1898**

a serious interest in the archipelago. The Dutch held military bases in the country in the 1600s, the British briefly controlled Manila in the 1800s, and the Japanese overran the entire country in the 1940s. But it was Spain, in control of most of the country for more than 300 years (1565–1898), that established the cultural base for the modern Philippines. Spain's interest in the islands—its only colony in Asia—was primarily material and secondarily spiritual. It wanted to take part in the lucrative spice trade and fill its galleon ships each year with products from Asia for the benefit of the Spanish Crown. It also wanted (or, at least, Rome wanted) to convert the so-called heathens (that is, nonbelievers) to Christianity. The friars were particularly successful in winning converts to Roman Catholicism because, despite some local resistance, there were no competing Christian denominations in the Philippines, and because the Church quickly gained control of the resources of the island, which it used to entice converts. Resisting conversion were the Muslims of the island of Mindanao, a group that continues to remain on the fringe of Philippine society but which signed a cease-fire with the government in 1994, after 20 years of guerrilla warfare (although sporadic violence continues, as in 1995, when 200 armed Muslims attacked and burned the town of Ipil on Mindanao). Eventually, a Church-dominated society was established that mirrored in structure—social-class divisions as well as religious and social values—the mother cultures of Spain and Mexico.

Spanish rule in the Philippines came to an inglorious end in 1898, at the end of the Spanish-American War. Spain granted independence to Cuba and ceded the Philippines, Guam, and Puerto Rico to the United States. Filipinos hoping for independence were disappointed to learn that yet another foreign power had assumed control of their lives. Resistance to American rule cost several thousand lives in the early years, but soon Filipinos realized that the U.S. presence was fundamentally different from that of Spain. The United States was interested in trade, and it certainly could see the advantage of having a military presence in Asia, but it viewed its primary role as one of tutelage. American officials believed that the Philippines

should be granted independence, but only when the nation was sufficiently schooled in the process of democracy. Unlike Spain, the United States encouraged political parties and attempted to place Filipinos in positions of governmental authority.

Preparations were under way for independence when World War II broke out. The war and the occupation of the country by the Japanese undermined the economy, devastated the capital city of Manila, caused divisions among the political elite, and delayed independence. After Japan's defeat, the country was, at last, granted independence, on July 4, 1946. Manuel Roxas, a well-known politician, was elected president. Despite armed opposition from Communist groups, the country, after several elections, seemed to be maintaining a grasp on democracy.

MARCOS AND HIS AFTERMATH

Then, in 1965, Ferdinand E. Marcos, a Philippines senator and former guerrilla fighter with the U.S. armed forces, was elected president. He was reelected in 1969. Rather than addressing the serious problems of agrarian reform and trade, Marcos maintained people's loyalty through an elaborate system of patronage, whereby his friends and relatives profited from the misuse of government power and money. Opposition to his corrupt rule manifested itself in violent demonstrations and in a growing Communist insurgency. In 1972, Marcos declared martial law, arrested some 30,000 opponents, and shut down newspapers as well as the National Congress. Marcos continued to rule the country by personal proclamation until 1981. He remained in power thereafter, and he and his wife, Imelda, and their extended family and friends increasingly were criticized for corruption. Finally, in 1986, after nearly a quarter-century of his rule, an uprising of thousands of dissatisfied Filipinos overthrew Marcos, who fled to Hawaii. He died there in 1990.

Taking on the formidable job of president was Corazon Aquino, the widow of murdered opposition leader Benigno Aquino. Aquino's People Power revolution had a heady beginning. Many observers believed that at last Filipinos had found a democratic leader around whom they could unite and who would end corruption and put the persistent Communist

insurgency to rest. Aquino, however, was immediately beset by overwhelming economic, social, and political problems.

Opportunists and factions of the Filipino military and political elite still loyal to Marcos attempted numerous coups d'état in the years of Aquino's administration. Much of the unrest came from within the military, which had become accustomed to direct involvement in government during Marcos's martial-law era. Some Communist separatists turned in their arms at Aquino's request, but many continued to plot violence against the government. Thus, the sense of security and stability that Filipinos needed in order to attract more substantial foreign investment and to reestablish the habits of democracy continued to elude them.

Nevertheless, the economy showed signs of improvement. Some countries, particularly Japan and the United States and, more recently, Hong Kong, invested heavily in the Philippines, as did half a dozen international organizations. In fact, some groups complained that further investment was unwarranted, because already-allocated funds had not yet been fully utilized. Moreover, misuse of funds entrusted to the government—a serious problem during the Marcos era—continued, despite Aquino's promise to eradicate corruption.

A 1987 law, enacted after Corazon Aquino assumed the presidency, limited the president to one term in office. Half a dozen contenders vied for the presidency in 1992, including Imelda Marcos and other relatives of former presidents Marcos and Aquino; U.S. West Point graduate General Fidel Ramos, who had thwarted several coup attempts against Aquino and who thus had her endorsement, won the election. It was the first peaceful transfer of power in more than 25 years (although campaign violence claimed the lives of more than 80 people).

In the 1998 presidential campaign, some 83 candidates filed with the election commission, including Imelda Marcos. Despite the deaths of nearly 30 people and some bizarre moments, such as when a mayoral candidate launched a mortar attack on his opponents, the election was the most orderly in years. Former movie star Joseph Estrada won by a landslide. However, when he attempted to revise the law limiting presidents to one six-year

The Japanese attack the Philippines
1941

General Douglas MacArthur makes a triumphant return to Manila
1944

The United States grants complete independence to the Philippines
1946

Military-base agreements are signed with the United States
1947

Ferdinand Marcos is elected president
1965

Marcos declares martial law
1972

Martial law is lifted; Corazon Aquino and her People Power movement drive Marcos into exile
1980s

The United States closes its military bases in the Philippines; economic crisis
1990s

2000s

The crippling "Love Bug" computer virus emanates from the Philippines

President Estrada is ousted for "economic plunder"

term, some 100,000 people took to the streets in protest.

Estrada had to handle such problems as the Mayon Volcano eruption, which forced 66,000 people from their homes, and with the tension over the Spratly Islands. In 1998, and again in 1999, the Philippine Navy sank Chinese fishing vessels that were operating in the South China Sea in areas claimed by the Philippines. The Philippines demanded that China remove a pier that it had built on Mischief Reef. Due to the likelihood of large oil deposits in the area, the Philippines is likely to pursue this issue.

In January 2001, growing evidence of "economic plunder" led to popular demonstrations to remove Estrada from office. Investigators believe that Estrada stole $200 million to $300 million, hiding the money in banks under various aliases. Under pressure, Estrada left office and was replaced by Vice-President Gloria Macapagal-Arroyo.

SOCIAL PROBLEMS

Much of the foreign capital coming into the Philippines in the 1990s was invested in stock and real-estate speculation rather than in agriculture or manufacturing. Thus, with the financial collapse of 1997, there was little of substance to fall back on. Even prior to the financial crisis, inflation had been above 8 percent per year and unemployment was nearing 9 percent. And one problem never seemed to go away: extreme social inequality. As in Malaysia, where ethnic Malays have constituted a seemingly permanent class of poor peasants, Philippine society is fractured by distinct classes. Chinese and mestizos constitute the top of the hierarchy, while Muslims and most country dwellers form the bottom. About half the Filipino population of 76 million make their living in agriculture and fishing; but even in Manila, where the economy is stronger than any-

where else, thousands of residents live in abject poverty as urban squatters. Officially, one third of Filipinos live below the poverty line. Disparities of wealth are striking. Worker discontent has been such that the Philippines lost more work days to strikes between 1983 and 1987 than any other Asian country.

Adding to the country's financial woes was the sudden loss of income from the six U.S. military bases that closed in 1991 and 1992. The government had wanted the United States to maintain a presence in the country, but in 1991, the Philippine Legislature, bowing to nationalist sentiment, refused to renew the land-lease agreements that had been in effect since 1947. Occupying many acres of valuable land and bringing as many as 40,000 Americans at one time into the Philippines, the bases had come to be seen as visible symbols of American colonialism or imperialism. But they had also been a huge boon to the economy. Subic Bay Naval Base alone had provided jobs for 32,000 Filipinos on base and, indirectly, to 200,000 more. Moreover, the United States paid nearly $390 million each year to lease the land and another $128 million for base-related expenses. Base-related monies entering the country amounted to 3 percent of the entire Philippines economy. After the base closures, the U.S. Congress cut other aid to the Philippines, from $200 million in 1992 to $48 million in 1993. To counterbalance the losses, the Philippines accepted a $60 million loan from Taiwan to develop 740 acres of the former Subic Bay Naval Base into an industrial park. The International Monetary Fund also loaned the country $683 million—funds that have been successfully used to transform the former military facilities into commercial zones.

Other efforts to revitalize the economy are also under way. In 1999, the Legisla-

ture passed a law allowing more foreign investment in the retail sector; and, in the most surprising development, the government approved resumption of large-scale military exercises with U.S. forces.

CULTURE

Philippine culture is a rich amalgam of Asian and European customs. Family life is valued, and few people have to spend their old age in nursing homes. Divorce is frowned upon. Women have traditionally involved themselves in the worlds of politics and business to a greater degree than have women in other Asian countries. Educational opportunities for women are about the same as those for men; adult literacy in the Philippines is estimated at 95 percent. Unfortunately, many college-educated men and women are unable to find employment befitting their skills. Discontent among these young workers continues to grow, as it does among the many rural and urban poor.

Nevertheless, many Filipinos take a rather relaxed attitude toward work and daily life. They enjoy hours of sports and folk dancing or spend their free time in conversation with neighbors and friends, with whom they construct patron/client relationships. In recent years, the growing nationalism has been expressed in the gradual replacement of the English language with Pilipino, a version of the Malay-based Tagalog language.

DEVELOPMENT

The Philippines has more than $50 billion in foreign debt. Payback from development projects has been so slow that about half of the earnings from all exports has to be spent just to service the debt. The Philippines sells most of its products to the United States, Japan, Hong Kong, Great Britain, Singapore, and the Netherlands.

FREEDOM

Marcos's one-man rule meant that both the substance and structure of democracy were ignored. The Philippine Constitution is similar in many ways to that of the United States. President Aquino attempted to adhere to democratic principles; her successors pledged to do the same. The Communist Party was legalized in 1992.

HEALTH/WELFARE

Quality of life varies considerably between the city and the countryside. Except for the numerous urban squatters, city residents generally have better access to health care and education. Most people still do not have access to safe drinking water. The gap between the upper-class elite and the poor is hugely pronounced, and growing.

ACHIEVEMENTS

Filipino women often run businesses or hold important positions in government. Folk dancing is very popular, as is the *kundiman,* a unique blend of music and words found only in the Philippines.

Singapore (Republic of Singapore)

GEOGRAPHY
Area in Square Miles (Kilometers):
250 (648) (about 3½ times the
size of Washington, D.C.)
Capital (Population): Singapore
(3,737,000)
Environmental Concerns: air and
industrial pollution; limited
fresh water; waste-disposal
problems
Geographical Features: lowlands;
gently undulating central
plateau; many small islands
Climate: tropical; hot, humid,
rainy

PEOPLE

Population
Total: 4,151,3000
Annual Growth Rate: 3.54%
Rural/Urban Population (Ratio):
0/100
Major Languages: Malay; Man-
darin Chinese; Tamil; English
Ethnic Makeup: 77% Chinese;
14% Malay; 7% Indian; 2%
others
Religions: 42% Buddhist and
Taoist; 18% Christian; 16%
Muslim; 5% Hindu; 19% others

Health
Life Expectancy at Birth: 77
years (male); 83 years (female)
Infant Mortality Rate (Ratio):
3.65/1,000
Physicians Available (Ratio):
1/667

Education
Adult Literacy Rate: 91%

COMMUNICATION
Telephones: 1,778,000 main lines
Daily Newspaper Circulation:
340 per 1,000 people
Televisions: 218 per 1,000 people
Internet Service Providers: 8

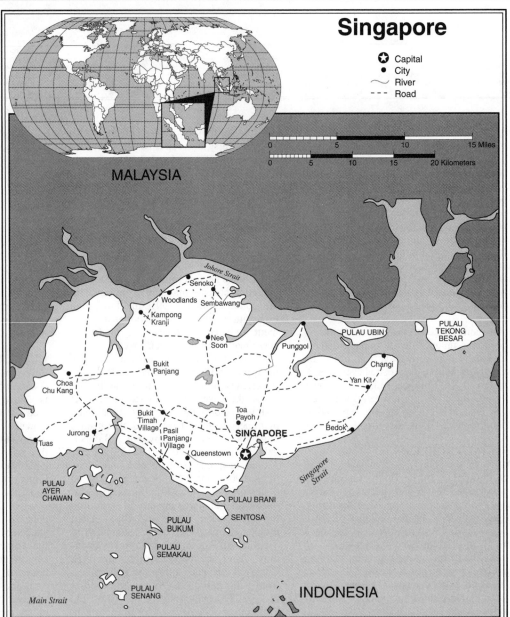

TRANSPORTATION
Highways in Miles (Kilometers): 1,936
(3,122)
Railroads in Miles (Kilometers): 23 (38)
Usable Airfields: 9
Motor Vehicles in Use: 521,000

GOVERNMENT
Type: parliamentary republic
Independence Date: August 9, 1965
(from Malaysia)
Head of State/Government: President
Sellapan Rama Nathan; Prime Minister
Goh Chok Tong

Political Parties: People's Action Party;
Workers' Party; Singapore Democratic
Party; National Solidarity Party;
Singapore People's Party
Suffrage: universal and compulsory at 21

MILITARY
Military Expenditures (% of GDP): 4.9%
Current Disputes: territorial dispute with
Malaysia

ECONOMY
Currency ($ U.S. Equivalent): 1.77
Singapore dollars = $1
Per Capita Income/GDP: $27,800/$98
billion

GDP Growth Rate: 6.2% (est.)
Inflation Rate: 0.4%
Unemployment Rate: 3.2%
Labor Force: 1,932,000
Natural Resources: fish; deepwater ports
Agriculture: rubber; copra; fruit; vegetables
Industry: petroleum refining; electronics;
oil-drilling equipment; rubber process-
ing and rubber products; processed
food and beverages; ship repair; finan-
cial services; biotechnology
Exports: $114 billion (primary partners
United States, Malaysia, Hong Kong)
Imports: $111 billion (primary partners
United States, Malaysia, Japan)

SINGAPORE

It is often said that North Americans are well off because they inhabit a huge continent that abounds with natural resources. This explanation for prosperity does not fit even remotely the case of Singapore. The inhabitants of this tiny, flat, humid tropical island, located near the equator off the tip of the Malay Peninsula, must import even their drinking water. With only 250 square miles of land (including 58 mostly uninhabited islets), Singapore is just half the size of Hong Kong; however, it has one of the highest per capita incomes ($27,800) in Asia. With one of the highest population densities in the region, Singapore might be expected to have the horrific slums that characterize parts of other crowded urban areas. But unemployment in Singapore is only about 3 percent; inflation is less than 1 percent; and almost 85 percent of its 4.1 million people live in spacious and well-equipped, albeit government-controlled, apartments.

Imperialism, geography, and racism help to explain Singapore's unique characteristics. For most of its recorded history, beginning in the thirteenth century A.D., Singapore was controlled variously by the rulers of Thailand, Java, Indonesia, and even India. In the early 1800s, the British were determined to wrest control of parts of Southeast Asia from the Dutch and expand their growing empire. Facilitating their imperialistic aims was Sir Stamford Raffles, a Malay-speaking British administrator who not only helped defeat the Dutch in Java but also diminished the power of local elites in order to fortify his position as lieutenant governor.

Arriving in Singapore in 1819, Raffles found it to be a small, neglected settlement with an economy based on fishing. Yet he believed that the island's geographic location endowed it with great potential as a transshipment port. He established policies that facilitated its development just in time to benefit from the British exports of tin, rubber, and timber leaving Malaya. Perhaps most important was his declaration of Singapore as a free port. Skilled Chinese merchants and traders, escaping racist discrimination against them by Malays on the Malay Peninsula, flocked to Singapore, where they prospered in the free-trade atmosphere.

In 1924, the British began construction of a naval base on the island, the largest in Southeast Asia, which was nonetheless overcome by the Japanese in 1942. Returning in 1945, the British continued to build Singapore into a major maritime center. Today, oil supertankers from Saudi Arabia must exit the Indian Ocean through the Strait of Malacca and skirt Singapore to enter the South China Sea for deliveries to Japan and other Asian nations. Thus, Singapore has found itself in the enviable position of helping to refine and transship millions of barrels of Middle Eastern oil. Singapore's oil-refining capacities have been ranked the world's third largest since 1973.

Singapore is now the second-busiest port in the world (Rotterdam in the Netherlands is number one). It has become the largest shipbuilding and -repair port in the region and a major shipping-related financial center. During the 1990s, Singapore's economy grew at the astounding rates of between 6 and 12 percent a year, making it one of the fastest-growing economies in the world. The Asian financial crisis of the late 1990s temporarily slashed growth to 1.5 percent in 1998, but estimates put growth back to 6.2 percent for 2001. In recent years, the government has aggressively sought out investment from non-shipping–related industries in order to diversify the economy. In 1992, Singapore hosted a summit of the Association of Southeast Asian Nations in which a decision was made to create a regional common market by the year 2008. In order to compete with the emerging European and North American regional trading blocs, it was decided that tariffs on products traded within the ASEAN region would be cut to 5 percent or less.

A UNIQUE CULTURE

Britain maintained an active interest in Singapore throughout its empire period. At its peak, some 100,000 British military men and their dependents were stationed on the island. The British military remained until 1971. (The U.S. Navy's Seventh Fleet's logistics operations have recently been transferred from the Philippines to Singapore, thereby increasing the number of U.S. military personnel in Singapore to about 300 persons.) Thus, British culture, from the architecture of the buildings, to the leisure of a cricket match, to the prevalence of the English language, is everywhere present in Singapore. Yet, because of the heterogeneity of the population (77 percent Chinese, 14 percent Malay, and 7 percent Indian), Singapore accommodates many philosophies and belief systems, including Confucianism, Buddhism, Islam, Hinduism, and Christianity. In recent years, the government has attempted to promote the Confucian ethic of hard work and respect for law, as well as the Mandarin Chinese language, in order to develop a greater Asian consciousness among the people. But most Singaporeans seem content to avoid extreme ideology in favor of pragmatism; they prefer to believe in whatever approach works—that is, whatever allows them to make money and have a higher standard of living.

Their great material success has come with a price. The government keeps a firm hand on the people. For example, citizens can be fined as much as $250 for dropping a candy wrapper on the street or for driving without a seat belt. Worse offenses, such as importing chewing gum or selling

(UPI/Corbis-Bettmann photo by Paul Wedel)

Singapore, one of the most affluent nations in Asia, features a wealthy financial district that overlooks a harbor in the island state.

Singapore is controlled by several different nearby nations, including Thailand, Java, India, and Indonesia **A.D. 1200–1400**	British take control of the island **1800s**	The Japanese capture Singapore **1942**	The British return to Singapore **1945**	Full elections and self-government; Lee Kuan Yew comes to power **1959**	Singapore, now unofficially independent of Britain, briefly joins the Malaysia Federation **1963**	Singapore becomes an independent republic **1965**	Singapore becomes the second-busiest port in the world and achieves one of the highest per capita incomes in the Pacific Rim **1980s**	The U.S. Navy moves some of its operations from the Philippines to Singapore; the Asian financial crisis briefly slows Singapore's economic growth **1990s**

2000s

Singapore tries to position itself to become a regional financial center by liberalizing foreign investment in the banking sector

it, carry fines of $6,000 and $1,200 respectively. Death by hanging is the punishment for murder, drug trafficking, and kidnapping, while lashing is inflicted on attempted murderers, robbers, rapists, and vandals. Being struck with a cane is the punishment for crimes such as malicious damage, as an American teenager, in a case that became a brief international cause célèbre in 1994, found out when he allegedly sprayed graffitti on cars in Singapore. Later that year, a Dutch businessperson was executed for alleged possession of heroin. The death penalty is required when one is convicted of using a gun in Singapore. The United Nations regularly cites Singapore for a variety of human-rights violations, and the world press frequently makes fun of the Singapore government for such practices as giving prizes for the cleanest public toilet.

If restrictions are the norm for ordinary citizens, they are even more severe for politicians, especially those who challenge the government. In recent years, opposition politicians have been fined and sent to jail for speaking in public without police permission, or for selling political pamphlets on the street without permission. Such actions can potentially bar would-be politicians from running for office for up to five years. Foreign media coverage of opposition parties is allowed only under strict guidelines, and in 1998, Parliament banned all political advertising on television. Government leaders argue that order and hard work are necessities since, being a tiny island, Singapore could easily be overtaken by the envious and more politically unstable countries nearby; with few natural resources, Singapore must instead develop its people into disciplined, educated workers. Few deny that Singapore is an amazingly clean and efficient city-state; yet in recent years,

younger residents have begun to wish for a greater voice in government.

The law-and-order tone exists largely because after its separation from Malaysia in 1965, Singapore was controlled by one man and his personal hard-work ethic, Prime Minister Lee Kuan Yew, along with his Political Action Party (PAP). He remained in office for some 25 years, resigning in 1990, but his continuing role as "senior minister" gives him considerable clout. In 2000, for example, he was able to engineer the selection process for president in such a way that no election took place at all, so that his personal choice, S. R. Nathan, could be simply appointed by the election board. Similarly, he saw to it that his personal preference for prime minister, Goh Chok Tong, would, in fact, become his chosen successor. Goh had been the deputy prime minister and had been the designated successor-in-waiting since 1984. The transition has been smooth, and the PAP's hold on the government remains intact.

The PAP originally came into prominence in 1959, when the issue of the day was whether Singapore should join the proposed Federation of Malaysia. Singapore joined Malaysia in 1963, but serious differences persuaded Singaporeans to declare independence two years later. Lee Kuan Yew, a Cambridge-educated, ardent anti-Communist with old roots in Singapore, gained such strong support as prime minister that not a single opposition-party member was elected for more than 20 years. Only one opposition seat exists today.

The two main goals of the administration have been to fully utilize Singapore's primary resource—its deepwater port—and to develop a strong Singaporean identity. The first goal has been achieved in a way that few would have thought possible; the question of national identity, however,

continues to be problematic. Creating a Singaporean identity has been difficult because of the heterogeneity of the population, a situation that is likely to increase as foreign workers are imported to fill gaps in the labor supply resulting from a very successful birth-control campaign started in the 1960s. Identity formation has also been difficult because of Singapore's seesaw history in modern times. First Singapore was a colony of Britain, then it became an outpost of the Japanese empire, followed by a return to Britain. Next Malaysia drew Singapore into its fold, and finally, in 1965, Singapore became independent. All these changes transpired within the lifetime of many contemporary Singaporeans, so their confusion regarding national identity is understandable. Many still have a sense that their existence as a nation is tenuous, and they look for direction. In 1996, Singapore reaffirmed its support for a five-nation defense agreement among itself, Australia, Malaysia, New Zealand, and Great Britain, and it strengthened its economic agreements with Australia.

DEVELOPMENT

Development of the deepwater Port of Singapore has been so successful that at any single time, 400 ships are in port. Singapore has also become a base for fleets engaged in offshore oil exploration and a major financial center, the "Switzerland of Southeast Asia." Singapore has key attributes of a developed country.

FREEDOM

Under former prime minister Lee Kuan Yew, Singaporeans had to adjust to a strict regimen of behavior involving both political and personal freedoms. Citizens want more freedoms but realize that law and order have helped produce their high quality of life. Political opposition voices have largely been silenced since 1968, when the People's Action Party captured all the seats in the government.

HEALTH/WELFARE

About 85 percent of Singaporeans live in government-built dwellings. A government-created pension fund, the Central Provident Fund, takes up to one quarter of workers' paychecks; some of this goes into a compulsory savings account that can be used to finance the purchase of a residence. Other forms of social welfare are not condoned. Care of the elderly is the duty of the family, not the government.

ACHIEVEMENTS

Housing remains a serious problem for many Asian countries, but virtually every Singaporean has access to adequate housing. Replacing swamplands with industrial parks has helped to lessen Singapore's reliance on its deepwater port. Singapore successfully overcame a Communist challenge in the 1950s to become a solid home for free enterprise in the region.

South Korea (Republic of Korea)

GEOGRAPHY

Area in Square Miles (Kilometers):
38,013 (98,480) (about the size
of Indiana)

Capital (Population): Seoul
(10,232,000)

Environmental Concerns: air and
water pollution; overfishing

Geographical Features: mostly
hills and mountains; wide
coastal plains in west and
south

Climate: temperate, with rainfall
heaviest in summer

PEOPLE

Population

Total: 47,471,000

Annual Growth Rate: 0.93%

Rural/Urban Population (Ratio):
17/83

Major Language: Korean

Ethnic Makeup: homogeneous
Korean

Religions: 49% Christian; 47%
Buddhist; 3% Confucian; 1%
Shamanist and Chondogyo

Health

Life Expectancy at Birth: 71 years
(male); 79 years (female)

Infant Mortality Rate (Ratio):
7.8/1,000

Physicians Available (Ratio):
1/784

Education

Adult Literacy Rate: 98%

Compulsory (Ages): 6–12; free

COMMUNICATION

Telephones: 20,422,000 main lines

Daily Newspaper Circulation: 404
per 1,000 people

Televisions: 233 per 1,000 people

Internet Service Providers: 11
(1999)

TRANSPORTATION

Highways in Miles (Kilometers): 53,940
(87,000)

Railroads in Miles (Kilometers): 3,869
(6,240)

Usable Airfields: 103

Motor Vehicles in Use: 10,420,000

GOVERNMENT

Type: republic

Independence Date: August 15, 1948
(from Japan)

Head of State/Government: President
Kim Dae Jung; Prime Minister Lee
Han Dong

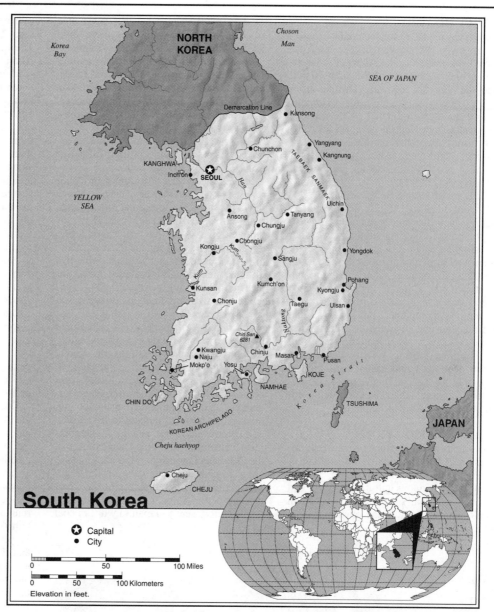

South Korea

★ Capital
• City

0   50   100 Miles

0   50   100 Kilometers

Elevation in feet.

Political Parties: Grand National Party;
United Liberal Democratic Party;
Millennium Democratic party

Suffrage: universal at 20

MILITARY

Military Expenditures (% of GDP): 3.2%

Current Disputes: Demarcation Line
disputed with North Korea; Liancourt
Rocks, claimed by Japan

ECONOMY

Currency ($ U.S. Equivalent): 1,286 won = $1

Per Capita Income/GDP: $13,300/$625.7
billion

GDP Growth Rate: 10%

Inflation Rate: 0.8%

Unemployment Rate: 6.3%

Labor Force: 22,000,000

Natural Resources: coal; tungsten;
graphite; molybdenum; lead; hydropower

Agriculture: rice; root crops; barley;
vegetables; fruit; livestock; fish

Industry: textiles; clothing; footwear; food
processing; chemicals; steel; electronics;
automobile production; shipbuilding

Exports: $144 billion (primary partners
United States, Japan, China)

Imports: $116 billion (primary partners
United States, Japan, China)

http://memory.loc.gov/frd/cs/
krtoc.html
http://www.skas.org

EARLY HISTORY

Korea was inhabited for thousands of years by an early people who may or may not have been related to the Ainus of northern Japan, the inhabitants of Sakhalin Island, and the Siberian Eskimos. Distinct from this early civilization are today's Koreans whose ancestors migrated to the Korean Peninsula from Central Asia and, benefiting from close contact with the culture of China, established prosperous kingdoms as early as 1000 B.C. (legends put the date as early as 2333 B.C.).

The era of King Sejong, who ruled Korea from 1418 to 1450, is notable for its many scientific and humanistic accomplishments. Ruling his subjects according to neo-Confucian thought, Sejong taught improved agricultural methods; published books on astronomy, history, religion, and medicine (the Koreans invented movable metal type); and was instrumental in the invention of sundials, rain gauges, and various musical instruments. Of singular importance was his invention of *han-qul*, a simplified writing system that uneducated peasants could easily learn. Before *han-qul*, Koreans used the more complicated Chinese characters to represent sounds in the Korean language.

REGIONAL RELATIONS

For most of its history, Korea has remained at least nominally independent of foreign powers. China, however, always wielded tremendous cultural influence and at times politically dominated the Korean Peninsula. Similarly, Japan often cast longing eyes toward Korea but was never able to control affairs successfully there until the beginning of the twentieth century.

Korean influence on Japanese culture was pronounced in the 1400s and 1500s, when, through peaceful trade as well as forced labor, Korean artisans and technicians taught advanced skills in ceramics, textiles, painting, and other arts to the Japanese. (Historically, the Japanese received most of their cultural influence from China via Korea.)

In this century, the direction of influence reversed—to the current Japan-to-Korea flow—with the result that the two cultures share numerous qualities. Ironically, cultural closeness has not eradicated emotional distance: Modern Japanese continue to discriminate against Koreans who live in Japan, and Japanese brutality during Japan's of occupation of Korea (1905–1945) remains a frequent topic of conversation among Koreans.

Japan achieved its desire to rule Korea in 1905, when Russia's military, along with its imperialistic designs on Korea, was soundly defeated by the Japanese in the Russo–Japanese War; Korea was granted to Japan as part of the peace settlement. Unlike other expansionist nations, Japan was not content to rule Korea as a colony but, rather, attempted the complete cultural and political annexation of the Korean Peninsula. Koreans had to adopt Japanese names, serve in the Japanese Army, and pay homage to the Japanese emperor. Some 1.3 million Koreans were forcibly sent to Japan to work in coal mines or to serve in the military. The Korean language ceased to be taught in school, and more than 200,000 books on Korean history and culture were burned.

Many Koreans joined clandestine resistance organizations. In 1919, a Declaration of Korean Independence was announced in Seoul by resistance leaders, but the brutally efficient Japanese police and military crushed the movement. They killed thousands of demonstrators, tortured and executed the leaders, and set fire to the homes of those suspected of cooperating with the movement. Despite suppression of this kind throughout the 40 years of Japanese colonial rule, a provisional government was established by the resistance in 1919, with branches in Korea, China, and Russia. However, a very large police force—one Japanese for every 40 Koreans—kept resistance in check.

One resistance leader, Syngman Rhee, vigorously promoted the cause of Korean independence to government leaders in the United States and Europe. Rhee, supported by the United States, became the president of South Korea after the defeat of the Japanese in 1945.

Upon the surrender of Japan, the victorious Allied nations decided to divide Korea into two zones of temporary occupation, for the purposes of overseeing the orderly dismantling of Japanese rule and establishing a new Korean government. The United States was to occupy all of Korea south of the 38th Parallel of latitude (a demarcation running east and west across the peninsula, north of the capital city of Seoul), while the Soviet Union was to occupy Korea north of that line. The United States was uneasy about permitting the Soviets to move troops into Korea, as the Soviet Union had entered the war against Japan just eight days before Japan surrendered, and its commitment to the democratic intentions of the Allies was questionable. Nevertheless, it was granted occupation rights.

Later, the United Nations attempted to enter the zone occupied by the Soviet Union in order to oversee democratic elections for all of Korea. Denied entry, UN advisers proceeded with elections in the South, which brought Syngman Rhee to the presidency. The North created its own government, with Kim Il-Song at the head. Tensions between the two governments resulted in the elimination of trade and other contacts across the new border. This was difficult for each side, because the Japanese had developed industries in the North while the South had remained primarily agricultural. Each side needed the other's resources; in their absence, considerable civil unrest occurred. Rhee's

(Reuters/Bettmann photo by Tony Chung)

South Korea became an economic powerhouse, following the Japanese model of development. Workers in the Hyundai shipyards are pictured above.

government responded by suppressing dissent, rigging elections, and using strong-arm tactics on critics. Autocratic rule, not unlike that of the colonial Japanese, remained the norm in South Korea, and citizens, particularly university students, have been quick to take to the streets in protest of human-rights violations by the various South Korean governments. Equally stern measures were instituted by the Communist government in the North, so that despite a half-century of Korean rule, the repressive legacy of the Japanese police state remained.

AN ECONOMIC POWERHOUSE
Upon the establishment of two separate political entities in Korea, the North pursued a Communist model of economic restructuring. South Korea, bolstered by massive infusions of economic and military aid from the United States, pursued a decidedly capitalist strategy. The results of this choice have been dramatic. For many years, South Korea's economic growth has been one of the fastest in the world; before the financial crisis of the late 1990s, it had been predicted that South Korea's per capita income would rival that in European countries by the year 2010. Predictions have since been revised downward, but South Korea's success in improving the living standards of its people has been phenomenal. About 75 percent of South Korean people live in urban centers, where they have access to good education and jobs. Manufacturing accounts for 30 percent of the gross domestic product. Economic success and recent improvements in the political climate seem to be slowing the rate of outward migration. In recent years, some Koreans have even returned home after years abroad.

North Koreans, on the other hand, are finding life unbearable. Hundreds have defected, some via a "safe house" system through China (similar to the famous Underground Railroad of U.S. slavery days). Some military pilots have flown their jets across the border to South Korea. Food shortages are increasingly evident, and some reports indicate that as many as 2,000 hungry North Koreans attempt to flee into China each month.

Following the Japanese model, South Korean businesspeople work hard to capture market share rather than to gain immediate profit—that is, they are willing to sell their products at or below cost for several years in order to gain the confidence of consumers, even if they make no profit. Once a sizable proportion of consumers buy their product and trust its reliability, the price is raised to a profitable level.

During the 1980s and much of the 1990s, South Korean businesses began in-vesting in other countries, and South Korea became a creditor rather than a debtor member of the Asian Development Bank, putting it in a position to loan money to other countries. There was even talk that Japan (which is separated from Korea by only 150 miles of ocean) was worried that the two Koreas would soon unify and thus present an even more formidable challenge to its own economy—a situation not unlike some Europeans' concern about the economic strength of a reunified Germany.

The magic ended, however, in late 1997, when the world financial community would no longer provide money to Korean banks. This happened because Korean banks had been making questionable loans to Korean *chaebol,* or business conglomerates, for so long, that the banks' creditworthiness came into question. With companies unable to get loans, with stocks at an 11-year low, and with workers eager to take to the streets in mass demonstrations against industry cutbacks, many businesses went under. One of the more well-known firms that went into receivership in 1998 was Kia Motors Corporation. In 2000, the French automaker Renault bought the failed Samsung Motors company. By 1999, the unemployment rate had reached 7 percent, and South Korea had applied for a financial bailout—with all of its restrictions and forced closures of unprofitable businesses—from the International Monetary Fund. Workers deeply resented the belt-tightening required by the IMF (shouting "No to layoffs!" thousands of them threw rocks at police who responded with tear gas and arrests), but IMF funding probably prevented the entire economy from collapsing.

SOCIAL PROBLEMS
Economically, South Korea was once, and likely will be again, an impressive showcase for the fruits of capitalism. Politically, however, the country has been wracked with problems. Under Presidents Syngman Rhee (1948–1960), Park Chung Hee (1964–1979), and Chun Doo Hwan (1981–1987), South Korean government was so centralized as to constitute a virtual dictatorship. Human-rights violations, suppression of workers, and other acts incompatible with the tenets of democracy were frequent occurrences. Student uprisings, military revolutions, and political assassinations became more influential than the ballot box in forcing a change of government. President Roh Tae-woo came to power in 1987, in the wake of a mass protest against civil-rights abuses and other excesses of the previous government. Students began mass protests against various candidates long before the 1992 elections that brought to office the first civilian

president in more than 30 years, Kim Young-sam. Kim was once a dissident himself and was victimized by government policies against free speech; once elected, he promised to make major democratic reforms. The reforms, however, were not good enough for thousands of striking subway workers, farmers, or students whose demonstrations against low pay, foreign rice imports, or the placement of Patriot missiles in South Korea sometimes had to be broken up by riot police.

Replacing Kim Young-sam as president in 1998 was opposition leader Kim Dae Jung. Kim's election was a profound statement that the Korean people were tired of human-rights abuses, because Kim himself had once been a political prisoner. Convicted of sedition by a corrupt government and sentenced to die, Kim had spent 13 years in prison or house arrest, and then, like Nelson Mandela of South Africa, rose to defeat the system that had abused him. That the Korean people were ready for real democratic reform was also revealed in the 1996 trial of former president Chun Doo Hwan, who was sentenced to death (later commuted) for his role in a 1979 coup.

A primary focus of the South Korean government's attention at the moment is the several U.S. military bases in South Korea, currently home to approximately 43,500 U.S. troops. The government (and apparently most of the 47.4 million South Korean people), although not always happy with the military presence, believes that the U.S. troops are useful in deterring possible aggression from North Korea, which, despite an enfeebled economy, still invests massive amounts of its budget in its military. Many university students, however, are offended by the presence of these troops. They claim that the Americans have suppressed the growth of democracy by propping up authoritarian regimes—a claim readily admitted by the United States, which believed during the cold war era that the containment of communism was a higher priority. Strong feelings against U.S. involvement in South Korean affairs have precipitated hundreds of violent demonstrations, sometimes involving as many as 100,000 protesters. The United States' refusal to withdraw its forces from South Korea left an impression with many Koreans that Americans were hard-line, cold war ideologues who were unwilling to bend in the face of changing international alignments.

In 1990, U.S. officials announced that in an effort to reduce its military costs in the post–Cold War era, the United States would pull out several thousand of its troops from South Korea and close three of its five air bases. The United States also

The Yi dynasty begins a 518-year reign over Korea
A.D. 935

Korea pays tribute to Mongol rulers in China
1637

Korea opens its ports to outside trade
1876

Japan formally annexes Korea at the end of the Russo-Japanese War
1910

Korea is divided into North and South
1945

North Korea invades South Korea: the Korean War begins
1950

Cease-fire agreement; the DMZ is established
1953

Democratization movement; the 1988 Summer Olympic Games are held in Seoul
1980s

Reunification talks; a nonaggression pact is signed with North Korea; cross-border exhanges begin; The economy suffers a major setback
1990s

2000s

The first ever summit between South and North Korea in 2000 holds promise for improved peace on the peninsula

South and North Korean athletes march together in the 2000 Summer Olympics in Sydney

declared that it expected South Korea to pay more of the cost of the U.S. military presence, in part as a way to reduce the unfavorable trade balance between the two countries. The South Korean government agreed to build a new U.S. military base about 50 miles south of the capital city of Seoul, where current operations would be relocated. South Korea would pay all construction costs—estimated at about $1 billion—and the United States would be able to reduce its presence within the Seoul metropolitan area, where many of the anti-U.S. demonstrations take place. Although the bases have been a focus of protest by students, surveys now show that about 90 percent of South Koreans want the U.S. military to remain in the country.

The issue of the South's relationship with North Korea has occupied the attention of every government since the 1950s. The division of the Korean Peninsula left many families unable to visit or even communicate with relatives on the opposite side. Moreover, the threat of military incursion from the North forced South Korea to spend huge sums on defense. Both sides engaged in spying, counter-spying, and other forms of subversive activities. Most worrisome of all was that the two antagonists would not sign a peace treaty, meaning that they were technically still at war—since the 1950s.

In 2000, just at the moment when the North was once again engaged in saber-rattleing, an amazing breakthrough occurred: The two sides agreed to hold a summit in P'yongyang, the North Korean capital. The South Korean president was greeted with cheering crowds and feted at state dinners. His counterpart, the reclusive Kim Jong-Il, appeared to be ready to make substantial concessions, including opening the border for family visits, halting the nonstop broadcasting of anti-South

propaganda, and seeking a solution for long-term peace and reunification. The impetus behind the dramatic about-face appears to be North Korea's dire economic situation and its loss of solid diplomatic and economic partners, now that the Communist bloc of nations no longer exists. Reunification will, of course, take many years to realize, and some estimate that it would cost close to a trillion dollars. Still, it is a goal that virtually every Korean wants.

South Korean government leaders have to face a very active, vocal, and even violent populace when they initiate controversial policies. Among the more vocal groups for democracy and human rights are the various Christian congregations and their Westernized clergy. Other vocal groups include the college students who hold rallies annually to elect student protest leaders and to plan antigovernment demonstrations. In addition to the military-bases question, student protesters are angry at the South Korean government's willingness to open more Korean markets to U.S. products. The students want the United States to apologize for its alleged assistance to the South Korean government in violently suppressing an antigovernment demonstration in 1981 in Kwangju, a southern city that is a frequent locus of antigovernment as well as labor-related demonstrations and strikes. Protesters were particularly angered by then–president Roh Tae-woo's silencing of part of the opposition by convincing two opposition parties to merge with his own to form a large Democratic Liberal Party, not unlike that of the Liberal Democratic Party that governed Japan almost continuously for more than 40 years.

Ironically, demands for changes have increased at precisely the moment that the government has been instituting changes de-

signed to strengthen both the economy and civil rights. Under Roh's administration, for example, trade and diplomatic initiatives were launched with Eastern/ Central European nations and with China and the former Soviet Union. Under Kim Youngsam's administration, 41,000 prisoners, including some political prisoners, were granted amnesty, and the powerful chaebol business conglomerates were brought under a tighter rein. Similarly, relaxation of the tight controls on labor-union activity gave workers more leverage in negotiating with management. Unfortunately, union activity, exploding after decades of suppression, has produced crippling industrial strikes—as many as 2,400 a year—and the police have been called out to restore order. In fact, since 1980, riot police have fired an average of more than 500 tear-gas shells a day, at a cost to the South Korean government of tens of millions of dollars.

The sense of unease in the country was tempered, until the late 1990s, by the dynamism of the economy. What will happen with the economy in difficult straits is not easy to predict, but economic recovery and democratization will likely continue to be high on the government's agenda. South Korea recently established unofficial diplomatic ties with Taiwan in order to facilitate freer trade. It then signed an industrial pact with China to merge South Korea's technological know-how with China's inexpensive labor force. Relations with Japan have also improved since Japan's apology for atrocities during World War II.

DEVELOPMENT

The South Korean economy was so strong in the 1980s and early 1990s that many people thought Korea was going to be the next Japan of Asia. The standard of living was increasing for everyone until a major slowdown in 1997–1998. The resulting difficulties forced companies to abandon plans for wage increases or to decrease work hours.

FREEDOM

Suppression of political dissent, manipulation of the electoral process, and restrictions on labor union activity have been features of almost every South Korean government since 1948. Martial law has been frequently invoked, and governments have been overthrown by mass uprisings of the people. Reforms have been enacted under Presidents Roh Tae-woo, Kim Young-sam, and Kim Dae Jung.

HEALTH/WELFARE

Korean men usually marry at about age 27, women at about 24. In 1960, Korean women, on average, gave birth to 6 children; in 1990, the expected births per woman were 1.6. The average South Korean baby born today can expect to live well into its 70s.

ACHIEVEMENTS

In 1992, Korean students placed first in international math and science tests. South Korea achieved self-sufficiency in agricultural fertilizers in the 1970s and continues to show growth in the production of grains and vegetables. The formerly weak industrial sector is now a strong component of the economy.

Taiwan (Republic of China)

GEOGRAPHY

Area in Square Miles (Kilometers): 22,320 (36,002) (about the size of Maryland and Delaware combined)

Capital (Population): T'aipei (2,596,000)

Environmental Concerns: water and air pollution; poaching; contamination of drinking water; radioactive waste

Geographical Features: mostly rugged mountains in east; flat to gently rolling plains in west

Climate: tropical; marine

PEOPLE

Population

Total: 22,191,000

Annual Growth Rate: 0.81%

Rural/Urban Population Ratio: 25/75

Major Languages: Mandarin Chinese; Taiwanese and Hakka dialects also used

Ethnic Makeup: 84% Taiwanese; 14% Mainlander Chinese; 2% aborigine

Religions: 93% mixture of Buddhism, Confucianism, and Taoism; 4.5% Christian; 2.5% others

Health

Life Expectancy at Birth: 74 years (male); 80 years (female)

Infant Mortality Rate (Ratio): 7.06/1,000

Physicians Available (Ratio): 1/867

Education

Adult Literacy Rate: 94%

Compulsory (Ages): 6–15; free

COMMUNICATION

Telephones: 11,500,400 main lines

Televisions: 327 per 1,000 people

Internet Service Providers: 15 (1999)

TRANSPORTATION

Highways in Miles (Kilometers): 20,940 (34,901)

Railroads in Miles (Kilometers): 1,488 (2,481)

Usable Airfields: 38

Motor Vehicles in Use: 5,300,000

GOVERNMENT

Type: multiparty democratic regime

Head of State/Government: President Chen Shui-bian; Vice President Annette Lu

Political Parties: Nationalist Party (Kuomintang); Democratic Progressive Party; Chinese New Democratic Party; Labour Party; New KMT Alliance

Suffrage: universal at 20

MILITARY

Military Expenditures (% of GDP): 2.8%

Current Disputes: territorial disputes with various countries

ECONOMY

Currency ($ U.S. Equivalent): 32.61 Taiwan dollars = $1

Per Capita Income/GDP: $16,100/$357 billion

GDP Growth Rate: 5.5%

Inflation Rate: 0.4%

Unemployment Rate: 2.9%

Labor Force: 9,700,000

Natural Resources: coal; natural gas; limestone; marble; asbestos

Agriculture: rice; tea; fruit; vegetables; corn; livestock; fish

Industry: steel; iron; chemicals; electronics; cement; textiles; food processing; petroleum refining

Exports: $121.6 billion (primary partners United States, Hong Kong, Europe)

Imports: $101.7 billion (primary partners Japan, United States, Europe)

http://www.gio/gov.tw/
http://www.cia.gov/cia/publications/factbook/geos/tw.html

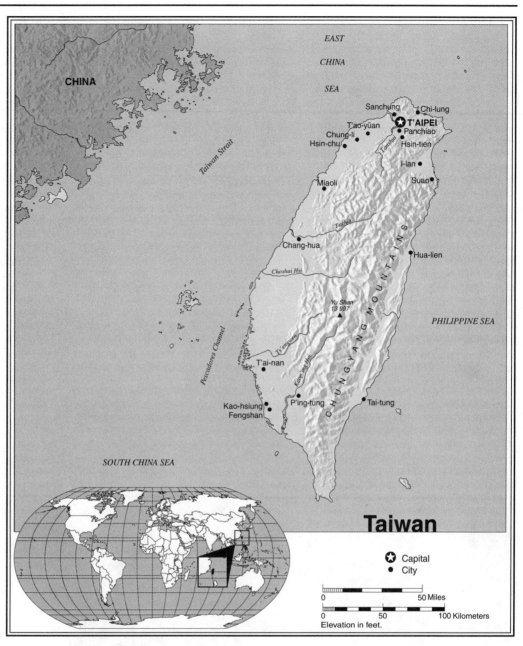

Taiwan

★ Capital
● City

0 ——— 50 Miles
0 — 50 — 100 Kilometers
Elevation in feet.

A LAND OF REFUGE

It has been called "beautiful island," "treasure island," and "terraced bay island," but to the people who have settled there, Taiwan (formerly known as Formosa) has come to mean "refuge island."

Typical of the earliest refugees of the island were the Hakka peoples of China, who, tired of persecution on the mainland, fled to Taiwan (and to Borneo) before A.D. 1000. In the seventeenth century, tens of thousands of Ming Chinese soldiers, defeated at the hands of the expanding Manchu Army, sought sanctuary in Taiwan. In 1949, a third major wave of immigration to Taiwan brought thousands of Chinese Nationalists, retreating in the face of the victorious Red Chinese armies. Hosting all these newcomers were the original inhabitants of the islands, various Malay–Polynesian-speaking tribes. Their descendants live today in mountain villages throughout the island.

Since 1544, other outsiders have shown interest in Taiwan, too: Portugal, Spain, the Netherlands, Britain, and France have all either settled colonies or engaged in trade along the coasts. But the non-Chinese power that has had the most influence is Japan. Japan treated parts of Taiwan as its own for 400 years before it officially acquired the entire island in 1895, at the end of the Sino–Japanese War. From then until 1945, the Japanese ruled Taiwan with the intent of fully integrating it into Japanese culture. The Japanese language was taught in schools, college students were sent to Japan for their education, and the Japanese style of government was implemented. Many Taiwanese resented the harsh discipline imposed, but they also recognized that the Japanese were building a modern, productive society. Indeed, the basic infrastructure of contemporary Taiwan—roads, railways, schools, and so on—was constructed during the Japanese colonial era (1895–1945). Japan still lays claim to the Senkaku Islands, a chain of uninhabited islands, which the Taiwanese say belong to Taiwan.

After Japan's defeat in World War II, Taiwan became the island of refuge of the anti-Communist leader Chiang Kai-shek and his 3 million Kuomintang (KMT, or Nationalist Party) followers, many of whom had been prosperous and well-educated businesspeople and intellectuals in China. These Mandarin-speaking mainland Chinese, called Mainlanders, now constitute about 14 percent of Taiwan's people.

During the 1950s, Mao Zedong, the leader of the People's Republic of China, planned an invasion of Taiwan. However, Taiwan's leaders succeeded in obtaining military support from the United States to prevent the attack. They also convinced the United States to provide substantial amounts of foreign aid to Taiwan (the U.S. government saw the funds as a way to contain communism) as well as to grant it diplomatic recognition as the only legitimate government for all of China.

China was denied membership in the United Nations for more than 20 years because Taiwan held the "China seat." World opinion on the "two Chinas" issue began to change in the early 1970s. Many countries believed that a nation as large and powerful as the People's Republic of China should not be kept out of the United Nations nor out of the mainstream of world trade in favor of the much smaller Taiwan. In 1971, the United Nations withdrew its China seat from Taiwan and gave it to the P.R.C. Taiwan has consistently reapplied for membership, arguing that there is nothing wrong with there being either two China seats or one China seat and one Taiwan seat; its requests have been denied. The United States and many other countries wished to establish diplomatic relations with China but could not get China to cooperate as long as they recognized the sovereignty of Taiwan. In 1979, desiring access to China's huge market, the United States, preceded by many other nations, switched its diplomatic recognition from Taiwan to China. Foreign-trade offices in Taiwan remained unchanged but embassies were renamed; the U.S. Embassy was called the American Institute in Taiwan. As far as official diplomacy with the United States was concerned, Taiwan became a non-nation, but that did not stop the two countries from engaging in very profitable trade, including a controversial U.S. agreement in 1992 to sell $4 billion to $6 billion worth of F-16 fighter jets to Taiwan. Similarly, Taiwan has refused to establish diplomatic ties, yet continues to trade, with nations that recognize the mainland Chinese authorities as a legitimate government. In 1992, for instance, when South Korea established ties with mainland China, Taiwan immediately broke off formal relations with South Korea and suspended direct airline flights. However, Taiwan continued to permit trade in many commodities. Recognizing

(UN photo by Chen Jr.)

Taiwan has one of the highest population densities in the world, but it has been able to expand its agricultural output rapidly and efficiently through utilization of a number of practices. By terracing, using high-yield seeds, and supplying adequate irrigation, Taiwanese can grow a succession of crops on the same piece of land throughout the year.

(UN photo by Chen Jr.)

After World War II, Taiwan emerged as a tremendous source for labor-intensive industries, such as electronics and clothing.

a potentially strong market in Vietnam, Taiwan also established air links with Vietnam in 1992, links that had been broken since the end of the Vietnam War. In 1993, a Taiwanese company collaborated with the Vietnamese government to construct a $242 million highway in Ho Chi Minh City (formerly Saigon). Just over 30 states formally recognize Taiwan today, but Taiwan nevertheless maintains close economic ties with more than 140 countries.

AN ECONOMIC POWERHOUSE

Diplomatic maneuvering has not affected Taiwan's stunning postwar economic growth. Like Japan, Taiwan has been described as an economic miracle. In the past two decades, Taiwan has enjoyed more years of double-digit economic growth than any other nation. With electronics leading the pack of exports, a substantial portion of Taiwan's gross domestic product comes from manufacturing. Taiwan has been open to foreign investment and, of course, to foreign trade. However, for many years, Taiwan insisted on a policy of no contact and no communication with mainland China. Private enterprises eventually were allowed to trade with China—as long as the products were transshipped through a third country, usually Hong Kong. In 1993, government-owned enterprises such as steel and fertilizer plants were allowed to trade with China, on the same condition. In 2000, the Legislature lifted the ban on direct trade, transportation, and communications with China from some of the Taiwanese islands nearest to the mainland. Little by little, the door toward China is opening. By the early 1990s, Taiwanese trade with China had exceeded $13 billion a year and China had become Taiwan's seventh-largest trading partner. The liberalization of trade between China (especially its southern and coastal provinces), Taiwan, and Hong Kong has made the region, now known as Greater China, an economic dynamo. Economists predict that Greater China will someday bypass Japan's economy.

As one of the newly industrializing countries of Asia, Taiwan certainly no longer fits the label "underdeveloped." Taiwan holds large stocks of foreign reserves and carries a trade surplus with the United States (in Taiwan's favor) far greater than Japan's, when counted on a per capita basis. The Taipei stock market has been so successful—sometimes outperforming both Japan and the United States—that a number of workers reportedly have quit their jobs to play the market, thereby exacerbating Taiwan's already serious labor shortage. (This shortage has led to an influx of foreign workers, both legal and illegal.)

Successful Taiwanese companies have begun to invest heavily in other countries where land and labor are plentiful and less expensive. In 1993, the Philippines accepted a $60 million loan from Taiwan to build an industrial park and commercial port at Subic Bay, the former U.S. naval base; and Thailand, Australia, and the United States have also seen inflows of Taiwanese investment monies. By the early 1990s, some 200 Taiwanese companies had invested $1.3 billion in Malaysia alone (Taiwan supplanted Japan as the largest outside investor in Malaysia). Taiwanese investment in mainland China has also increased.

Taiwan's economic success is attributable in part to its educated population, many of whom constituted the cultural and economic elite of China before the Communist revolution. Despite resentment of the mainland immigrants by native-born Taiwanese, everyone, including the lower classes of Taiwan, has benefited from this infusion of talent and capital. Yet the Taiwanese people are beginning to pay a price for their sudden affluence. It is said that T'aipei, the capital city of Taiwan (and one of the most expensive cities in the world for foreigners), is awash in money, but it is also awash in air pollution and traffic congestion. Traffic congestion in T'aipei is rated near the worst in the world. Concrete high-rises have displaced the lush greenery of the mountains. Many residents spend their earnings on luxury foreign cars and on cigarettes and alcohol, the consumption rate of which has been increasing by about 10 percent a year. Many Chinese traditions—for instance, the roadside restaurant serving noodle soup—are giving way to 7-Elevens selling Coca-Cola and ice cream.

Some Taiwanese despair of ever turning back from the growing materialism; they wish for the revival of traditional Chinese (that is, mostly Confucian) ethics. They doubt that it will happen. Still, the government, which has been dominated since 1949 by the conservative Mandarin migrants from the mainland, sees to it that Confucian ethics are vigorously taught in school. And there remains in Taiwan more of traditional China than in China itself, because, unlike the Chinese Communists, the Taiwanese authorities have had no reason to attempt an eradication of the values of Buddhism, Taoism, or Confucianism. Nor has grinding poverty—often the most serious threat to the cultural arts—negatively affected literature and the fine arts, as it has in China. Parents, with incense sticks burning before small religious altars,

Portuguese sailors are the first Europeans to visit Taiwan
A.D. 1544

Taiwan becomes part of the Chinese Empire
1700s

The Sino-Japanese War ends; China cedes Taiwan to Japan
1895

Taiwan achieves independence from Japan
1945

Nationalists, under Chiang Kai-shek, retreat to Taiwan
1947–49

A de facto separation of Taiwan from China; Chinese aggression is deterred with U.S. assistance
1950s

China replaces Taiwan in the United Nations
1971

Chiang Kai-shek dies and is succeeded by his son, Chiang Ching-Kuo
1975

The first two-party elections in Taiwan's history are held; 38 years of martial law end
1980s

Relations with China improve; the United States sells F-16 jets to Taiwan
1990s

China conducts military exercises to intimidate Taiwanese voters

2000s

Trade and communication with China continue to improve

The opposition Democratic Progessive Party wins the presidency

still emphasize respect for authority, the benefits of harmonious cooperative effort, and the inestimable value of education. Traditional festivals dot each year's calendar, among the most spectacular of which is Taiwan's National Day parade. Marching bands, traditional dancers, and a huge dragon carried by more than 50 young men please the crowds lining the streets of Taipei. Temples are filled with worshipers praying for health and good luck.

But the Taiwanese will need more than luck if they are to escape the consequences of their intensely rapid drive for material comfort. Some people contend that the island of refuge is being destroyed by success. Violent crime, for instance, once hardly known in Taiwan, is now commonplace. Six thousand violent crimes, including rapes, robberies, kidnappings, and murder, were reported in 1989—a 22 percent increase over the previous year, and the upward trend has continued since then. Extortion against wealthy companies and abductions of the children of successful families have created a wave of fear among the rich.

Like other Asian countries, Taiwan was affected by the Asian financial crisis of 1997 and 1998. But Taiwan remains in a strong growth mode. Labor shortages have forced some companies to operate at only 60 percent of capacity, and low-interest loans are hard to get because the government fears that too many people will simply invest in get-rich stocks instead of in new businesses.

POLITICAL LIBERALIZATION
These disturbing trends notwithstanding, in recent years, the Taiwanese people have had much to be grateful for in the political sphere. Until 1986, the government, dominated by the influence of the Chiangs, had permitted only one political party, the Na-

tionalists, and had kept Taiwan under martial law for nearly 4 decades. A marked political liberalization began near the time of Chiang Ching-Kuo's death, in 1987. The first opposition party, the Democratic Progressive Party, was formed; martial law (officially, the "Emergency Decree") was lifted; and the first two-party elections were held, in 1986. In 1988, for the first time a native-born Taiwanese, Lee Teng-hui, was elected to the presidency. He was reelected in 1996 in the first truly democratic, direct presidential election ever held in Taiwan. Although Lee has never promoted the independence of Taiwan, his high-visibility campaign raised the ire of China, which attempted to intimidate the Taiwanese electorate into voting for a more pro-China candidate by conducting military exercises and firing missiles just 20 miles off the coast of Taiwan. As expected, the intimidation backfired, and Lee soundly defeated his opponents.

It is still against the law for any group or person to advocate publicly the independence of Taiwan—that is, to advocate international acceptance of Taiwan as a sovereign state, separate and apart from China. When the opposition Democratic Progressive Party (DPP) resolved in 1990 that Taiwan should become an independent country, the ruling Nationalist government immediately outlawed the DPP platform.

In 2000, after 50 years of Nationalist Party rule in Taiwan, the DPP was able to gain control of the presidency. Chen Shui-bian, a native-born Taiwanese, was elected. Both Chen and his running rate, Annette Lu, had spent time in prison for activities that had angered the ruling Nationalists. Although his party had openly sought independence, Chen toned down that rhetoric and instead invited the mainland to begin talks for reconciliation. As usual, China had threatened armed intervention if Taiwan declared independence;

but after the vote for Chen, the mainland seemed to moderate its position. The bilateral talks that were initiated in 1998 continue to hold promise. Some believe that such talks will eventually result in Taiwan being annexed by China, just as in the cases of Hong Kong and Macau (although from a strictly legalistic viewpoint, Taiwan has just as much right to annex China). Others believe that dialogue will eventually diminish animosity, allowing Taiwan to move toward independence without China's opposition. In one sense, China has already won the world-opinion battle, since virtually everyone knows that China regards Taiwan as a rogue state that belongs back in the fold.

Opinion on the independence issue is clearly divided. Even some members of the anti-independence Nationalist Party have bolted and formed a new party (the New KMT Alliance, or the New Party) to promote closer ties with China. As opposition parties proliferate, the independence issue could become a more urgent topic of political debate. In the meantime, contacts with the P.R.C. increase daily; Taiwanese students are now being admitted to China's universities, and Taiwanese residents by the thousands now visit relatives on the mainland. Despite complaints from China, Taiwanese government leaders have been courting their counterparts in the Philippines, Thailand, Indonesia, and South Korea. But China has vowed to invade Taiwan if it should ever declare independence. Under these circumstances, many—probably most—Taiwanese will likely remain content to let the rhetoric of reunification continue while enjoying the reality of de facto independence.

DEVELOPMENT

Taiwan has vigorously promoted export-oriented production, particularly of electronic equipment. In the 1980s, manufacturing became a leading sector of the economy, employing more than one third of the workforce. Virtually all Taiwanese households own color televisions, and other signs of affluence are abundant.

FREEDOM
For nearly 4 decades, Taiwan was under martial law. Opposition parties were not tolerated, and individual liberties were limited. A liberalization of this pattern began in 1986. Taiwan now seems to be on a path toward greater democratization. In 1991, 5,574 prisoners, including many political prisoners, were released in a general amnesty.

HEALTH/WELFARE

Taiwan has one of the highest population densities in the world. Education is free and compulsory to age 15, and the country boasts more than 100 institutions of higher learning. Social programs, however, are less developed than those in Singapore, Japan, and some other Pacific Rim countries.

ACHIEVEMENTS

From a largely agrarian economic base, Taiwan has been able to transform its economy into an export-based dynamo with international influence. Today, only about 10 percent of the population work in agriculture, and Taiwan ranks among the top 20 exporters in the world.

Thailand (Kingdom of Thailand)

GEOGRAPHY

Area in Square Miles (Kilometers): 198,404 (514,000) (about twice the size of Wyoming)

Capital (Population): Bangkok (6,547,000)

Environmental Concerns: air and water pollution; poaching; deforestation; soil erosion

Geographical Features: central plain; Khorat Plateau in the east; mountains elsewhere

Climate: tropical monsoon

PEOPLE

Population

Total: 61,231,000

Annual Growth Rate: 0.93%

Rural/Urban Population Ratio: 80/20

Major Languages: Thai; English; various dialects

Ethnic Makeup: 75% Thai; 14% Chinese; 11% Malay and others

Religions: 95% Buddhist; 4% Muslim; 1% others

Health

Life Expectancy at Birth: 65 years (male); 72 years (female)

Infant Mortality Rate (Ratio): 31.4/1,000

Physicians Available (Ratio): 1/4,165

Education

Adult Literacy Rate: 94%

Compulsory (Ages): 6–15

COMMUNICATION

Telephones: 4,827,000 main lines

Daily Newspaper Circulation: 47 per 1,000 people

Televisions: 56 per 1,000 people

Internet Service Providers: 13 (1999)

TRANSPORTATION

Highways in Miles (Kilometers): 38,760 (64,600)

Railroads in Miles (Kilometers): 2,364 (3,940)

Usable Airfields: 106

Motor Vehicles in Use: 5,650,000

GOVERNMENT

Type: constitutional monarchy

Independence Date: founding date 1238; never colonized

Head of State/Government: King Bhumibol Adulyadej; Prime Minister Chuan Leekpai

Political Parties: Thai Nation Party; Democratic Party; National Development Party; many others

Suffrage: universal and compulsory at 18

MILITARY

Military Expenditures (% of GDP): 1.3%

Current Disputes: boundary disputes with Laos, Cambodia, and Myanmar

ECONOMY

Currency ($ U.S. Equivalent): 44.0 baht = $1

Per Capita Income/GDP: $6,400/$388.7 billion

GDP Growth Rate: 4%

Inflation Rate: 2.4%

Unemployment Rate: 4.5%

Labor Force: 32,600,000

Natural Resources: tin; rubber; natural gas; tungsten; tantalum; timber; lead; fish; gypsum; lignite; fluorite; arable land

Agriculture: rice; cassava; rubber; corn; sugarcane; coconuts; soybeans

Industry: tourism; textiles and garments; agricultural processing; beverages; tobacco; cement; electric appliances and components; electronics; furniture; plastics

Exports: $58.5 billion (primary partners United States, Japan, Singapore)

Imports: $45 billion (primary partners Japan, United States, Singapore)

 http://www.mahidol.ac.th/Thailand/Thailand-main.html

THAILAND'S ANCIENT HERITAGE

The roots of Thai culture extend into the distant past. People were living in Thailand at least as early as the Bronze Age. By the time Thai people from China (some scholars think from as far away as Mongolia) had established the first Thai dynasty in the Chao Phya Valley, in A.D. 1238, some communities, invariably with a Buddhist temple or monastery at their centers, had been thriving in the area for 600 years. Early Thai culture was greatly influenced by Buddhist monks and traders from India and Sri Lanka (Ceylon).

By the seventeenth century, Thailand's ancient capital, Ayutthaya, boasted a larger population than that of London. Ayutthaya was known around the world for its wealth and its architecture, particularly its religious edifices. Attempts by European nations to obtain a share of the wealth were so inordinate that in 1688, the king expelled all foreigners from the country. Later, warfare with Cambodia, Laos, and Malaya yielded tremendous gains in power and territory for Thailand, but it was periodically afflicted by Burma (present-day Myanmar), which briefly conquered Thailand in the 1760s (as it had done in the 1560s). The Burmese were finally defeated in 1780, but the destruction of the capital required the construction of a new city, near what, today, is Bangkok.

Generally speaking, the Thai people have been blessed over the centuries with benevolent kings, many of whom have been open to new ideas from Europe and North America. Gathering around them advisers from many nations, they improved transportation systems, education, and farming while maintaining the central place of Buddhism in Thai society. Occasionally royal support for religion overtook other societal needs, at the expense of the power of the government.

The gravest threat to Thailand came during the era of European colonial expansion, but Thailand—whose name means "Free Land"—was never completely conquered by European powers. Today, the country occupies a land area about the size of France.

MODERN POLITICS

Since 1932, when a constitutional monarchy replaced the absolute monarchy, Thailand (formerly known as Siam) has weathered 17 attempted or successful military or political coups d'état, most recently in 1991. The Constitution has been revoked and replaced numerous times; governments have fallen under votes of no-confidence; students have mounted violent demonstrations against the government; and the military has, at various

(Photo credit United Nations/Prince)

Buddhism is an integral part of the Thai culture. Six hundred years ago, Buddhist monks traveled from India and Ceylon (present-day Sri Lanka) and built temples and monasteries throughout Thailand. These newly ordained monks are meditating in the courtyard of a temple in Bangkok.

times, imposed martial law or otherwise curtailed civil liberties.

Clearly, Thai politics are far from stable. Nevertheless, there is a sense of stability in Thailand. Miraculously, its people were spared the direct ravages of the Vietnam War, which raged nearby for 20 years. Despite all the political upheavals, the same royal family has been in control of the Thai throne for nine generations, although its power has been severely delimited for some 60 years. Furthermore, before the first Constitution was enacted in 1932, the country had been ruled continuously, for more than 700 years, by often brilliant and progressive kings. At the height of Western imperialism, when France, Britain, the Netherlands, and Portugal were in control of every country on or near Thailand's borders, Thailand remained free of Western domination, although it was forced—sometimes at gunpoint—to relinquish sizable chunks of

its holdings in Cambodia and Laos to France, and holdings in Malaya to Britain. The reasons for this singular state of independence were the diplomatic skill of Thai leaders, Thai willingness to Westernize the government, and the desire of Britain and France to let Thailand remain interposed as a neutral buffer zone between their respective armies in Burma and Indochina.

The current king, Bhumibol Adulyadej, born in the United States and educated in Switzerland, is highly respected as head of state. The king is also the nominal head of the armed forces, and his support is critical to any Thai government. Despite Thailand's structures of democratic government, any administration that has not also received the approval of the military elites, many of whom hold seats in the Senate, has not prevailed for long. The military has been a rightist force in Thai politics, resisting reforms from the left

(Photo by Lisa Clyde)

Bangkok is one of the largest cities in the world. The city is interlaced with canals, and the population crowds along river banks. With the enormous influx of people who are lured by industrialization and economic opportunity, the environment has been strained to the limit.

that might have produced a stronger labor union movement, more freedom of expression (many television and radio stations in Thailand are controlled directly by the military), and less economic distance between the social classes. Military involvement in government increased substantially during the 1960s and 1970s, when a Communist insurgency threatened the government from within and the Vietnam War destabilized the external environment.

Until the February 1991 coup, there had been signs that the military was slowly withdrawing from direct meddling in the government. This may have been because the necessity for a strong military appeared to have lessened with the end of the Cold War. In late 1989, for example, the Thai government signed a peace agreement with the Communist Party of Malaya, which had been harassing villagers along the Thai border for more than 40 years. Despite these political/military improvements, Commander Suchinda Kraprayoon led an army coup against the legally elected government in 1991 and, notwithstanding promises to the contrary, promptly had himself named prime minister. Immediately, Thai citizens, tired of the constant instability in government occasioned by military meddling, began staging mass demonstrations against Suchinda. The protesters were largely middle-class office workers who used their cellular telephones to communicate from one protest site to another. The demonstrations were the largest in 20 years, and the military responded with violence; nearly 50 people were

killed and more than 600 injured. The public outcry was such that Suchinda was forced to appear on television being lectured by the king; he subsequently resigned. An interim premier dismissed several top military commanders and removed military personnel from the many government departments over which they had come to preside. Elections followed in 1992, and Thailand returned to civilian rule, with the military's influence greatly diminished.

The events of this latest coup show that the increasingly educated and affluent citizens of Thailand wish their country to be a true democracy. Still, unlike some democratic governments that have one dominant political party and one or two smaller opposition parties, party politics in Thailand is characterized by diversity. Indeed, so many parties compete for power that no single party is able to govern without forming coalitions with others. Parties are often founded on the strength of a single charismatic leader rather than on a distinct political philosophy, a circumstance that makes the entire political setting rather volatile. The Communist Party remains banned. Campaigns to elect the 360-seat Parliament often turn violent; in recent elections, 10 candidates were killed when their homes were bombed or sprayed with rifle fire, and nearly 50 gunmen-for-hire were arrested or killed by police, who were attempting to protect the candidates of the 11 political parties vying for office.

In 1999, corruption charges by the New Aspiration Party against the ruling Democrat Party failed to topple the government.

Many citizens seemed to credit the government with bringing Thailand out of economic recession (but the International Monetary Fund loan of $17.2 billion probably helped more). Still, corruption was everywhere evident in 2000, when massive election fraud was uncovered in Senate elections. It was the first time that senators had been directly elected instead of appointed, and many of them resorted to wholesale vote-buying and ballot-tampering. Out of 200 senators, 78 were disqualified as a result of election fraud, and the elections had to be held again.

FOREIGN RELATIONS
Thailand is a member of the United Nations, the Association of Southeast Asian Nations, and many other regional and international organizations. Throughout most of its modern history, Thailand has maintained a pro-Western political position. During World War I, Thailand joined with the Allies; and during the Vietnam War, it allowed the United States to stage air attacks on North Vietnam from within its borders, and it served as a major rest and relaxation center for American soldiers. During World War II, Thailand briefly allied itself with Japan but made decided efforts after the war to reestablish its former Western ties.

Thailand's international positions have seemingly been motivated more by practical need than by ideology. During the colonial era, Thailand linked itself with Britain because it needed to offset the influence of France; during World War II, it joined with Japan in an apparent effort to prevent its country from being devastated by Japanese troops; during the Vietnam War, it supported the United States because the United States seemed to offer Thailand its only hope of not being directly engaged in military conflict in the region.

Thailand now seems to be tilting away from its close ties with the United States and toward a closer relationship with Japan. In the late 1980s, disputes with the United States over import tariffs and international copyright matters cooled the countries' warm relationship (the United States accused Thailand of allowing the manufacture of counterfeit brand-name watches, clothes, computer software, and many other items, including medicines). Moreover, Thailand found in Japan a more ready, willing, and cooperative economic partner than the United States.

During the Cold War and especially during the Vietnam War era, the Thai military strenuously resisted the growth of Communist ideology inside Thailand, and the Thai government refused to engage in normal diplomatic relations with the Communist regimes on its borders. Because of military pressure, elected officials refrained from advocating improved relations with the Communist governments. However, in 1988, Prime Minister Prem Tinslanond, a former general in the army who had been in control of the government for eight years, stepped down from office, and opposition to normalization of relations seemed to mellow. The subsequent prime minister, Chatichai Choonhavan, who was ousted in the 1991 military coup, invited Cambodian leader Hun Sen to visit Thailand; he also made overtures to Vietnam and Laos. Chatichai's goal was to open the way for trade in the region by helping to settle the agonizing Cambodian conflict. He also hoped to bring stability to the region so that the huge refugee camps in Thailand, the largest in the world, could be dismantled and the refugees repatriated. Managing regional relations will continue to be difficult: Thailand fought a brief border war with Communist Laos in 1988. The influx of refugees from the civil wars in adjacent Cambodia and Myanmar continues to

strain relations. Currently some 100,000 Karen refugees live precariously in 20 camps in Thailand along the border with Myanmar. The Karens, many of whom practice Christianity and are the second-largest ethnic group in Myanmar, have fought the various governments in their home country for years in an attempt to create an independent Karen state. Despite the patrol efforts of Thai troops, Myanmar soldiers frequently cross into Thailand at night to raid, rape, and kill the Karens. Thailand has tolerated the massive influx of war refugees from Myanmar, but its patience seems to have been wearing thin in recent years.

THE ECONOMY

Part of the thrust behind Thailand's diplomatic initiatives is the changing needs of its economy. For decades, Thailand saw itself as an agricultural country; indeed, more than half of the laborforce work in agriculture today, with rice as the primary commodity. Rice is Thailand's single most important export and a major source of government revenue. Every morning, Thai

families sit on the floor of their homes around bowls of hot and spicy *tom yam goong* soup and a large bowl of rice; holidays and festivals are scheduled to coincide with the various stages of planting and harvesting rice; and, in rural areas, students are dismissed at harvest time so that all members of a family can help in the fields. So central is rice to the diet and the economy of the country that the Thai verb equivalent of "to eat," translated literally, means "to eat rice." Thailand is the fifth-largest exporter of rice in the world.

Unfortunately, Thailand's dependence on rice subjects its economy to the cyclical fluctuations of weather (sometimes the monsoons bring too little moisture) and market demand. Thus, in recent years, the government has invested millions of dollars in economic diversification. Not only have farmers been encouraged to grow a wider variety of crops, but tin, lumber, and offshore oil and gas production have also been promoted. Thailand is the world's largest rubber-producing country, but with prices at 30-year lows, that industry is struggling to survive. Foreign investment

(UN/photo by Saw Lwin)

Rice is Thailand's most important export. Its production utilizes a majority of the agricultural workforce. Today, the government is attempting to diversify this reliance on rice, encouraging farmers to grow a wider variety of crops that are not so dependent on world markets and the weather.

The formal beginning of Thailand as a nation A.D. 1200s	King Rama I ascends the throne, beginning a nine-generation dynasty 1782	Coup; constitutional monarchy 1932	The country's name is changed from Siam to Thailand 1939	Thailand joins Japan and declares war on the United States and Britain 1942	Thailand resumes its historical pro-Western stance 1946	Communist insurgency threatens Thailand's stability 1960s–1970s	Student protests usher in democratic reforms 1973	Currency decisions in Thailand precipitate the Southeast Asian financial crisis; mass demonstrations force a return to civilian rule 1990s

2000s

Thailand rebounds from the economic crisis of the late 1990s

Bangkok looks for ways to reduce pollution and traffic congestion

in export-oriented manufacturing has been warmly welcomed. Japan in particular benefits from trading with Thailand in food and other commodities, and it sees Thailand as one of the more promising places to relocate smokestack industries. For its part, Thailand seems to prefer Japanese investment over that from the United States, because the Japanese seem more willing to engage in joint ventures and to show patience while enterprises become profitable. Indeed, economic ties with Japan are very strong. For instance, in recent years Japan has been the largest single investor in Thailand and has accounted for more than 40 percent of foreign direct investment (Taiwan, Hong Kong, and the United States each have accounted for about 10 percent). About 24 percent of Thai imports come from Japan, and approximately 14 percent of its exports go to Japan.

Thailand's shift to an export-oriented economy paid off until 1997, when pressures on its currency, the baht, required the government to allow it to float instead of having it pegged to the U.S. dollar. That action triggered the Southeast Asian financial crisis. Until that time, Thailand's gross domestic product growth rate had averaged about 10 percent a year—one of the highest in the world, and as high, or higher than, all the newly industrializing countries of Asia (Hong Kong, South Korea, Singapore, Taiwan, and China). Furthermore, unlike the Philippines and Indonesia, Thailand was able to achieve this incredible growth without very high inflation. The 1997–1998 financial crisis hit Thailand very hard, but projections show that economic growth should soon return to a respectable 5 percent.

SOCIAL PROBLEMS

Industrialization in Thailand, as everywhere, draws people to the cities. Bangkok is one of the largest cities in the world. Numerous problems, particularly air pollution, traffic congestion, and overcrowding, complicate life for Bangkok residents. An international airport that opened near Bangkok in 1987 was so overcrowded just four years later that a new one had to be planned, and new harbors had to be constructed south of the city to alleviate congestion in the main port. Demographic projections indicate that there will be a decline in population growth in the future as the birth rate drops and the average Thai household shrinks from the six people it was in 1970 to only three people by 2015. This will alter the social structure of urban families, especially as increased life expectancy adds older people to the population and forces the country to provide more services for the elderly. Today, however, many Thai people still make their living on farms, where they grow rice, rubber, and corn, or tend chickens and cattle, including the ever-present water buffalo. Thus, it is in the countryside (or "upcountry," as everywhere but Bangkok is called in Thailand) that the traditional Thai culture may be found. There, one still finds villages of typically fewer than 1,000 inhabitants, with houses built on wooden stilts alongside a canal or around a Buddhist monastery. One also finds, however, unsanitary conditions, higher rates of illiteracy, and lack of access to potable water. Of increasing concern is deforestation, as Thailand's growing population continues to use wood as its primary fuel for cooking and heat. The provision of social services does not meet demand even in the cities, but rural residents are particularly deprived.

Culturally, Thai people are known for their willingness to tolerate (although not necessarily to assimilate) diverse lifestyles and opinions. Buddhist monks, who shave their heads and make a vow of celibacy, do not find it incongruous to beg for rice in districts of Bangkok known for prostitution and wild nightlife. And worshippers seldom object when a noisy, congested highway is built alongside the serenity of an ancient Buddhist temple. (However, the mammoth scale of the proposed $3.2 billion, four-level road-and-railway system in the city and its likely effect on cultural and religious sites prompted the Thai cabinet to order the construction underground; but the cabinet had to recant, when the Hong Kong firm designing the project announced that it was technically impossible to build it underground.)

Relative tolerance has mitigated ethnic conflict among Thailand's numerous minority groups. The Chinese, for instance, who are often disliked in other Asian countries because of their dominance of the business sectors, are able to live with little or no discrimination in Thailand; indeed, they constitute the backbone of Thailand's new industrial thrust.

DEVELOPMENT

Many Thais are small-plot or tenant farmers, but the government has energetically promoted economic diversification. Despite high taxes, Thailand has a reputation as a good place for foreign investment. Electronics and other high-tech industries from Japan, the United States, and other countries have been very successful in Thailand.

FREEDOM

Since 1932, when the absolute monarchy was abolished, Thailand has endured numerous military coups and countercoups, most recently in February 1991. Combined with the threat of Communist insurgencies, these have resulted in numerous declarations of martial law, press censorship, and suspensions of civil liberties. Thai censors banned the movie *Anna and the King* in 1999, just as they had done *The King and I* in 1956.

HEALTH/WELFARE

About 2,000 Thais out of every 100,000 inhabitants attend college (as compared to only 200 per 100,000 Vietnamese). Thailand has devoted substantial sums to the care of refugees from Cambodia and Vietnam. The rate of nonimmigrant population growth has dropped substantially since World War II. AIDS has emerged as a significant problem in Thailand.

ACHIEVEMENTS

Thailand is the only Southeast Asian nation never to have been colonized by a Western power. It was also able to remain detached from direct involvement in the Vietnam War. Unique among Asian cultures, Thailand has a large number of women in business and other professions. Thai dancing is world-famous for its intricacy. In 1996, boxer Somluck Khamsing became the first Thai to win an Olympic gold medal.

Vietnam (Socialist Republic of Vietnam)

GEOGRAPHY

Area in Square Miles (Kilometers):
121,278 (329,560) (about the size of New Mexico)

Capital (Population): Hanoi (1,236,000)

Environmental Concerns: deforestation; soil degradation; overfishing; water and air pollution; groundwater contamination

Geographical Features: low, flat delta in south and north; central highlands; hilly and mountainous in far north and northwest

Climate: tropical in south; monsoonal in north

PEOPLE

Population
Total: 78,774,000
Annual Growth Rate: 1.49%
Rural/Urban Population Ratio: 81/19
Major Languages: Vietnamese; French; Chinese; English; Khmer; tribal languages
Ethnic Makeup: 90% Vietnamese; 7% Muong, Thai, Meo, and other mountain tribes; 3% Chinese
Religions: Buddhists, Confucians, and Taoists most numerous; Roman Catholics; Cao Dai; animists; Muslims; Protestants

Health
Life Expectancy at Birth: 67 years (male); 72 years (female)
Infant Mortality Rate (Ratio): 31.1/1,000
Physicians Available (Ratio): 1/2,444

Education
Adult Literacy Rate: 94%
Compulsory (Ages): 6–11; free

COMMUNICATION
Telephones: 775,000 main lines
Daily Newspaper Circulation: 8 per 1,000 people
Televisions: 43 per 1,000 people
Internet Service Providers: 5 (1999)

TRANSPORTATION
Highways in Miles (Kilometers): 63,629 (106,048)
Railroads in Miles (Kilometers): 1,701 (2,835)
Usable Airfields: 48
Motor Vehicles in Use: 177,000

GOVERNMENT
Type: Communist state

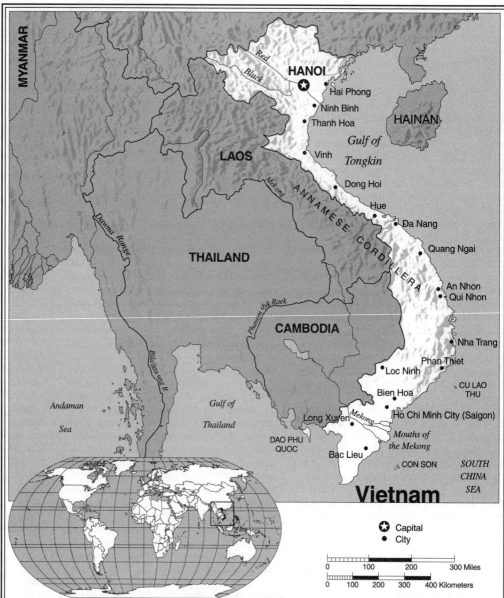

Independence Date: September 2, 1945 (from France)
Head of State/Government: Chairman, State Council (President) Tran Duc Luong; Chairman, Council of Ministers (Premier) Phan Van Khai
Political Party: Communist Party of Vietnam
Suffrage: universal at 18

MILITARY
Military Expenditures (% of GDP): 2.5%
Current Disputes: boundary disputes with Cambodia; other border disputes

ECONOMY
Currency ($ U.S. Equivalent): 14,603 dong = $1
Per Capita Income/GDP: $1,850/$143.1 billion
GDP Growth Rate: 4.8%
Inflation Rate: 4%
Unemployment Rate: 25%
Labor Force: 38,200,000
Natural Resources: phosphates; coal; manganese; bauxite; chromate; oil and gas deposits; forests; hydropower
Agriculture: rice; corn; potatoes; rubber; soybeans; coffee; tea; animal products; fish
Industry: food processing; textiles; machine building; mining; cement; chemical fertilizer; glass; tires; petroleum; fishing; shoes; steel; coal; paper
Exports: $11.5 billion (primary partners Japan, Singapore, Germany, Taiwan)
Imports: $11.6 billion (primary partners Singapore, South Korea, Japan)

FOREIGNERS IN VIETNAM

Foreign powers have tried to control Vietnam for 2,000 years. Most of that time it has been the Chinese who have had their eye on control—specifically of the food and timber resources of the Red River Valley in northern Vietnam.

Most of the northern Vietnamese are ethnically Chinese themselves; but over the years, they forged a separate identity for themselves and came to resent Chinese rule. Vietnam was conquered by China as early as 214 B.C. and again in 111 B.C., when the Han Chinese emperor Wu Ti established firm control. For about 1,000 years (until A.D. 939, and sporadically thereafter by the Mongols and other Chinese), the Chinese so thoroughly dominated the region that the Vietnamese people spoke and wrote in Chinese, built their homes like those of the Chinese, and organized their society according to Confucian values. In fact, Vietnam (*viet* means "people" and *nam* is Chinese for "south") is distinct among Southeast Asian nations because it is the only one whose early culture—in the north, at least—was influenced more by China than by India.

The Chinese did not, however, directly control all of what constitutes modern Vietnam. Until the late 1400s, the southern half of the country was a separate kingdom known as Champa. It was inhabited by the Chams, who originally came from Indonesia. For a time Champa was annexed by the north. However, between the northern region called Tonkin and the southern Chams-dominated region was a narrow strip of land occupied by Annamese peoples (a mixture of Chinese, Indonesian, and Indian ethnic groups), who eventually overthrew the Cham rulers and came to dominate the entire southern half of the country. In the 1500s, the northern Tonkin region and the southern Annamese region were ruled separately by two Vietnamese family dynasties. In the 1700s, military generals took power, unifying the two regions and attempting to annex or control parts of Cambodia and Laos as well.

In 1787, Nguyen-Anh, a general with imperial ambitions, signed a military-aid treaty with France. The French had already established Roman Catholic missions in the south, were providing mercenary soldiers for the Vietnamese

generals, and were interested in opening up trade along the Red River. The Vietnamese eventually came to resent the increasingly active French involvement in their internal affairs and took steps to curtail French influence. The French, however, impressed by the resources of the Red River Valley in the north and the Mekong River Delta in the south, were in no mood to pull out. Vietnam's geography contains rich tropical rain forests in the south, valuable mineral deposits in the north, and oil deposits offshore.

War broke out in 1858, and by 1863, the French had won control of many parts of the country, particularly in the south around the city of Saigon. Between 1884 and 1893, France solidified its gains in Southeast Asia by taking the northern city of Hanoi and the surrounding Tonkin region and by putting Cambodia, Laos, and Vietnam under one administrative unit, which it named *Indochina*.

Ruling Indochina was not easy for the French. For one thing, the region comprised hundreds of different ethnic groups, many of whom had been traditional enemies long before the French arrived.

(UN/DPI photo by Evan Scheider)

Addressing the UN General Assembly in 1998, Pham Gia Khiem, Deputy Prime Minister of Vietnam, expresses Vietnam's concern regarding the escalating worldwide drug problem.

Within the borders of Vietnam proper lived Thais, Laotians, Khmers, northern and southern Vietnamese, and mountain peoples whom the French called Montagnards. Most of the people could not read or write—and those who could wrote in Chinese, because the Vietnamese language did not have a writing system until the French created it. Most people were Buddhists and Taoists, but many also followed animist beliefs.

In addition to the social complexity, the French had to contend with a rugged and inhospitable land filled with high mountains and plateaus as well as lowland swamps kept damp by yearly monsoon rains. The French were eager to obtain the abundant rice, rubber, tea, coffee, and minerals of Vietnam, but they found that transporting these commodities to the coast for shipping was extremely difficult.

VIETNAMESE RESISTANCE

France's biggest problem, however, was local resistance. Anti-French sentiment began to solidify in the 1920s; by the 1930s, Vietnamese youths were beginning to engage in open resistance. Prominent among these was Nguyen ai Quoc, who founded the Indochinese Communist Party in 1930 as a way of encouraging the Vietnamese people to overthrow the French. He is better known to the world today as Ho Chi Minh, meaning "He Who Shines."

Probably none of the resisters would have succeeded in evicting the French had it not been for Adolf Hitler's overrunning of France in 1940 and Japan's subsequent military occupation of Vietnam. These events convinced many Vietnamese that French power was no longer a threat to independence; the French remained nominally in control of Vietnam, but everyone knew that the Japanese had the real power. In 1941, Ho Chi Minh, having been trained in China by Maoist leaders, organized the League for the Independence of Vietnam, or Viet Minh. Upon the defeat of Japan in 1945, the Viet Minh assumed that they would take control of the government. France, however, insisted on reestablishing a French government. Within a year, the French and the Viet Minh were engaged in intense warfare, which lasted for eight years.

The Viet Minh initially fought the French with weapons supplied by the United States when that country was helping local peoples to resist the Japanese. Communist China later became the main supplier of assistance to the Viet Minh. This development convinced U.S. leaders that Vietnam under the Viet Minh would very likely become another Communist state. To prevent this occurrence, U.S. president Harry S. Truman decided to

(United Nations/J. K. Isaac)

Many thousands of Vietnamese fled the country during the past 30 years. The Vietnamese, seen here in a Hong Kong refugee camp, have been either repatriated or absorbed into Hong Kong society.

back France's efforts to recontrol Indochina (although the United States had originally opposed France's desire to regain its colonial holdings). In 1950, the United States gave $10 million in military aid to the French—an act that began a long, costly, and painful U.S. involvement in Vietnam.

In 1954, the French lost a major battle in the north of Vietnam, at Dien Bien Phu, after which they agreed to a settlement with the Viet Minh. The country was to be temporarily divided at the 17th Parallel (a latitude above which the Communist Viet Minh held sway and below which non-Communist Vietnamese had the upper hand), and countrywide elections were to be held in 1956. The elections were never held, however; and under Ho Chi Minh, Hanoi became the capital of North Vietnam, while Ngo Dinh Diem became president of South Vietnam, with its capital in Saigon.

THE UNITED STATES
ENTERS THE WAR

Ho Chi Minh viewed the United States as yet another foreign power trying to control the Vietnamese people through its backing of the government in the South. The United States, concerned about the continuing attacks on the south by northern Communists and by southern Communist sympathizers known as Viet Cong, increased funding and sent military advisers to help prop up the increasingly fragile southern government. By 1963, President John F. Kennedy had sent 12,000 military "advisers" to Vietnam. In 1964, an American destroyer ship was attacked in the Gulf of Tonkin by North Vietnam. The U.S. Congress responded by giving then-president Lyndon Johnson a free hand in ordering U.S. military action against the north; before this time, U.S. troops had not been involved in direct combat.

By 1969, some 542,000 American soldiers and nearly 66,000 soldiers from 40 other countries were in battle against North Vietnamese and Viet Cong troops. Despite unprecedented levels of bombing and use of sophisticated electronic weaponry, U.S. and South Vietnamese forces continued to lose ground to the Communists, who used guerrilla tactics and built their successes on latent antiforeign sentiment among the masses as well as extensive Soviet military aid. At the height of the war, as many as 300 U.S. soldiers were being killed every week.

Watching the war on the evening television news, many Americans began to withdraw support. Anti-Vietnam rallies became a daily occurrence on American university campuses, and many people began finding ways to protest U.S. involvement: dodging the draft by fleeing to Canada, burning down ROTC buildings, and publicly challenging the U.S. government to withdraw. President Richard Nixon had once declared that he was not going to be the first president to lose a war, but, after his expansion of the bombing into Cambodia to destroy Communist supply lines and after significant battlefield losses, domestic resistance became so great that an American withdrawal seemed inevitable. The U.S. attempt to "Vietnamize" the war by training South Vietnamese troops and supplying them with advanced weapons did little to change South Vietnam's sense of having been sold out by the Americans.

Secretary of State Henry Kissinger negotiated a cease-fire settlement with the North in 1973, but most people believed that as soon as the Americans left, the North would resume fighting and would probably take control of the entire country. This indeed happened, and in April 1975, under imminent attack by victorious North Vietnamese soldiers, the last Americans lifted off in helicopters from the grounds of the U.S. Embassy in Saigon, the South Vietnamese government surrendered, and South Vietnam ceased to exist.

The war wreaked devastation on Vietnam. It has been estimated that nearly 2 million people were killed during just the American phase of the war; another 2.5 million were killed during the French era. In addition, 4.5 million people were wounded, and nearly 9 million lost their homes. U.S. casualties included more than 58,000 soldiers killed and 300,000 wounded.

A CULTURE, NOT
JUST A BATTLEFIELD

Because of the Vietnam War, many people think of Vietnam as if it were just a battlefield. But Vietnam is much more than that. It is a rich culture made up of peoples representing diverse aspects of Asian life. In good times, Vietnam's dinner tables are supplied with dozens of varieties of fish and the ever-present bowl of rice. Sugarcane and bananas are also favorites. Because about 80 percent of the people live in the countryside, the population as a whole possesses a living library of practical know-how about farming, livestock raising, fishing, and home manufacture. Today, only about 200 out of every 100,000 Vietnamese people attend college, but most children attend elementary school and nearly 94 percent of the adult population can read and write.

Literacy was not always so high; much of the credit is due the Communist government, which, for political-education reasons, has promoted schooling throughout the country. Another thing that the government has done, of course, is to unify the northern and southern halves of the country. This has not been an easy task, for, upon the division of the country in 1954, the North followed a socialist route of economic development, while in the South, capitalism became the norm.

Religious belief in Vietnam is an eclectic affair and reflects the history of the nation; on top of a Confucian and Taoist foundation, created during the centuries of Chinese rule, rests Buddhism (a modern version of which is called Hoa Hao and claims 1 million believers); French Catholicism, which claims about 15 percent of the population; and a syncretist faith called Cao Dai, which claims about 2 million followers. Cao Dai models itself after Catholicism in terms of hierarchy and religious architecture, but it differs in that it accepts many Gods—Jesus, Buddha, Mohammed, Lao-Tse, and others—as part of its pantheon. Many Vietnamese pray to their ancestors and ask for blessings at small shrines located inside their homes. Animism, the worship of spirits believed to live in nature, is also practiced by many of the Montagnards. (About 400 Christianized Montagnards, incidentally, fought the Communists continually since 1975 and have only recently taken refuge outside of Vietnam.)

Freedom of religious worship has been permitted, and church organizational hierarchies have not been declared illegal. In fact, the government, sensing the need to solicit the support of believers (especially the Catholics), has been careful in its treatment of religions and has even avoided collectivizing farms in areas known to have large numbers of the faithful.

THE ECONOMY

When the Communists won the war in 1975 and brought the capitalist South under its jurisdiction, the United States imposed an economic embargo on Vietnam, which most other nations honored and which remained in effect for 19 years, until President Bill Clinton ended it in 1994. As a consequence of war damage as well as the embargo and the continuing military involvement of Vietnam in the Cambodian War and against the Chinese along their mutual borders, the first decade after the end of the Vietnam War saw the entire nation fall into a severe economic slump. Whereas Vietnam had once been an exporter of rice, it now had to import rice from abroad. Inflation raged at 250 percent a year, and the government was hard-pressed to cover its debts. Many South

China begins 1,000 years of control or influence over the northern part of Vietnam **214 B.C.**	Northern and southern Vietnam are ruled separately by two Vietnamese families **A.D. 1500s**	Military generals overthrow the ruling families and unite the country **1700s**	General Nguyen-Anh signs a military-aid treaty with France **1787**	After 5 years of war, France acquires its first holdings in Vietnam **1863**	France establishes the colony of Indochina **1893**	Ho Chi Minh founds the Indochinese Communist Party **1930**	The Japanese control Vietnam **1940s**	France attempts to regain control **post-1945**

(United Nations/Photo by J. K. Isaac)

As a result of the 1994 removal of a U.S.-imposed economic embargo on Vietnam, several foreign countries found Vietnam a good place to invest funds and to start businesses. Following a 1998 slowdown in the economy, and under the continued control of a Communist government, Vietnam continues to struggle with its future, and the plight of many of Vietnam's peasants is dependent upon the success of the economy and relations with surrounding neighbors.

Vietnamese were, of course, opposed to Communist rule and attempted to flee on boats—but, contrary to popular opinion, most refugees left Vietnam because they could not get enough to eat, not because they were being persecuted.

Beginning in the mid-1980s, the Vietnamese government began to liberalize the economy. Under a restructuring plan called *doi moi* (similar in meaning to Soviet *perestroika*), the government began to introduce elements of free enterprise into the economy. Moreover, despite the Communist victory, the South remained largely capitalist; today, with wages lower than al-most every other country in Asia, an infrastructure built by France and the United States, and laborers who can speak at least some English or French, foreign nations are finding Vietnam a good place to invest funds. In 1991 alone, Australian, Japanese, French, and other companies spent $3 billion in Vietnam. After the embargo ended, firms poured into Vietnam to do business. By the end of 1994, more than 540 firms, especially from Singapore, Hong Kong, Japan, and France, were doing business in Vietnam, and some 70 other countries had expressed interest. The government's move toward privatization and capitalism, begun in 1986, was so successful that by the early 1990s, nearly 90 percent of the workforce were in the private sector, and loans from the World Bank, the Asian Development Bank, and the International Monetary Fund were flowing in to jumpstart national development.

Unfortunately, the financial crises in nearby Malaysia, Indonesia, and other countries caused Vietnam to devalue its currency in 1998. Combined with El Niño–caused drought, forest fires (some 900 in 1998), and floods (1999 floods killed more than 700 people), the currency devaluation slowed economic growth

| The United States begins to aid France to contain the spread of communism **1950s** | Geneva agreements end 8 years of warfare with the French; Vietnam is divided into North and South **1954** | South Vietnam's regime is overthrown by a military coup **1961** | The United States begins bombing North Vietnam **1965** | The United States withdraws its troops and signs a cease-fire **1973** | North Vietnamese troops capture Saigon and reunite the country; a U.S. embargo begins **1975** | Vietnamese troops capture Cambodia; China invades Vietnam **1979** | Communist Vietnam begins liberalization of the economy **1980s** | The U.S. economic embargo of Vietnam is lifted; the United States establishes full diplomatic relations **1990s** |

2000s

The Vietnamese economy shows encouraging growth

Relations with China improve with the resolution of a border dispute

and gave government leaders an excuse to trumpet the failures of capitalism. Leaders declared that despite some liberalization, the government is committed to communism.

Much to the worry of government traditionalists, the Vietnamese people seem fascinated with foreign products. They want to move ahead and put the decades of warfare behind them. Western travelers in Vietnam are treated warmly, and the Vietnamese government has cooperated with the U.S. government's demands for more information about missing U.S. soldiers. In 1994, after a 40-year absence, the United States opened up a diplomatic mission in Hanoi as a first step toward full diplomatic recognition. So eager are the Vietnamese to reestablish economic ties with the West that the Communist authorities have even offered to allow the U.S. Navy to lease its former port at Cam Ranh Bay (the offer has not yet been accepted). Diplomatic bridge-building between the United States and Vietnam increased in the 1990s, when a desire to end the agony of the Cambodian conflict created opportunities for the two sides to talk together. Telecommunications were established in 1992, and in the same year, the United States gave $1 million in aid to assist handicapped Vietnamese war veterans. Finally, in 1995, some two decades after the end of the Vietnam War, the United States established full diplomatic relations with Vietnam.

Despite this gradual warming of relations, however, anti-Western sentiment remains strong in some parts of the population, particularly the military. As recently as 1996, police were still tearing down or covering up signs advertising Western products, and anti–open-door policy editorials were still appearing in official newspapers. The situation was not helped by the visit, in 2000, of U.S. senator John McCain, a former prisoner of war in Vietnam, during which he publicly declared that the wrong side had won the war. Still, the Vietnamese gave a formal state welcome to U.S. defense secretary William Cohen, who discussed military and other topics, especially the missing-in-action, with Vietnamese leaders.

HEARTS AND MINDS
As one might expect, resistance to the current Vietnamese government comes largely from the South Vietnamese, who, under both French and American tutelage, adopted Western values of capitalism and consumerism. Many South Vietnamese had feared that after the North's victory, South Vietnamese soldiers would be mercilessly killed by the victors; some were in fact killed, but many former government leaders and military officers were instead sent to "reeducation camps," where, combined with hard labor, they were taught the values of socialist thinking. Several hundred such internees remain incarcerated two decades after the end of the war. Many of the well-known leaders of the South fled the country when the Communists arrived and have long since made new lives for themselves in the United States, Canada, Australia, and other Western countries. Those who have remained—for example, Vietnamese members of the Roman Catholic Church—have occasionally resisted the Communists openly, but their protests have been silenced. Hanoi continues to insist on policies that remove the rights to which the South Vietnamese had become accustomed. For instance, the regime has halted publication and dissemination of books that it judges to have "harmful contents." There is not much that average Vietnamese can do to change these policies except passive obstruction, which many are doing even though it damages the efficiency of the economy.

Vietnam has made progress with some of its neighbors in recent years. In 1999, it settled a long-standing, 740-mile border dispute with China; and in 2000, it accepted the last repatriated Vietnamese from refugee camps in Hong Kong. Also in 2000, the government announced that it would build a two-lane highway along the route of the famous Ho Chi Minh Trail. It will border Cambodia and Laos, and, at a cost of $375 million, will be completed in 2003.

DEVELOPMENT

Vietnam is again a major exporter of rice. It also produces cement, fertilizer, steel, and coal. Aid and loans from other Asian nations are helping with the construction of roads and other infrastructure, but the per capita income based on gross domestic product is still only $1,850 a year.

FREEDOM

Vietnam is nominally governed by an elected National Assembly. Real power, however, resides in the Communist Party and with those military leaders who helped defeat the U.S. and South Vietnamese armies. Civil rights, such as the right of free speech, are curtailed. Private-property rights are limited. In 1995, Vietnam adopted its first civil code providing property and inheritance rights for citizens.

HEALTH/WELFARE

Health care has been nationalized and the government operates a social-security system, but the chronically stagnant economy has meant that few Vietnamese receive sufficient health care or have an adequate nutritional intake. The World Health Organization has been involved in disease-abatement programs since reunification of the country in 1975.

ACHIEVEMENTS

Vietnam provides free and compulsory schooling for all children. The curricular content has been changed in an attempt to eliminate Western influences. New Economic Zones have been created in rural areas to try to lure people away from the major cities of Hanoi, Hue, and Ho Chi Minh City (formerly Saigon).

Annotated Table of Contents for Articles

Regional Articles

Topic Guide to Articles

TOPIC AREA	TREATED IN	TOPIC AREA	TREATED IN
Foreign Trade	8. Japan: A Rising Sun? 26. Dying for Rice	Natural Resources	14. State of the Staple 21. Indonesia: Starting Over
Health and Welfare	11. Who Wants to Be a Volunteer? 13. Arthritic Nation 14. State of the Staple 15. Continental Divide: Who Owns Aboriginal Lands? 18. Solving the Tibetan Problem 20. Sticking Point 21. Indonesia: Starting Over 26. Dying for Rice 29. Little House on the Paddy: Life in a Northern Vietnam Village	Philippines	27. Roots of Poverty
		Political Development	1. Asia, a Civilization in the Making 7. Reinventing Japan . . . Again 25. Future of Korea: Chronology
		Political Reform	3. South-East Asia: The Tigers That Changed Their Stripes 23. Future of Korea: Background 24. Relations Between North and South Korea: Unification of Our Homeland
History	1. Asia, a Civilization in the Making 6. Japanese Roots 11. Who Wants to Be a Volunteer? 12. Difficulty of Apology: Japan's Struggle With Memory and Guilt 15. Continental Divide: Who Owns Aboriginal Lands? 19. Capital Idea 22. Jakarta's Shame 25. Future of Korea: Chronology	Political Unrest	15. Continental Divide: Who Owns Aboriginal Lands? 28. Uncivil Society
		Politics	10. Privilege of Choosing: The Fallout From Japan's Economic Crisis 17. Living With the Colossus: How Southeastern Asian Countries Cope With China 30. Vietnam's Communists Eye New Vices as Market Worries Rise
Human Rights	5. Cradling Commerce 13. Arthritic Nation 17. Living With the Colossus: How Southeastern Asian Countries Cope With China 18. Solving the Tibetan Problem 22. Jakarta's Shame 26. Dying for Rice	Poverty	27. Roots of Poverty
		Religion	1. Asia, a Civilization in the Making
		Social Reform	10. Privilege of Choosing: The Fallout From Japan's Economic Crisis 11. Who Wants to Be a Volunteer? 13. Arthritic Nation
Medicine	20. Sticking Point		
Minorities	18. Solving the Tibetan Problem	Social Unrest	15. Continental Divide: Who Owns Aboriginal Lands?
Natives	6. Japanese Roots 15. Continental Divide: Who Owns Aboriginal Lands? 16. Burma: Constructive Engagement in Cyberspace? 18. Solving the Tibetan Problem 28. Uncivil Society 29. Little House on the Paddy: Life in a Northern Vietnam Village	Standard of Living	10. Privilege of Choosing: The Fallout From Japan's Economic Crisis 21. Indonesia: Starting Over 26. Dying for Rice 29. Little House on the Paddy: Life in a Northern Vietnam Village 30. Vietnam's Communists Eye New Vices as Market Worries Rise
		Women	5. Cradling Commerce

Article 1   *Foreign Affairs*, July/August, 1996

Asia, a Civilization in the Making

Masakazu Yamazaki

MASAKAZU YAMAZAKI is a playwright and Professor of comparative Studies on Cultures at East Asia University. Mask and Sword collects two of his plays in English translations.

EAST ASIA, THE PACIFIC, AND THE MODERN AGE

AS THE specter of communism fades, some warn of a new East-West confrontation. The remarkable rise of East Asia in recent decades, they say, has been fostered by a civilization very different from the West's, and this poses dangers for international relations. Such thinking, however, is based on Kiplingesque assumptions about an Asian civilization whose existence it fails to demonstrate. At no time in history has an Asian or Eastern civilization arisen over and above the many national and ethnic civilizations and cultures found in that vast region.

Much writing from the West on the purported divide is economically or militarily alarmist, focusing on huge trade deficits with East Asian countries, China's flexing of military muscle, and a few cases in which Chinese or North Korean arms were reportedly sold to Iraq or Iran. Some go so far as to predict that what they see as East Asian civilization may cozy up to Islamic civilization and make common cause against Western power and values. East Asian writers, on the other hand, tend to be extremely sanguine about their region's recent development and its future, contrasting these with Europe's economic plight and the West's social problems. All participants in the debate, however, emphatically affirm the existence of a distinctive East Asian frame of mind, even if they describe it only by saying that it, unlike its Western counterpart, subscribes to no shared value system like democracy or capitalism.

This very diversity and flexibility, some in East Asia argue, will smooth the way for the integration of their region; even North Korea and Myanmar may be brought in. But such integration requires a binding force capable of overriding the logically incompatible value systems the people of the region espouse. That force could only be a tacitly shared psychology or style of life. Some of the thinkers lined up along the artificial East-West divide have noted common features among cities all around the Pacific Basin and even speculated about a melding there of Western and what they call Eastern civilization. What few have seen clearly, however, is that the force behind the convergence observable in the region today is modernity, which was born in the West but has radically transformed both East and West in this century.

AMBIGUOUS ASIA

IN TREATING the question of civilization in Asia, one must first deal with the ambiguity of Asia as a concept. This ambiguity is an irritant to Asians and non-Asians alike and the source of a more than semantic problem in international diplomacy. From around 130 B.C. "Asia" was the name of a province of the Roman Empire on the eastern shore of the Aegean. Today it refers to a sweeping stretch of land and sea from the Middle East to the South Pacific islands—an area too broad to make any sense as a geographical unit. The 1994 Asian Sports Festival in Hiroshima saw Kyrgyz and Tajik athletes from the former Soviet Union in action, but no Hawaiians, Siberians, Australians, or New Zealanders were invited because of the host organization's uncertainty about what constituted Asia. At times, admittedly, countries exploit the confusion over the region's boundaries for political purposes. Many nations along the Pacific Rim—including the United States, Canada, and Chile—participate in the Asia-Pacific Economic Cooperation forum, organized on Australia's initiative, but the white-dominated nations are denied membership in the East Asian Economic Caucus envisaged by Malaysian Prime Minister Mahathir bin Mohamad. And if delineating Asia is a problem, East Asia poses even greater difficulties. This region's energy is palpable but its identity is elusive. Is it a geographical area, an agglomeration of ethic populations, or a civilization in the making?

One thing is certain: the region the West disdained for its "Asiatic stagnation" and whose people suffered because of its lack of economic growth is no more. Flush with Western and Japanese capital and technology, Asian nations are growing vigorously, supplying the rest of the world with products and workers and opening their own markets. Riding the global tide of modernization and industrialization, the region at long last has been integrated into the world economic system. This, however, does not mean that the development that has occurred has been "Asiatic," or that an Asia once seen as dormant is now wide awake.

CIVILIZATION AS UMBRELLA

TO REPEAT: there has never been an Asian, let alone East Asian, sphere of civilization. Western civilization is dominant in Europe and North America, but Asia has known only the individual national and ethnic cultures and civilizations that have arisen in areas of the region.

Western civilization, whose beginnings I place toward the end of the eighth century A.D., created a world that contained different nationalities while transcending national identity. Earlier civilizations, by contrast, whether Greek, Judaic, or Chinese, were essentially ethnic or national and maintained their identity through unity. Customs and forms adopted from the outside were fused with traditional patterns, never acknowledged as a foreign presence. Everyone and everything outside the group was relegated to the realm of the "barbarous," beyond the civilized pale.

Asia has known diverse civilizations, never an Asian civilization.

From Constantine until the latter part of the eighth century, the dominant force in the West was Christianity, which fused the Judaic and Hellenic traditions and, thanks to extensive trade and the use of Latin as the official language, constituted a unified sphere of civilization. But toward the end of the eighth century, as Charlemagne consolidated his empire, Islamic control of Mediterranean trade routes forced fundamental changes in the West. Denied any chance at prosperity through commerce, the West became an agricultural society based on large landholdings. This system of land ownership gave rise to decentralization, leading to dual rule by powerful princes and the Catholic Church. Latin's status gradually eroded, allowing local vernaculars to assert themselves as national languages.

The rise of duality in both rule and language marked the beginning of the Western world civilization. Under the civilizational umbrella dating back to the Roman Empire, and within the unifying framework of Christian civilization, the West set out on its journey toward a world civilization that would encompass national and ethnic civilizations and cultures alien to one another. The crucial factor in the process was that no single nation claimed the supranational umbrella as its own. The Greeks had been debilitated, while the Romans had turned Ital-

ian and Latin remained the common language only for writing. The Jews preserved their identity but were driven to the bottom of the social scale, with Hebrew consigned to libraries and Yiddish and Ladino taking its place. Westerners, whether English, German, or French, could and still can talk about Judeo-Hellenistic civilization on an equal footing.

Asia has never had a comparable superstructure of civilization. Asians lack an experience of political unification like the West's under the Roman Empire, nor do they possess a common tradition in language, currency, laws, roads, or architecture. In the absence of an overall, if loose, religious framework such as Christianity provided for the West, Confucianism, Buddhism, Taoism, Islam, Christianity, and a variety of indigenous religions have coexisted in Asia. There was no writing system like the alphabet that could spell words from different national languages. There was no universal system of musical notation, nor contemporaneous development of artistic styles as in the West's Romanesque, Gothic, and Renaissance periods. Far larger than Europe, Asia stretches from the Arctic to the tropics, and one cannot find in that swath any fundamental similarity in mores, manners, or customs.

CHINESE AND BARBARIANS

SOME WOULD contend that Chinese civilization is the basis of an Asian civilization, and China's influence has indeed been extensive. But the Chinese Empire differed greatly from the Roman. It was the homogeneous empire of the Han, conquering the Manchurians, to be sure, but failing to bring the Mongolians, Vietnamese, Koreans, or Japanese under its control. China exported its laws, religions, art forms, and ideographic writing, but their impact was on the same order as, say, French civilization's on the Germans, in no way tantamount to the framework a world civilization provides. Although the use of Chinese ideograms is widespread in neighboring nations, it failed to progress beyond mimicry into the universalization of the civilization; even today, Japanese politicians are reportedly embarrassed when they sign Sino-Japanese diplomatic agreements with brush and ink, as their ancestors learned to do from the Chinese.

The Chinese, for their part, were generally allergic to outside cultural influences and were particularly reluctant to credit alien contributions to the development of their culture. For them, the Japanese and the Vietnamese were always the "eastern barbarians" and the "southern savages." The Italian descendants of the Romans recognize that they can learn something from the English about Latin language and literature, whereas the Chinese have never turned an attentive ear to Japanese interpretations of Confucius. German directors have impressed and moved Englishmen with their productions of Shakespeare, but the Japanese calligrapher Sugawara Michizane's distinguished work has never had the slightest impact on Chinese practitioners of the art.

The primary reason for Chinese civilization's unusual exclusivity is that the Han have endured, for good or for ill, for 4,000 years. Through the Mongol invasion and Manchu domination, the Han preserved their ethnic civilization as a badge of their identity; "Down with the Manchu, long live the Han!" was their motto as late as the end of the last century. And in the eyes of surrounding peoples, Chinese civilization was simply the source of their borrowings; the civilization never suggested that they had claims on it.

That the Chinese strand has dominated so large an area for so long has inhibited the development of an Asian civilization. The dynamism of a civilization derives from mutual influence, intermixture, and the friendly rivalry of different peoples, but no such chemistry has been at work in Asia. National or ethnic civilizations can undergo such changes only under the umbrella of a world civilization, and Asia has never known such a dual structure.

Buddhist civilization could have become Asia's world civilization. Born in India but disowned there, Buddhism spread to China, northeastern Asia, and Southeast Asia, establishing itself as a religion shared by many ethnic groups. But it has left no indelible mark in the Malay Peninsula or Indonesia, and has been emaciated in China and Korea under the Confucian onslaught that began in the fifteenth century. Buddhism has managed to retain some hold on Japan and part of Southeast Asia, but the two centers have little contact, and the faith survives in Asia at large only as a localized religion. The history of Buddhism, in fact, illustrates how difficult it is for any civilization without an ethnic proprietor to attain dominance and for any dual structure of civilization to take root in Asian soil.

Strangely enough, a prototype of a dual structure was once firmly in place in the early monoethnic Japanese civilization. From time immemorial into the modern era, the Japanese regarded Chinese civilization not as another national civilization but as a world civilization and were painfully conscious that their own civilization occupied a subsidiary position. Few, however, had set foot in China, and their knowledge of the civilization was limited to Chinese characters and other imported traits and institutions. They failed to appreciate that Chinese civilization was a living national civilization, mistaking it for a supranational world civilization. Thus they yielded tamely to Chinese influences, and saw themselves as an alien presence tolerated within the supposedly universal civilization. This mindset may well have facilitated Japanese acceptance of Western civilization in the nineteenth century. If exposure to a strange civilization does not set off alarms warning of imminent clashes but is instead taken as an invitation to share in common property, the recipient nation will naturally be more open and tolerant than it would otherwise be.

The dual structure of rule and language in the West significantly aided the acceptance of Arab civilization that started the West on the path of modernization as far back as the twelfth century. When Spaniards and Italians first encountered Arab civilization, they would have subconsciously placed it on the same level as Western world civilization—which would make it common property that they were encouraged to share in. Since the Arabs in real life were regarded as a great peril, how else could the West have accepted their insights on such fundamental subjects as mathematics, science and technology, and even—if Arab mysticism indeed influenced the twelfth-century troubadors, as some scholars believe—love?

Asia, unfortunately, possessed no such dual structure of civilization or the dynamism it generates. In Japan and a few other nations on the periphery, there was some notion of an Eastern world civilization encompassing all of Asia, but in actuality no such thing existed. This absence ensured that the seeds of modernization in Asia would fail to sprout but would lie dormant until the encounter with the West.

MORNING IN ASIA

MODERN WESTERN civilization has brought the world umbrella to Asia for the first time, and a dual structure of civilization is now taking shape in the region. The Asian world and Asian civilization cited so often of late have their origins not deep in the past but in modernization this century in an Asia in contact with the West.

The entire fabric of society is being geared toward modernization.

In the past 100 years or so, East Asian nations as a group have set out to modernize, and they have been fairly successful in the endeavor. Progress has extended beyond economic development; the entire fabric of society is being geared to modernization, more rapidly in some fields than in others. The formation of a nation-state under the rule of law and legitimate institutions, the secularization of ethics and mores, the rise of industry, and the growth of market economies integrated into the global economy all have been or soon will be attained in virtually all countries of the region except North Korea.

The world over, as education is extended, mass media grow, and leisure activities and consumer goods gain popularity, a middle class arises that favors democratic development. Although each country in East Asia defines and protects human rights and democratic principles differently, no national leader except perhaps North Korea's Kim Jung Il would deny their legitimacy. Members of the Association of Southeast Asian Nations have nearly reached consensus on such fundamentals as the separation of politics from religion, one man—one vote representation, and public trial. When it comes to social welfare, women's liberation, freedom of conscience, access to modern health care, and other social policies, almost all the countries of the region now speak the same language as the West.

In city after large city in East Asia, one finds glass-and-steel towers soaring, the metric system in use, and intellectuals employing American English as the lingua franca. People drive cars, wear Western-style clothes to work, have electric appliances at home, and enjoy jazz, motion pictures, and soap operas. Often television programs are broadcast across the Pacific Basin. It is getting so that one feels at home on both sides of the Pacific.

The secular tolerance of Asian religions has been very good for business.

These changes began in the early 1900s in Japan and in mid-century elsewhere in the region, with all countries going through the same process, experiencing its drawbacks as well as rewards, in the space of a single century. Nothing comparable has ever occurred in Africa, the Middle East, or Russia. It is this contemporaneous experience that is the driving force behind East Asia's integration as a region.

THE BUSINESS OF RELIGION

LOOKING AT the region for common factors that might have made such a transformation possible, the secular tolerance of Asian religions, or the weakness of what is fashionably called fundamentalism, stands out. Asia has had its share of ascetics and spiritual disciplinarians, but they have never joined the establishment. Religions that developed elsewhere tend to slacken in their precepts when they arrive in East Asia. Hinduism as practiced in Bali has reduced the caste system to a mere skeleton, and farmers are permitted to raise hogs for food. Islamic strictures against images and public entertainment, which have led to the closing of movie theaters in Saudi Arabia, are breezily dispensed with in Indonesia, and shadow puppet shows and traditional *gamelan* orchestra music are all the rage.

During the Middle Ages Europeans and Asians alike looked down on commercial profits, and ascetic renunciation of the world was the ideal. But an emphasis on diligence, if not financial gain, is detectable in East Asian religions. By the sixteenth century commerce and its profits were seen as legitimate in Japan and China, and a "secular asceticism" entailing hard work and thrift became established. In his *Religious Ethics and the Merchant Spirit in Early-Modern China,* Ying-shi Yu, a professor of Chinese history at Princeton University, calls this ethos precisely analogous with the Protestant ethic that Max

Weber saw as leading to the rise of capitalism and industrialization in Europe.

According to Ying-shi Yu, the notion of secular asceticism originated in China as early as the ninth century in the reforms of Zen Buddhism, then a new sect. The farm and domestic work required of Zen novices came to be equated with prescribed ascetic practices, and the Zen precept, "No eating without producing," was quoted and put into practice in society at large as well as the monasteries. Confucian scholars of the Sung Dynasty (960–1279) came to interpret the ancient ethic of character-building—"Work hard, be frugal, save time"—in terms of whatever daily work one did in the secular world.

In the sixteenth century, with the policies of the latter Ming Dynasty threatening to impoverish them, intellectuals moved away from the classic interpretation of Confucianism and embraced commerce. Business activity took off nationwide, with merchant cliques in Guangxi and Zhejiang provinces in the vanguard. Merchants' social status improved, and they became conscious of their own power. The insight of the neo-Confucian scholar Wang Yangming—"Though their walks of life are different, all four classes of people are on the same road"—became firmly established. His followers acknowledged that hard work and frugality were virtues on the same order as study. After the merchants agreed to high tax rates, the emperor opened the prestigious profession of government service to them. Scholars made themselves available to pen the epitaphs of magnates.

Merchants, for their part, committed themselves to diligence and thrift and sought to earn "profits controlled by justice." The moral code of merchants of the late Ming Dynasty and Ching Dynasty (1644–1912) boiled down to honest dealings, as the merchants took to heart the tenth-century saying, "In sincerity lies the passage to Heaven." Ying-shi Yu equates this animating principle with the old Protestant belief that worldly work crowned by material success is a sign of redemption: for the Chinese merchant, the secular moral value of open and fair dealings with customers and suppliers became a transcendental passage to heaven. The modern character of Japanese merchants of the period was even more pronounced that that of their Chinese counterparts. They strove to gain a reputation for honesty and trustworthiness, lived frugally, regarded their calling as given by Providence, and took pride in their business because it benefited the nation.

Why East Asians nurtured religious tolerance of the secular and a view of secular activity as akin to religious is not easy to explain. One possibility is that East Asia, along with the Protestant West, which underwent an almost identical ideological evolution, is located far from the centers where the ancient religions were born, and that the religions grew less dogmatic as they spread. In any case, when modern Western civilization encountered East Asia, it found civilizations with which it had a strong affinity. Little wonder, then, that it could serve as the framework for the integration of those civilizations.

CULTURE AND CIVILIZATION

INTEGRATION UNDER Western auspices, however, does not imply the wholesale Westernization of East Asian national civilizations, let alone an East-West fusion of cultures. Culture is a way of life, a conventional order, physically acquired and rooted in subliminal consciousness. Civilization, in contrast, is a consciously recognized ideational order. There is a gray area between the two, but they are distinct. Handiness with machines, for example, is part of culture, while mechanized industry is an aspect of civilization. The performing styles of individual musicians and idiosyncrasies of composers belong to the former, while the diatonic scale and rhythmic system of Western music belong to the latter. Cultures die hard, but their spheres of dominance are limited. Civilizations can become widespread, but they may be deliberately abandoned.

Failure to distinguish clearly between culture and civilization marks the thought of the prophets of the clash of civilizations. The thesis is predicated on the mistaken notions that a civilization can be as predetermined a property of an ethnic group as its culture and that a culture can be as universal and expansive as a civilization. Working from these misconceptions, it follows that a stubborn and irrational culture posing as a civilization could assert itself politically, stirring up conflict.

The rule of culture extends at most from the family, village, or circle of social acquaintances to the tribe or nation. Civilization, in contrast, encompasses different tribes and nations and creates a world. Ancient civilizations, however, had a limited sphere of dominance; in Greece, China, Judea, and elsewhere in the ancient world, the ethnic-national culture covered the same area as the civilization. After Western world civilization arose in the eighth century, the correspondence between cultures and nations still obtained, but civilization assumed a two-level structure: Western world civilization arching over distinct national civilizations. In twentieth-century Britain, a member of Parliament's oratorical style is part of culture, constitutional monarchy is part of national civilization, and democracy is part of Western civilization.

The peoples of East Asia today can be said to partake of modern Western civilization at the topmost stratum of their world, to retain their national civilizations and nation-states in the middle stratum, and to preserve their traditional cultures in their day-to-day lives. In political affairs, human rights and democratic principles belong to the first stratum, distinct bodies of law and political institutions to the second, and political wheeling and dealing to the third. In theater, the dramaturgy common to modern drama is at the topmost stratum, the national languages in which characters' lines are spoken are in the middle, and at ground level are distinctive ethnic styles and figures of speech.

Under the umbrella of modernization, traditional ethnic cultures are being revived with new elements of universality. The Korean agrarian folk music known as *samulnori* attracts percussion aficionados worldwide in the jazz-influenced version popularized by the musician Kim Deoksoo. The Japanese dance troupe Sankaijuku, currently popular in Europe, incorporates steps from Balinese *kechak* dancing, which in turn draws on steps learned from Germans in Bali at the beginning of the century. East Asia has also become a center for cinema, nurturing some promising young filmmakers who bring to their twentieth-century medium exquisite touches of ethnic aesthetics. These developments suggest the imminent birth in the region of what may be called the Pacific-International style.

Traditionalists fail to understand that a world civilization belongs not to one group but to all.

Charging the West with cultural imperialism or deploring the loss of traditional Asian cultures is the height of foolishness. Under the influence of the reigning world civilization, cultures inevitably change and may lose this or that, since they are living organisms. But some portion of their identity is always kept intact. Traditionalists of a nationalistic bent decry the changes, depicting them as impositions from abroad or trappings of a borrowed civilization. They fail to understand that a world civilization belongs not to any one group but to all.

THE MODERN MODE

MODERN CIVILIZATION originated in the West, but it is not an evolutionary phase of Western civilization. To the contrary, modernization began in the twelfth century with the rejection of the Western civilization born four centuries before and can be thought of as an 800-year-long progressive denial of Western civilization.

During the Renaissance the West was deeply influenced by Arab civilization and shaken by underground and local folk cultures that it had deemed heretical and had repressed. The investigations of alchemists led to scientific experimentation, and the grotesque pushed the limits of artistic taste. The seventeenth century witnessed the revival of animistic sensitivity as the West rediscovered and sometimes well-nigh worshiped Nature. Romanticism in the late eighteenth and nineteenth centuries fired the imaginations of the era's artist-exiles, and the West felt the impact first of eastern Europe and Russia and then of the East, as evidenced by the flood of chinoiserie and japonisme.

In this century modernization has driven Western cultures to transform themselves as rapidly as Asian ones. American puritanism has declined to the point that homosexuality is widely tolerated, French cuisine is cutting down on fat and alcohol, German has lost its fraktur script, and the British have abandoned their shillings and tuppence for the more rational decimal system. If the social scientist David Riesman, author of *The Lonely Crowd,* is right, self-centrism, once said to be the core of Western culture, is giving way among the masses to a group-oriented culture. And the revolution in information and communications is changing the West and East Asia at the same time and at about the same speed.

As its Latin etymology suggests, modernity (from *modo,* now) is the spirit of living in constant contrast to the past. Despite the conventional wisdom, it does not necessarily have anything to do with progressivism, which sets goals in pursuit of a future utopia. The essence of modernity is not programmed; there is only a patchwork of trial and error and changes in the status quo. Modernity casts a glance back and extrapolates in different directions. In its willingness to reject all previous values and systems, including itself, modernity verges on nihilism but differs from it in its deep faith in *élan vital.*

If a new sphere of East Asia civilization is in the making today, modernity is the topmost stratum of its "world." The most positive outcome for the region would be not merely diversity but an orderly, widely agreed-on framework encompassing a well-regulated market, human rights, and democratic principles. While narrower political considerations will inevitably affect the civilizational process, an East Asian sphere that defied these fundamental values is inconceivable.

But then, Asian peoples no longer need think in terms of an East Asian framework. In view of the prevailing economic, defense, and political relations in the region, it would seem reasonable to take the entire Pacific Basin as the sphere of the emerging civilization. In East Asia as in North America, Mexico, Australia, and New Zealand, the experience of the twentieth century is of crucial significance, which is why one can feel at home traveling between their cities.

The Pacific sphere should not and will not remain closed to the rest of the world for long. As a civilization-in-progress incorporating continually advancing industrial and communications technologies and unfolding mass societies, it will have to collaborate with the Atlantic sphere of civilization that is sharing the experience. As the 21st century begins, humankind must overcome fanatic nationalism and fundamentalism in all their forms. If it is to have historical relevance, the Pacific sphere of civilization must serve as a transitional stronghold in that struggle.

Article 2

USA Today Magazine, May 1998

USA LOOKS AT THE WORLD

Controlling Economic Competition in the
PACIFIC RIM

The Big Three economic powers in Asia—China, Japan, and the U.S.— must learn to cooperate economically, politically, and militarily if prosperity is to succeed in this burgeoning marketplace.

by Charles W. Kegley, Jr.

Dr. Kegley, Associate American Thought Editor of USA Today, is Pearce Professor of International Relations, University of South Carolina, Columbia.

AS ASTONISHED lenders and investors surveyed the wreckage from the chain-reaction near-collapse of Asia's currencies and markets, their former smug self-confidence began to fade. The spread of Asia's economic flu throughout the globalized marketplace has provoked widespread anxiety about the prospects for a 21st century of peace bolstered by prosperity through free trade in a borderless, integrated international economy. The financial typhoon amidst contagious currency devaluations, budget deficits, and bankruptcies has exposed the logical flaws underlying the megalomania during the 1990s' string of boom years. Now, in the face of crisis, the basic fault lines and vulnerabilities in the unmanaged trade and monetary system have been exposed. Policymakers worldwide have ample reasons to fear the future. The economic foundations of peace in the Pacific Rim are precarious and, without Asian regional stability, there exists little chance for the interdependent international political economy to remain stable.

The challenge of building a more stable institutional architecture for order must be faced if the future is to be prosperous and peaceful. The obstacles to cooperation must be confronted collectively to overcome nations' natural temptation to bolster exports at one another's expense. To avert the disaster of another competitive round of "beggar they neighbor" policies that was the tinderbox for the 1930s Great Depression which ignited World War II, the leading powers must overcome their divisions and work together or their national economies will implode. In a globalized economy where states only can help themselves if they help their rivals as well, *all* have a clear stake in the collective management of collective problems. Still, history does not inspire much confidence that parochial impulses can be resisted. How, then, should the pitfalls and prospects for a prosperous and peaceful future be envisioned?

Any understanding of the prospects for the Pacific Rim, and the globe in general, must begin with a sense of history. As British statesman Winston Churchill once observed, in looking to the future, the further back one looks, the further ahead one can see. To find an informed basis for grasping the major challenges that should be faced by the Big Three economic powers in Asia (China, Japan, and the U.S.), one should follow his advice by looking at long-term historical trends and learning from their lessons.

A second analytical axiom for viewing the future is that there exists a tight interface between the political and military preconditions for continued economic growth in the Pacific Rim. A synergistic, interactive interrelationship has developed among economics, politics, and military security, which can not be meaningfully separated. Peace is a precondition for prosperity, and inspired diplomacy to manage economic competitors' trade relations in the global market is, in turn, a precondition for lasting peace.

What are the long-term problems of the past and the core contemporary circumstances that most will affect the future? First, the U.S. still is a military superpower and an economic giant. It once was said that when the American economy sneezed, the rest of the world caught a cold. That assertion no longer is quite true, but the globalization of trade means that the impact of changes in the American economy continue to reverberate throughout the world.

A second feature of today's global realities lies in how the law of gravity applies in international relations. In the historical evolution of the global political economy, as in nature, what goes up must come down. That natural phenomenon pertains to the U.S. It now is recognized widely that, although the recent Japanese economic upheaval has interrupted that nation's ascendancy, Japan and China's economic growth will resume over the long term. China and Japan enjoy massive trade surpluses and hold vast reserves of foreign currency. Their own currencies are not challenged immediately, and, banking crises in Japan notwithstanding, both nations continue to cut into the U.S. share of world output. The American share peaked after World War II at almost 50%, but steadily has slid since.

Simply put, the U.S. is in decline relative to its two Pacific Rim economic rivals. Without a doubt, the next superpower contest will be between the U.S. and China. The question is not if, but when, China will surpass the U.S. as the leading economic power in the world. More problematic is if and how long it might take a recovering, restructuring Japan to challenge U.S. economic supremacy as well. Whatever the timetable, current economic trajectories suggest that a major power transition likely will occur in the early 21st century, with

three great powers wealthy enough to project military might throughout the region.

This leads to a third unfolding trend. In the likely new 21st century global distribution of power, military and economic might increasingly will be diffused. In contrast to bipolarity, where two superpowers held a preponderance of strength compared to all other countries, the multipolar state system of the future appears destined to contain three roughly equal great powers: China, the U.S., and Japan. The changes from this power transition promise to be profound, creating a new hierarchy with the Big Three at the top of a tripolar regional pyramid.

Such a reordering of the global economic pecking order raises important questions about future military and political order in the Pacific Rim and the world at large. Which powers might align with one another? Will these alignments be seen as a threat by others and stimulate the formation of counteralliances? Can a security regime be built to prevent the rise of such rival trade blocs and military coalitions primed for war on the economic and, perhaps, even military battlefield? To frame an answer, let us return to Churchill's advice that prophets need good memories and look at some additional facts surrounding the advent of a new multipolar Pacific Rim.

The diffusion of strength among the world's three economic powers demands attention because, in history, some forms of multipolarity have been more peaceful than others. For instance, the multipolar system of antagonistic blocs that developed on the eve of World War I proved particularly explosive. When many great powers split into rival camps, there is little chance that competitors in one policy arena (for example, trade) will emerge as partners somewhere else (say, defense), so as to mitigate the competition. Rather, the gains made by one side will be seen as losses by the other, ultimately causing minor disagreements to grow into larger face-offs from which neither coalition is willing to retreat. Since the international system of the early 21st century probably will be dominated by the Big Three extremely powerful states whose commercial and security interests are global, it is important that they do not become segregated into rival blocs.

Aside from the danger of trade wars and armed conflict, the security threats of the Pacific Rim future will include such challenges as interdependent monetary affairs, environmental degradation, resource depletion, internal rebellion by minority ethnic populations, the rising tide of refugees, and the cross-border spread of contagious diseases through increased contact, ranging from AIDS to multi-drug-resistant strains of tuberculosis. None of these regional problems can be met without substantial Big Three cooperation; they are transnational issues that necessitate not national, but global solutions. The threats on the horizon require a collective approach, at the very time when the accelerating erosion of national sovereignty is reducing states' control over their national fates and forcing them to confront common problems through multilateral diplomacy.

> "Conflict over political and territorial issues remains much in evidence. While there have been mutual gains made through foreign trade, geostrategic military differences in the interests of the Big Three have not disappeared."

However, whereas the impact of these traditional non-military threats to regional welfare and stability promises to be potent, they do not necessarily mean geo-economics or ecopolitics will replace geopolitics. Conflict over political and territorial issues remains much in evidence. While there have been mutual gains made through foreign trade, geostrategic military differences in the interests of the Big Three have not disappeared. As former U.S. Secretary of State Lawrence S. Eagleburger has pointed out, the world is "returning to a more traditional and complicated time of multipolarity, with a growing number of countries increasingly able to affect the course of events." The primary issues are how well the U.S. can adjust to its decline from overwhelming preponderance, and how well China and Japan will adapt to their new-found importance. "The change will not be easy for any of the players, as such shifts in power relationships have never been easy," Eagleburger cautions. The challenge to be confronted is ensuring that Big Three cooperation, not competition, becomes institutionalized. At issue is whether the traditional politico-military and non-traditional economic security threats that collectively face the greater Asian Pacific will be managed through concerted Big Three action, to preserve the order that has made the extraordinary progress of the past three decades possible.

Multipolar future options

As power in the Asian Pacific becomes more dispersed, what can be done to prevent the re-emergence of an unstable form of multipolarity? How can the Big Three avoid becoming polarized into antagonist trade and military blocs and avert the use of currency devaluations and protective tariffs as competitive weapons? Three general courses of action exist: they can act unilaterally; they can develop specialized bilateral alliances with another state; or they can engage in some form of collective collaboration with each other.

Of course, each option has many possible variations, and the foreign policies of most great powers contain a mix of acting single-handedly, joining with a partner, and cooperating globally. History shows that what has mattered most for the stability of past multipolar systems has been the relative emphasis placed on "going it alone" vs. "going it with others," and whether joint action was defined in inclusive or exclusive terms.

Unilateral policies, though attractive because they symbolize the nostalgic pursuit of national autonomy, are unlikely to be viable in a multipolar future. The end of the Cold War has reduced public anxieties about foreign dangers. In the U.S., the collapse of the communist threat has led to calls for a reduction in the scale of foreign commitments. In Japan, the Asian financial crisis has prompted the declaration at the December, 1997, Asian Summit that its neighbors could not count on Japan to cushion the cascading economic collapse by opening its market for their exports. In China, echoes of enthusiasm for the Middle Kingdom assertively to chart an independent path can be heard growing more vocal. However, an isolationistic retreat from world affairs by any or all would imperil efforts to deal with the many transnational threats to regional security that require global activity and engagement, and this decreases the incentives for a neo-isolationist withdrawal from existing involvements abroad.

On the other hand, a surge of unilateral activism by any of the Big Three powers would be equally harmful. None of them holds unquestioned hegemonic status with enough power to override all others. Although the U.S. is unrivaled in military might, its offensive capability and unsurpassed military technology is not paralleled by unrivaled financial clout. The U.S. economy faces problems that constrain and inhibit the projection of American power on a global scale. (The U.S. remains the biggest debtor country in history and suffers from extremely low savings rates. Moreover, the American investment in infrastructure is the lowest among all the G-7 industrialized economies.) Given the prohibitive costs of shouldering the financial burden of acting alone, and given the probability that other great powers would be unlikely to accept subordinate positions, unilateralism will be problematic in a multipolar future. The Big Three countries will have to accommodate themselves to the need for internationalist roles so they can coexist and interact with each other.

An alternative to acting unilaterally is joining with selected states in a series of special relationships. On the surface, this option appears attractive. Yet, in a world lacking the stark simplicities of obvious allies and adversaries, differentiating friend from foe is exceedingly difficult, particularly when allies in the realm of military security are the most likely to be trade competitors in a cutthroat global marketplace. Instead of adding restrictive predictability to international affairs, a network of special bilateral relationships and internally active restrictive trade blocs would foster a fear of encirclement among those who perceive themselves as the targets of these combinations.

Whether they entail informal understandings or formal treaties of alliance, all bilateral partnerships have a common drawback: they promote a politics of exclusion that can lead to dangerously polarized forms of multipolarity, in which the competitors align by forming countercoalitions. For example, the formation of new Russo-Sino, Japanese-American, and Sino-American mutual defense agreements have elevated others' security fears. In much the same way, construction of a U.S.-Russian-European Union axis stretching from the Atlantic to the Urals, if institutionalized, greatly would alarm both China and Japan. The trouble with bilateral alliances and multiparty defense and trade coalitions is that they inevitably shun those standing outside the charmed circle. Resentment, revenge, and revisionist efforts to overturn the *status quo* are the predictable consequence. The freewheeling dance of balance-of-power politics typically produces much switching of partners, with some cast aside and thus willing to break-up the whole dance. Arms races fed by beggar-thy-neighbor trade and monetary competition almost always result.

Beyond forming special bilateral alliances, the Big Three have the option of cementing their mutual financial fate in the strengthening of the broad, multilateral associations and cooperative institutions they have created. Two common variants of this option are concerts and collective security organizations. The former involves regularized consultation among those at the top of the global hierarchy; the latter, full participation by all states in the region. A concert offers the benefit of helping control the great-power rivalries that often spawn polarized blocs, though at the cost of ignoring the interests of those not belonging to the group.

> ## "Creating a new Pacific Rim security structure that enlarges the circle of participation to include all the Big Three in collective decision-making will not be easy, especially in a climate of financial crisis and fear."

Alternatively, the all-inclusive nature of collective security allows every voice to be heard, but makes more problematic providing a timely response to threatening situations. Consensus-building is both difficult and delayed, especially in identifying a party to the regime aiming at overturning its rules or preparing for either parochial economic protectionism or, even worse, armed conquest. Collective security mechanisms are deficient in choosing an appropriate response to a challenger to order and in implementing the selected course of deterrent action as well. Since a decision-making body can become unwieldy as its size expands, what is needed to make multilateralism a viable option for the Pacific Rim multipolar future looming on the political horizon is a hybrid that combines elements of a Big Three concert with elements of collective security.

Assuring collective security

Throughout history, different types of multipolar systems have existed. Some of these systems of diffused power were unstable because they contained antagonistic blocs poised on the brink of trade or military warfare. History shows that the key to the stability of any future multipolar system in Asia lies in the inclusiveness of multilateralism. While not a panacea for all of the region's security problems, it offers the best chance to avoid the kinds of polarized alignment patterns that have proven so destructive in the past.

Creating a new Pacific Rim security structure that enlarges the circle of participation to include all the Big Three in collective decision-making will not be easy, especially in a climate of financial crisis and fear. When seen from today's perspective, however, there is no alternative but to try. The other options— unilateralism or bilateralism—have severe costs, because any of the Big Three is certain to look disfavorably on any great power's hegemonic effort to bully the others, or on an alliance between the other pair that defines its purpose as the third's containment. Restricting security protection and free-trade zones in a way that ostracizes and encircles one of the Big Three is a path to division and destruction of the very pillars of prosperity provided through open regional trade and agreements to facilitate exports.

The Big Three must not ever return to the days of a world divided in separate blocs, each seeking to contain the expansion of the other. Such a *realpolitik* response is likely to produce the very kind of polarization into competing coalitions that would benefit no one and corrode the cooperative trade links that provide the basis for lasting friendship. As countries connected by a web of economic linkages, there are material incentives for the Big Three to avoid policies that will rupture profitable business transactions, such as those that stigmatize any core player in the game as an enemy.

This is not to ignore the stubborn fact that trading relationships involve both costs and benefits. The rewarding aspects of commerce likely will be offset by fierce competition, breeding irritation, disputes, and hostility between winners and losers. In view of the differential growth rates among the Big Three and their anxiety about trade competitiveness in an interdependent global marketplace, their major battles of the future indeed are likely to be clashes on the economic front. Still, the Big Three have it within their power to avoid armed combat among soldiers if they keep their aim set on the business of managing business instead of warfare.

A full-fledged, comprehensive Pacific Rim collective security system, dedicated to containing aggression anywhere at any time, may be too ambitious and doomed to failure. A restricted, concert-based collective security mechanism, though, could bring a modicum of order in a fragile and disorderly new Pacific Rim multipolar system, and provide the umbrella needed to allow the regional marketplace to contribute to continuing prosperity.

Whether the actions taken by a Pacific Rim concert-based collective security organization can succeed will hinge on how such a body is perceived and the ways members in it are treated. A Big Three consensus on the rules of trade and security regimes is imperative. It also is vital that each of the Big Three be accorded the equal status it deserves in such multilateral institutions, and none be deprived of membership or equal power over decision-making in such organizations.

In this light, China needs to be included in the World Trade Organization and Group of Seven (and agree to abide by its rules for membership as Russia has done in its participation as the G-7's eighth member). Japan has to be seated on the United Nations' Security Council, and the enlarged NATO and European Union must define their agendas with greater sensitivity to the fears enlargement provoked in the minds of leaders in Tokyo, Beijing, and Moscow. Unity for a globalized, increasingly borderless world is the *sine qua non* to future prosperity and peace. The Big Three have special responsibilities to lead in fostering a unified collective spirit, not only in the Pacific Rim, but in the 21st-century global system.

Article 3   *The Economist,* February 12, 2000

SOUTH-EAST ASIA
The tigers that changed their stripes

Now that East Asia is roaring again, economic reform may be postponed, writes Paul Markillie. But the biggest change of all could be the growth of democracy

---

THIS is supposed to be the "Asian century", and already many people in South-East Asia are of good cheer. Among them is Goh Chok Tong, Singapore's prime minister. The doom and gloom of the financial crisis which began to engulf the region in 1997 has given way to renewed optimism. The decade of despondency that many predicted now seems unlikely to materialise. But it's not quite business as usual. The crisis, says Mr Goh, has produced "four positive outcomes": it has speeded up the opening of economies, forced Asians to be more aware of good corporate governance, made the region concentrate on its real competitive strengths, and provided a hard lesson about globalisation. If so, that is all for the good. But this survey is mostly about another positive outcome, the one that Mr Goh left out: the emergence of more open and democratic government.

This is also the one that will make the most difference to the future of South-East Asia. Most people believed that a big reason for the region's decades of rapid economic growth were its tough, often authoritarian leaders. These strongmen tolerated little dissent, but delivered increased wealth and stability. It was a bargain many South-East Asians were prepared to accept as millions of them were lifted out of the poverty that still haunts many developing countries. But the bargain was also abused by political and business elites, nowhere more so than in Indonesia during 32 years of repressive rule by President Suharto. By the time he was forced to step down in May 1998, a new acronym had entered the lexicon of South-East Asian politics: KKN. In Indonesian Bahasa, it stands for

Sources: EIU; UN   1,000 km

0.0 — Population, m
0.0 — GDP per head, $
Figures for all countries are
1999 estimates except *1998

korupsi, kolusi and nepotisme. It hardly requires translation, but corruption, collusion and nepotism are the evils against which all governments in the region are now being judged.

The complexion of those governments is as diverse as the ten countries this survey covers. They are all members of the regional club, the Association of South-East Asian Nations (ASEAN). They range from Indonesia, the world's biggest Muslim country, to Vietnam, one of the last bastions of communism, Brunei, a small oil-rich Islamic sultanate, and the Philippines, the region's most raucous democracy. In-between come Cambodia, Laos, Malaysia, Myanmar, Singapore and Thailand. And now there is another place close to becoming a country: East Timor. Amid horrendous violence last year, the East Timorese opted for independence from Indonesia. The territory is now under the protection of the United Nations while it prepares for statehood—although some East Timorese do not want to join ASEAN.

The countries of South-East Asia are home to some 500m people and have a combined GDP of more than $700 billion (see map). Their largely young populations, with large numbers of well-educated and hard-working people, helped to make the region one of the fastest-growing in the world. But rampant KKN, many now think, made the financial slump inevitable. True, there were other flaws, such as poorly developed financial systems, but nobody seemed to notice these as long as South-East Asia's tiger economies looked like a one-way bet on rapid growth. Besides, until 1997 there was a sense that the economic order was changing; that the West seemed to be in decline, and that the Asian century was about to dawn. All the lines on the flip-charts about business prospects in South-East Asia pointed one way: upwards. No one believed that the boom could stop.

When it did, the region's self-confidence was shattered. As the value of South-East Asian currencies tumbled, foreigners and locals alike tried to pull their money out, causing fragile financial systems to collapse. Far from being a little local difficulty—a "few small glitches along the road", as President Clinton initially described it—the trouble spread well beyond

East Asia, to Russia and the Americas. For the people of South-East Asia, the worst part of it was western triumphalism. Their economies, which had once been held up as models, were now depicted as fundamentally flawed. Indeed, many in the West seemed to revel in the victory of American capitalism over so-called "Asian values". Seen from within the region, the crisis looked extremely dangerous, having brought a number of countries to the point of economic collapse and perhaps social chaos. Yet the West seemed to be offering help only grudgingly, showing little concern for growth and stability in South-East Asia. This is a memory that is likely to stick.

In the end the bust proved to be overhyped, just like the boom before it. Last year the region staged a sharp recovery, with average growth of about 3.4%, and some countries doing far better than that (see chart 1). This swift upturn contains its own risk: that many of the promised reforms which governments said they would undertake will not now be completed. This is particularly worrying in the financial sector, with its mountains of debt. Paul Krugman, an American economist and one of the few who were sceptical about the "Asian miracle", even when Asia was fashionable, reckons there is now no prospect of serious financial reforms. "This crisis has come to an end too soon," he told a group of Thai bankers last December.

The speed of the recovery has been such that growth is likely to slow, perhaps later this year. If it turns out that countries have indeed failed to put their houses in order, growth will flatten further. Whether the region then heads back into another crisis could hinge on external factors, such as Japan's ability to revive its own sickly economy or, more important, whether the American economy continues to thrive and suck in imports from the region. The chances are that growth will settle down to a slower pace than the breakneck speed of much of the 1990s. That may be no bad thing.

Nevertheless, the possibility of another financial shock is not lost on South-East Asia's leaders. They are now trying harder to work together, and to co-operate with their powerful northern neighbours, China, South Korea and Japan. Should there be another crisis, East Asia might well try to deal with it by itself. Eventually such regional co-operation could lead to the emergence of a trading block with enough clout to rival America and the European Union.

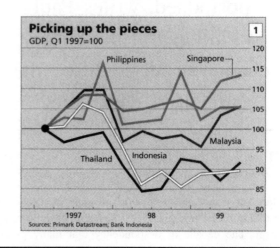

Picking up the pieces   1
GDP, Q1 1997=100

Philippines   Singapore

Thailand   Indonesia

Malaysia

1997   98   99

Sources: Primark Datastream; Bank Indonesia

More immediately, South-East Asia's leaders will be trying to ensure that their economic recovery is sustainable. Some of these leaders are not the same ones who held power when the region got into its mess. Indonesia, for the first time in more than 30 years, is now run by a freely elected government, and new governments have taken over in Thailand and the Philippines. Even where there has been no change in leadership, as in Malaysia, there is clamour for more accountable government. Yet the new, more liberal leadership that is emerging will still have to deliver what the old order did: social harmony and increased wealth. The biggest test case will be Indonesia, long the local giant with a defining influence on the whole region. If Indonesia can thrive as a democracy, then South-East Asia's tigers will really have changed their stripes. . . .

A prickly pair

How will Malaysia and Singapore respond to greater openness in the region?

THE Hard Rock Café in Bangkok has one of Thailand's ubiquitous tuk-tuks (a type of motorised tricycle-taxi) hanging over its door. Its sister restaurant in the Malaysian capital, Kuala Lumpur, boasts a Formula One racing car. Something symbolic here? Malaysia, and especially the tiny island-nation of Singapore off its southern tip, are South-East Asia's economic superstars and enjoy the highest living standards in the region. After gaining independence, both pulled ahead of their neighbours with strong men at the wheel: Mahathir Mohamad, Malaysia's prime minister since, 1981, and Lee Kuan Yew, Singapore's founding father. Yet these veteran leaders are not the best of friends.

Mr Lee was swept to power as Singapore's prime minister in 1959, when the island was moving towards full independence from Britain to become part of a federation with newly independent Malaya. At that time Mr Lee's People's Action Party (PAP) had a fiery left-wing image, but once in power it became less radical and shed its communist members. That enabled it to get closer to the United Malays National Organisation (UMNO), the voice of growing Malay nationalism on the peninsula.

Singapore joined the Malayasian federation in 1963, but there were bitter divisions from the start. When the PAP campaigned in an election on the peninsula, it was seen as encroaching on UMNO's Malay heartland. Singapore refused Malays the special privileges they enjoyed on the peninsula, which contributed to bloody race riots on the island. In Kuala Lumper the attacks on Mr Lee and his party intensified. One of them was led by Dr Mahathir, then a radical UMNO member of parliament, who denounced the PAP as "pro-Chinese, communist-orientated and positively anti-Malay."

Mr Lee was distraught. "We had jumped out of the frying pan of the communists into the fire of the Malay communalists," he wrote. Singapore was expelled from the federation after only two years, and Singaporeans were left wondering how their tiny state, with no natural resources, was going to survive. Remarkably, it turned into one of the richest and most modern city-states in the world. "And that's the problem," says a senior Singaporean official, insisting on anonymity. "They [Malaysia] look at us and see what might have been." But the Malaysians can give as good as they get, jibing that Singaporeans, stuck in their tiny apartments, are jealous of Malaysia's open spaces. Relations between the two countries remain frosty, but never to the point of breaking office.

Malaysia and Singapore still have authoritarian leaders (Mahathir Mohamad remains in office; Goh Chok Tong, who took over as prime minister from Mr Lee in 1990, is no more of a softie than his predecessor), but their political landscapes are very different. In Singapore the ethnic Chinese are in the majority, accounting for over three-quarters of the population of 3.5m. In Malaysia they account for about 25% of the population of 22m, with Malays and other indigenous people making up around 60%. Much of the tension comes from a lingering fear of the Chinese, who are seen as dominating business, big and small, throughout much of South-East Asia.

One estimate being bandied about is that people of Chinese origin control up to 70% of private wealth in Indonesia, Malaysia, the Philippines and Singapore, even though they make up only 6% of the combined population. They control much of the corporate wealth too. Some no longer speak Chinese or use Chinese names, but even so they are often seen as different from the rest, and richer. This has made them kidnapping victims in Manila, and scapegoats during times of tension in Indonesia, though in Singapore and Malaysia such inter-racial strife now seems to be largely a thing of the past.

This is why both countries can claim success in delivering not only prosperity, but also social harmony. With a GDP per head of $25,500, Singaporeans are now among the world's richest people. In Malaysia, economic development has been spread much more evenly than in some of the neighbouring countries. But although the two countries have both been suc-

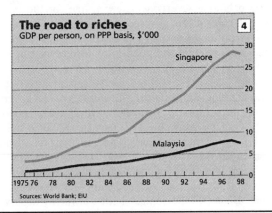

The road to riches
GDP per person, on PPP basis, $'000

4

Singapore

Malaysia

1975 76 78 80 82 84 86 88 90 92 94 96 98

Sources: World Bank; EIU

cessful, that success has been achieved through different political approaches. Singapore's PAP, a single multi-racial party, currently holds all but two of the elected seats in parliament. In Malaysia, UMNO dominates a coalition of 14 parties organised mainly along racial and religious lines, Both countries say they are democracies, although few voices are ever raised in opposition. Both the PAP and UMNO dominates a coalition of 14 parties organised mainly along racial and religious lines. Both countries say they are democracies, although few voices are ever raised in opposition. Both the PAP and UMNO tend to deal with potential opponents either by co-opting them or by cursing them. For most Singaporeans and Malaysians, that has been the price of prosperity and stability. But the financial whirlwind that engulfed the region has set some of them thinking.

That black eye

It was not only the PAP that the young Dr Mahathir attacked in the 1960s, but also the prime minister of the day, Tunku Abdul Rahman, for creating a political elite that abused power and became estranged from the people. The government, glorying in its massive strength, "became contemptuous of criticisms directed at it either from the opposition or its own supporters," Dr Mahathir wrote in his 1970 book "The Malay Dilemma". In November 1999, after 18 years as prime minister, Dr Mahathir led his ruling coalition into a snap election. He faced an opposition united as never before that criticised him in much the same terms as he had criticised the government over 30 years earlier.

What drew the opposition together was the sacking and prosecution of Anwar Ibrahim, Dr Mahathir's deputy and finance minister. This, too, had its roots in the financial crisis: Mr Anwar seemed prepared to work closely with the IMF to promote domestic reforms and tight monetary and fiscal policies, whereas Dr Mahathir blamed the crisis on all manner of things, mostly foreigners, hedge-fund managers and Jews. Mr Anwar appeared to be going against his boss. By the summer of 1998 he had started attacking cronyism, corruption and nepotism in government. He was asking for trouble, and trouble duly arrived. Malaysia refused IMF help and introduced selective currency controls. In September Mr Anwar was sacked, and subsequently arrested. He appeared in court with a black eye inflicted in a beating by police. Many Malaysians were appalled by his treatment. He was later convicted of conspiracy and is currently standing trial for sodomy, which is a crime in Malaysia. Mr Anwar maintains he is innocent.

With the economy rebounding and Mr Anwar in jail, there was one more thing Dr Mahathir needed: a victory at the polls. This would prove he was right about everything all along. To nobody's surprise, in the November election his coalition easily retained its two-thirds majority in parliament, which has a practical as well as a symbolic value because it allows the government to amend the constitution. The election gave Malaysia's 74-year-old leader a new five-year mandate. But his victory was not as sweet as it might have been because the opposition managed to split the Malay vote. Many Malays deserted the ruling coalition, chiefly for the opposition Muslim conservative Parti Islam se-Malaysia (PAS). This meant Dr Ma-

hathir had to rely more heavily than ever on the Chinese and Indian parties in his coalition.

PAS, which has long held the north-eastern state of Kelantan, also captured the oil-rich neighbouring state of Terengganu. By tripling its number of parliamentary seats, it became Malaysia's main opposition party. The opposition gained strength by forming a united front in the election and teaming up with other parties, including that established by Mr Anwar's wife, Wan Azizah Ismail.

As Mr Anwar's second trial continues, Dr Wan Azizah can use her seat in parliament to try to keep her husband's cause alive. But the fight against the ruling coalition is now being led by PAS, which could prove far more threatening to Dr Mahathir. UMNO officials like to portray PAS as a party of religious hardliners who intend to impose an intolerant form of Islam on secular Malaysia. Certainly there have been moves in that direction in the two states that PAS now controls: in Terengganu non-Muslims and other ethnic minorities have been angered by plans to impose Islamic taxes. But PAS is now a modern party whose parliamentarians and new members are mostly professionals keen to take up other social and economic issues, says Fadzil Noor, the party's president. One of those issues is KKN.

Growing support for PAS would widen the split among Malays, Dr Mahathir's own constituency. Many of them seem to have been upset by what they saw as the excessive use of the institutions of state against Mr Anwar, whether he was guilty or not. Many were also riled by the government's heavy-handiness in relentlessly showing Indonesian riots on television and in campaign material to frighten the ethnic Chinese, implying that such things would happen in Malaysia if the ruling party fell. Some also wondered about the wisdom of Dr Mahathir forging ahead with quite so many ambitious pet projects: the new motorway to the new airport has turnings to a new administrative capital, a new high-tech business corridor and a new Formula One motor-racing circuit, all recently built.

Try voting for the opposition

Meanwhile the mass media, with their government-biased reporting, have lost all credibility. Sales of PAS's newspaper *Harakah* have soared in the past year, even though government officials have lately been stopping news vendors from selling it because it is supposed to be distributed only to party members. Scores of Internet sites have also sprung up to distribute news and commentary. With the strength of the ruling party's machine behind it, everyone knew a vote for the opposition would not change the government, says Lim Guan Eng, an opposition politician recently released from jail after serving a sentence for sedition and publishing false news. But, he adds, those who voted for the opposition were making a protest which they hoped might check the powers of government.

It did not happen. In January, five critics of the government, including the editor of *Harakah,* were arrested and charged with sedition. Dr Mahathir also persuaded UMNO's Supreme Council to let him stand unopposed in the party leadership election due in May. Whoever heads UMNO runs Malaysia: it has been that way since independence. But at least Dr Mahathir has anointed someone to succeed him when the time comes: Abdullah Badawi, who replaced Mr Anwar as deputy prime minister. Mr Abdullah will also stand unopposed as vice-president of UMNO, or at least that is the plan: a challenge for either of the top jobs is still possible.

Mr Abdullah has spelled out clearly what he wants to happen: "In order to spread our message, UMNO must embark on recruiting more educated, young professionals and thinkers into our ranks so that the party remains committed to reform and is better able to empathise with the concerns of the younger electorate." He maintains that PAS will not be able to claim broad-based support because only the ruling coalition promotes religious moderation and tolerance, and that is what most Malays want. But they may want more. "Some sort of dislocation is coming because the process of democratisation is not taking place," says Hishamuddin Rais, a film maker and a veteran activist. "Malaysia is increasingly getting out of step with our neighbours."

Perhaps Dr Mahathir should have followed the example of his opposite number in Singapore and handed over the reins of power while remaining a spokesman for his country. Indeed, Dr Mahathir has been a vocal and often conscience-pricking champion for developing countries. Somehow, though, retiring does not seem to be his style. Those looking for change, provided they look hard enough, are more likely to find it in Singapore.

"It's a change in the generations," explains Singapore's Lee Kuan Yew when asked if the government really has begun a cautious dismantling of the country's nanny state. Younger leaders, mostly more widely travelled and often educated abroad, have different experiences and expectations, "so you have got to adjust a couple of things." Yet Mr Lee also talks of the danger that change might inflame racial and religious tensions: "It has to be a compromise between a cosmopolitan elite and a still very conservative base."

Letting go a little

A compromise it may be, but Singapore is loosening up. In the 35 years since its Malaysian divorce, the city-state has prospered by offering what the rest of South-East did not: an honest and professional administration; a modern, clean and efficient environment; a highly trained workforce; and a big welcome to foreign investors. But in embracing the global market, Singapore also has to move with it. Its leaders are well aware that as its neighbours are forced to put their houses in order, the island risks losing some of its competitive advantage.

The effort to keep Singapore ahead of the competition is being led by a younger generation, in particular Lee Hsien Loong, the deputy prime minister. He is also the veteran Mr Lee's son, so it is tempting to imagine him being advised by his father not to push change too far. But the younger Mr Lee is widely respected as his own man. He won praise for slashing business costs, not least by cutting everyone's benefits by 5%, which helped Singapore to weather the financial crisis with only the shallowest of recessions. Now he is trying to open up the economy to foreign competition, starting with the banking industry and telecoms. But his plan also calls for the creation of a more entrepreneurial workforce, able to think for itself, to innovate and to stay ahead in the knowledge-based, Internet-wired commercial future that is believed to lie ahead. For a government used to telling its citizens what they can and cannot do, that is quite a challenge.

Hence other boundaries are being pushed outwards too. A lot of censorship has gone, and political debate has become a little more open: some of Singapore's opposition leaders, for instance, have been allowed to address students instead of being locked up. But there is no sign that mechanisms of control such as the Internal Security Act will be removed, and until they are, critics doubt that reform will go very far. One of them observes: "This sort of loosening has happened before, only for it to be drawn back in again once elections approach."

Those elections are not due until mid-2002. Mr Goh, the present prime minister, remains popular among Singaporeans. He is expected to stay on until after the poll, and then to hand over to the younger Mr Lee. Like previous leaders, this Mr lee seems convinced that in the end PAP knows best. The ruling party is determined not to lose a single seat. Singapore, Mr Goh and his colleagues argue, is a small place well suited to a government that wins overwhelmingly—as long as it does a good job running the country. This, they say, is because local issues tend to be national issues too.

So loosening up in Singapore seems at best to offer a younger, more liberal nanny. Will that be enough for Singaporeans? Provided their government remains competent and honest, they seem unlikely to take to the streets demanding *reformasi*. But stranger things have happened in South-East Asia. Singaporeans who want more will probably pack their bags to work overseas. That means the island risks losing some of the home-grown talent it is desperately trying to cultivate. Today's entrepreneurs, after all, can increasingly choose where to live and work. If they don't like Singapore, they won't stay there. Yet Singapore is a paragon of virtue compared with some of its neighbours. . . .

Living together

South-East Asia looks north to buttress its future

ASKED about the prospects for a common East Asian currency, Rodolfo Severino, the secretary-general of the Association of South-East Asian Nations (ASEAN), looks around the room for the youngest person present. "Per-

haps in her lifetime," he says. A giant trading block that comprises South-East Asia, Japan, China and South Korea may be 30 years off, but it seems to be on its way. It would bring together one-third of the world's population with a combined

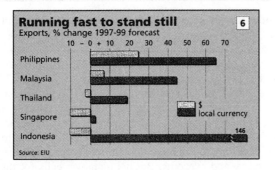

Running fast to stand still
Exports, % change 1997-99 forecast

Source: EIU

GDP of almost $7 trillion at present. That would make it a considerable counterweight to the world's two existing economic powerhouses, the United States and the European Union. In time it might outweigh them.

With so many historical animosities to overcome, progress will be slow. But as Edgardo Espiritu, until last month the Philippines' finance minister, points out, it took divided Europe many decades to move towards a common currency. "Perhaps it will also take us that time." Paradoxically, the economic crisis may have given the region a push towards greater unity. "The East Asian financial crisis should bring the East Asian nations closer together," says Daim Zainuddin, Malaysia's finance minister. Even countries that once saw themselves as somewhat apart from the rest have become more clubby. Kim Dae Jung, South Korea's president and its most democratic leader to date, now talks of the "mutual interests and concerns" that his country has discovered with others in the region.

The strongest evidence of this new enthusiasm for unity is that East Asia's leaders are planning to get together more often. When the ten members of ASEAN met in Manila last November, they were joined not only by Mr Kim but also by Zhu Rongji, China's prime minster, and Keizo Obuchi, Japan's prime minister. All of them agreed that meetings of this group, awkwardly named "ASEAN-plus-three", would become a regular event.

South-East Asia's three powerful neighbours have a strong commercial reason for looking south: they know that a bigger pie will be good for everybody. At the Manila meeting, the ASEAN members agreed to speed up the elimination of tariffs as part of their plan to form a free-trade area among themselves. The entry of China into the World Trade Organisation will make that country's economy more competitive and more open to investment from the region, not least from the many overseas Chinese. Japan and South Korea, which are among the biggest foreign investors in South-East Asia, will benefit from a single ASEAN market which their factories in the region can supply.

But not everything about such a single market is straightforward. Take the question of a financial life-raft. During the crisis Japan launched a temporary $30 billion fund to help Asian countries overcome their problems. The South-East Asians would like Japan to turn this into a permanent facility, a sort of lender of last resort should anyone get into trouble again. Although Japanese officials deny it, such a fund sounds remarkably like the "Asian Monetary Fund" that Japan and Malaysia proposed during the worst of the region's troubles, only to have the idea quickly shot down by America. The fear

then was that any bail-out fund led by Asia would not impose sufficiently tough conditions to ensure that the recipients of the money would carry out the necessary restructuring. Indeed, Japan's own sickly economy, caused by failure to reform its financial industry, bears witness to that.

Even so, America seems to be less bothered by the scheme in its latest guise. But what does Japan get out of it? Easy: the more finance Japan provides in the region, the better it will be for its campaign to internationalise the yen, which seems to have lost ground against the dollar and the euro of late. And at the same time as doling out money at the Manila meeting, Japanese officials were trying to elicit pledges of support from ASEAN leaders for their candidate for the vacant post of managing director of the IMF, Eisuke Sakakibara, a former top financial official.

The Japanese were pushing at open doors. The IMF had been the chief bogeyman during the financial crisis, mainly because it was seen to be too western-influenced and uncaring. All the same, some South-East Asian countries swallowed its medicine and got better. One refused, but still got better. "We have resorted to different ways to overcome our economic problems," says Malaysia's Mr Daim, referring to the imposition of selective capital controls against the advice of many financial commentators. "The results have been positive. The speed and extent of our recovery is comparable to any other crisis-hit economy in the region . . . But what makes our recovery outstanding is the fact that it was achieved without the painful social and political upheaval."

Really? Political life in Malaysia certainly seems to have its painful moments. As for the economy, it does look in better shape, and perhaps the capital controls really were a help. But if they have delayed corporate and financial restructuring, they may have been less helpful than they seemed.

Looking at the region as a whole, it is too early to say who has done the best job of putting his house in order. When the crisis struck, different countries were at different economic stages and consequently suffered different degrees of damage. For instance, whereas Thailand is concerned mainly about the effect of its high level of non-performing loans on the financial sector, the Philippines' Jose Pardo, who took over from Mr Espiritu as finance minister last month, says his top priority is to curb a ballooning budget deficit.

When South-East Asia's bounce-back starts to peter out, more problems may emerge. And although the renewed growth looks impressive in local-currency terms, it seems less wonderful when measured in dollars (for the effect on exports, see chart 6). "The crisis really has not changed the sort of things we have to look at," says Bruce Gale, of Political & Economic Risk Consultancy in Singapore. As the recovery continues, he will be watching two main indicators: how well countries are carrying out their economic reconstruction, and how strong their national institutions are. At present, few countries in South-East Asia can claim good marks for both.

The main political impetus behind a more united East Asia is the hope that it would provide greater security. The area contains some of the world's most dangerous flashpoints. A real war between nuclear-armed India and Pakistan over the disputed territory of Kashmir would greatly unsettle all of Asia. China has threatened to invade Taiwan should it ever declare independence from the motherland. And no one really knows what will happen to dangerously unstable North Korea.

China has long been suspected of wanting to increase its political hegemony in the region. Meanwhile, America's direct influence may decline, although Japan could increasingly pick up that role through a new defence pact with the Americans. For South-East Asian countries that have been variously colonised, invaded or pushed around by China or Japan in the past, the prospect of either country extending its influence in the region is a highly sensitive issue.

Trouble specks

The Spratly Islands are a case in point. China lays claim over all of the South China Sea (see map), which includes many of the world's busiest shipping lanes and is the main highway for Japan's trade. The Spratlys are no more than a collection of reefs, although the area may yet prove to be rich in oil and other minerals. But the islands are also claimed, in whole or in part, by Taiwan, the Philippines, Brunei, Malaysia and Vietnam. During the past year the Chinese have erected a building on one of the reefs which they claim is a shelter for fishermen, but which to all the world looks like a fort. This has increased tensions, and provoked several tussles with the Philippine navy.

Now that all the countries involved are pals in ASEAN-plus-three, the Philippines has tried to broker a peace plan. Unhelpfully, the Philippine navy is so ill-equipped that it can hardly put to sea, let alone defend a territorial claim. The Chinese have noted, but not gone along with, the Philippines' idea for a code of conduct for countries with competing claims over the Spratlys. However, Chinese officials have been reported as saying that although they would not drop claims of sovereignty, they might consider joint development of the islands. That could mean one less flashpoint, but others could still pop up. That is why South-East Asia's leaders have decided to form a troika of (yet-to-be-named) ministers who would respond to regional emergencies. Joseph Estrada, the president of the Philippines, says the team will be "more proactive" in dealing with urgent problems. In the past, ASEAN has maintained a policy of "non-interference" in members' internal affairs. That policy has not been scrapped, but the new troika points that way.

What triggered the latest move was the crisis in East Timor. The lack of a regional response to the killings and destruction unleashed when the East Timorese voted for independence from Indonesia left a vacuum which had to be filled by Australia and the West. Many East Timorese leaders are suspicious of their South-East Asian neighbours, viewing them as supporters of Indonesia's brutality against them, if only by omission. Many do not want an independent East Timor to join ASEAN. They see their country as a Pacific nation and consider Australia as one of their best friends.

Yet some in South-East Asia saw the events in East Timor almost as an attempt by the West to recolonise part of the region. The UN transitional authority on the island is keen to dispel that impression: the military forces there will soon pass from the command of an Australian to a Filipino. If an ASEAN troika had been in place at the time, would it have moved

swiftly enough to end the violence, even if backed by the UN? No one will know unless something like East Timor happens again. It might do, perhaps in the shape of Aceh in Indonesia breaking away, or of an uprising against the regime in Myanmar. Although many in South-East Asia would doubt it, western nations—with plenty of other trouble spots nearer home to worry about—would probably welcome ASEAN trying to take care of problems in the region on its own.

As for Asian values . . .

If it does, the last thing on its mind will be "Asian values". When the tiger economies were going strong, those values were praised to the skies by leaders such as Malaysia's Dr lease, though rather less is heard of them now. Thailand and the Philippines never really believed that there were some features of life in Asia, such as supportive families and greater levels of trust, that somehow set it apart from the West. Asian values are now widely seen as an invention to find common ground for countries with widely differing ethnic, religious and political backgrounds. "The campaign for Asian values will come to be seen in the years ahead as a pragmatic interlude during which Asian leaders briefly sought to justify authoritarian rule before losing power to the middle class they themselves had helped to create by managing their economies for so long with such success," says Victor Mallet in his book "The Trouble with Tigers."

But the issue has not disappeared. "It now goes under another name: 'western interference'," says Zaitun Kasim, a parliamentary candidate for a grassroots women's initiative in Malaysia. It still echoes in South-East Asian criticism—some of it legitimate—of the West's dominance of international organisations such as the IMF and the WTO. It is also heard when some of the regional leaders inveigh against the forces of globalisation. It is least credible when it takes the form of accusations that greater democratic ambitions, too, are a western-influenced evil. After all, the world's biggest democracy, India, is an Asian country. The world's third-biggest may well be emerging in the east too, in the battered and bloodied shape of Indonesia. Democratic values, just like Asian values, are universal.

Different countries may have different ways of getting there. "We want to develop home-grown democracies," says Charles Santiago, a Malaysian economist at Stamford College in Kuala Lumpur. Some things may be the same, he says: greater financial transparency, free media and the right of assembly. Other things may be different: for example, many South-East Asians see the lobby system in the United States as a type of KKN. "The exact form of the political systems that evolve will be differ-

ent," says Abhisit Vejjajiva, the young Thai minister. But "greater democracy will happen all over the region. Even the new members of ASEAN have seen it coming. To be part of the international community, at least economically, they will have to adapt."

There is a risk that the speed of the region's recovery may delay the necessary tasks of fixing broken financial systems, restructuring companies and reforming economies. But the financial crisis has helped with something that will make such reforms more likely to succeed: the move toward more plural, more democratic societies. The faster South-East Asia can get on with that, the better.

Books that were useful in writing this survey include: "Asian Eclipse: Exposing the Dark Side of Business in Asia", by Michael Backman. John Wiley & Sons, Singapore, 1999. "The Trouble with Tigers: the Rise and Fall of South-East Asia", by Victor Mallet. HarperCollins, London, 1999. "South-East Asian Affairs 1999", by various authors. Institute of South-East Asian Studies, Singapore, 1999. "The Years of Living Dangerously: Asia—from Financial Crisis to the New Millennium", by Stephen Vines. Orion Business, London, 1999.

Article 4   *The National Interest,* Winter 1999–2000

Self-Inflicted Wounds.

Hilton Root

Hilton Root is senior fellow and head of Global Studies at the Milken Institute, Santa Monica, CA.

THE GLOBAL financial crisis that began with the collapse of the Thai baht in mid-1997 was fundamentally one of information mismanagement—that is, banks failed and markets collapsed mainly because crucial information was not collected and distributed to investors. Once burned, those same investors have understandably proved reluctant to return to markets where they mistrust the available financial data and, more generally, the credibility of host governments. To lure back investment, armies of financial specialists have been dispatched to Asia by the International Monetary Fund (IMF) and other multilateral institutions to establish basic standards of accounting and disclosure requirements for banks and firms.

But what these specialists have discovered is that the mismanagement they seek to remedy is often the product not of inefficiency but of direct and willful government action. More precisely, governments in Asia routinely misuse financial information for corrupt purposes or unreasonable taxation. Firms then respond by withholding this information, thereby reducing potential trade and investment. If a firm in China, for example, were to maintain the same standards of accounting as its counterparts in the United States, it would soon find itself subjected to a wide range of capricious interventions by state officials.

By contrast, the transformation of the U.S. economy made possible by the emergence of the public corporation in the late

nineteenth century depended upon protection from arbitrary government. Without government financing, managers turned to the public and issued stock, the sale of which required the release of vital information. Because governmental opportunism did not jeopardize firms that accurately disclosed information, American businesses were able to create a thriving equity market. As firms grew by drawing capital from dispersed shareholders, enormous economies of scale developed. Municipalities learned that responsible public accounting permitted the financing of infrastructure by issuing bonds.

Private sector initiatives have not emerged, however, in present-day China, India, Russia and a host of smaller countries having their first serious encounter with a market system. This is not surprising: information about a firm's assets or ownership structure is simply not available in these places. Fearing confiscation of their assets, firms disguise their holdings as well as their management structure, which in turn constrains their ability to access capital effectively.

Yet merely insisting upon improved accounting or auditing will accomplish very little in the developing world. The problem here is one of governance—specifically, malign governance. Legislation that forces banks to reveal their assets, liabilities and loan-loss provisions misses the source of these information failures. The obstacle to transparency in the financial sector is not a dearth of legislation but the developing world's political leadership. Financial mismanagement at the government level, which culminates in the plundering of banks and the restriction of access to alternative capital markets, has become a means of political survival in many states, one not easily surrendered to market forces the leadership cannot control. It is, of course, hardly news that many of the world's leaders stay in power by mismanaging their nations' resources. But until the practice ceases, the logic of international financial rescue will remain dubious.

IN JULY 1998, defending a $4.8 billion IMF loan to Russia that subsequently evaporated, then-Undersecretary of the Treasury Larry Summers reasoned that the United States took a calculated risk "because it was vastly better that Russia succeed than not succeed." The IMF loan was intended to foster financial sector reform. Instead, according to the IMF official that brokered the deal, the $4.8 billion was wasted on propping up Russia's currency long enough to "let the oligarchs get their money out of the country." IMF funds likewise provided Indonesia's Suharto sufficient breathing space to designate a successor while his children were overseas directing the family's money out of the country. As a result, the Suharto clan now possesses the resources to buy back at discount prices enough assets to dominate the Indonesian economy for years, perhaps even generations.

Indeed, it is in East Asia that we find this pattern at its worst. Twenty years ago leaders in that region were busily channeling resources toward national development and growth. As wealth and prosperity increased, however, supervision of the financial system as a public good declined. Instead, Asian leaders today routinely provide their friends and supporters with bank loans, public funds and access to subsidized credit, disregarding the economic risks of doing so. Hence, politicians concerned primarily with political power for its own sake have looked the other way as domestic banking credit has outpaced GDP growth, with predictably ruinous consequences.

The transformation of Korea's banking sector as that country became wealthy and democratic illustrates the point. Because political access lowered their cost of capital, Korea's large *chaebols* set about acquiring new assets, many of them unproductive. In the process, they disregarded profitability, subsidized loss-making business ventures, and promiscuously invested in unrelated sectors of the Korean economy, from steel and automobile manufacturing to real estate and hotels. Even the financial crisis that now threatens the country's prosperity has done little to disrupt this expansion. Quite the contrary: economic collapse has enabled the top five *chaebols* to seize control of a still larger portion of the devastated economy. Loss-making industries are simply dislodged on the public, while the *chaebols* consolidate the profit-making jewels.

Japan, too, features a banking system in which loans are divorced from market considerations. As in Korea, government policy in Japan, which encourages informal banking arrangements, has reduced the cost of capital for firms, allowing them to measure performance on the basis of market share rather than on rates of return on investment. For years, Japanese officials have looked aside as firm managers created vast profits not reflected in corporate accounts. Many observers believe that the Japanese government rejects full disclosure of the banking system's assets and liabilities out of fear that political malfeasance will be exposed. And though international organizations promote the use of outside auditors to ensure that the business sector discloses its borrowings, the Japanese Finance

Ministry has simultaneously promoted a means of circumventing this practice. It simply places a cap on how much a private firm can spend on an audit, a cap established at a level sufficiently low to thwart an effective job being done. If fraud is later exposed, it is the accounting firm, handicapped from the outset, that is usually blamed. Accounting firms, in turn, accept this risk as the price of gaining new business.

Japan is not the only country in which politicians encumber full disclosure of assets and liabilities. Political leaders across the globe have designed governmental bureaucracies with the express purpose of hiding financial information. Banks that cannot, or do not, perform basic accounting functions are often part of a larger structure of opaque regulations, politicized judiciaries and capricious tax systems. The resulting budgetary chaos distorts economic performance, to be sure, but such corruption keeps politicians in power.

The banking crisis the world confronts today recalls an observation once made by Schumpeter: "The public finances are one of the best starting points for an investigation of society, especially of its political life." One of the most basic tests of accountability that a country must pass is the ability to identify and collect taxes. It should come as no surprise, then, that the largest debtor countries in the world collect only a small fraction of the taxes owed them. Of the estimated 73 million Brazilians in the work force, only 7.6 million pay income taxes. Out of a Pakistani population of 140 million, an estimated 1 million pay income taxes. Mexico reports about 19 million registered taxpayers in a country of nearly 100 million people, and its finance secretary reports the country has the highest rate of tax evasion in the world. In the Philippines, of 12 million potential taxpayers only 2 million contribute. The situation in Russia is better; collections there account for as much as 50 percent of what taxpayers owe the treasury. Citizens in many of these countries so distrust the government that they hide their fortunes under mattresses or stash them overseas.

IRONICALLY, international financial assistance often exacerbates mismanagement by insulating failed regimes and allowing them to continue to reward supporters. Investors, of course, view loan-making institutions as a source of credible information. Indeed, the mere willingness of these institutions to extend credit establishes certain expectations about a nation's economic performance. Unfortunately, their loan approval processes rely on information gleaned from a compromised source—namely, the same political leaders who wrecked their nations' economies in the first place. Being accountable to their membership, international organizations thus become vehicles of misinformation and lack the independence to play the role expected of them as a neutral source of analysis and information. Under Suharto, for example, the representative of the Indonesian government at the World Bank was able to prevent the dissemination of research that questioned the wisdom of Indonesia's economic policies.

Then, too, international loan packages, which are often negotiated through secret agreements with a single branch of government, actually weaken the main source of policy reform: the need for public accountability. When domestic financial crises broke out in England (1669), France (1789) and Japan (1868), no international donor existed to bail out political leaders. Bankrupt rulers were driven into the hands of their own citizens, who insisted on sound policies in exchange for the taxes they remitted. Indeed, international watchdogs can never substitute for accountability to citizens who pay taxes and thereby tie their fortunes to those of the state. Such organizations can at best empower local systems of accountability. However, when the deal is secret and conceals essential economic data, neither investors nor citizens can assess how a government is performing. And such agreements allow leaders, *ex post facto,* to claim a lack of public support and to renege on loan agreements once the funds have been dispensed. Recall how in early 1998 Suharto sought to use the riots that his policies induced in Indonesia as a pretext for evading the conditions of the deal he had signed with the IMF.

Contrary to prevailing wisdom, the serial fiscal failures of some of the world's best endowed nations are not the result of a mysterious contagion, nor of exploitation by developed countries or by the "international system." The blame lies closer to home. The afflicted nations—Brazil, Indonesia, Mexico, Pakistan and the Philippines, to name a few—have all become caught in binds of their own devising. And the financial architecture that has been engaged to surmount this global economic disaster has merely ended up compounding it. That architecture—which has allowed leaders to draw support from international organizations without engaging the will of the citizens who must pay these organizations back—has created a fundamental misalignment of economic incentives, the benefits of which have been largely private, while the debt has been public and sovereign. Fostering greater accountability where it is ultimately needed will require subjecting the leaders of sovereign states to the will of the peoples they govern. Only then can this global system of taxation without representation be reformed.

Article 5   *Far Eastern Economic Review,* June 10, 1999

Cradling Commerce

Southeast Asian women have proven to be key economic players throughout the millennium. Now, they are continuing that role by helping to propel the region through its current slump and beyond.

By Margot Cohen

Margot Cohen is a writer based in Jakarta.

Indonesia's economic crisis hasn't left the women of West Sumatra with idle hands. Take Santi Marnis, for example. A farmer's wife in the Alahan Panjang district, Santi spends her days hunched over a sewing machine in a small kiosk bursting with brightly coloured, intricately patterned handbags. She started the business in November 1997 and now boasts a hefty waiting list of customers. "I'm the only one in the whole district who can make these designs," beams the 25-year-old mother.

She'll pass on more than just seamstress skills to her daughter. The Minangkabau people in West Sumatra's highlands follow a matrilineal line of inheritance, leaving homes and fields in the hands of women. Men typically reside in the homes of their mothers-in-law, or leave the province to seek their fortunes. Today, while the population is devoutly Muslim, customary law still takes precedence over an Islamic tradition that would deprive women of such considerable inheritance. In shaping their own destinies, women throughout Southeast Asia have always been in a stronger position than most of their sisters to the north. This strength, spanning a millennium, is now coming to the region's aid in a time of crisis. As women paddle feverishly to keep their families financially afloat, the ripples they create are washing across local economies and providing much needed relief.

Admittedly, the feisty Minangkabau women are lucky compared to their peers elsewhere in the region. Most Southeast Asian cultures—including Malay, Javanese, Thai, Burmese and Vietnamese—practise a bilateral kinship system, in which property rights can be passed down to either sex. However, as in the case of Minangkabau women, these rights have their roots in customary law, which is pre-Islamic, pre-Confucian and pre-Christian. This also gives women more equal access to divorce and property ownership than is formally allowed by either the Islamic or Confucian legal codes.

The relative influence and economic drive of the "fairer sex" astounded many early European travellers to Southeast Asia. The same visitors had watched Chinese women hobble on bound feet and winced as Indian women cast themselves on the funeral pyres of their deceased husbands. Southeast Asia was an entirely different world. In 1820, for example, British observer John Crawfurd recorded his impressions of women in the archipelago that would later become known as Indonesia:

They eat with the men, and associate with them in all respects on terms of equality... This equality of the sexes... is perhaps most thoroughly recognized among the most warlike tribes. Among the nations of the Celebes [modern-day South Sulawesi], the most warlike of the Archipelago, the women appear in public without any scandal; they take an active concern in all the business of life; they are consulted by men on all public affairs, and are frequently raised to the throne...

That Southeast Asian women have enjoyed such a relatively favourable position is rooted in several factors, says Carol Warren of Murdoch University in Australia. Their key role in food production is one element. In traditional societies dependent on hunting and gathering, women provided one-half to three-quarters of the group's nutritional requirements.

Women also tend to excel in agrarian economies, depending on the ecology and technology of different farming systems. In cultures dependent on intensive wet-rice cultivation, as in Southeast Asia, the division of labour in food production was apportioned equally between the sexes.

Finally, the general availability of land and relatively low demographic pressure in the region may also explain the prevalence of bilateral kinship systems. These factors led to the lack of strong gender preferences in children. Female infanticide, once common in China and some parts of India, never gained a toehold in Southeast Asia.

In this climate, Southeast Asian women began turning their energies to commerce. In a 1727 account of Siam, precursor to Thailand, British observer Alexander Hamilton marvelled: "The women in Siam are the only merchants in buying goods, and some of them trade very considerably." Indeed, often the best way for a Chinese, Indian, Persian or Dutch merchant to advance his economic interests was to marry one of Siam's women traders, or marry his own daughter into a Siamese family. These days, foreigners still gain entry into the country's business world by marrying Thai women, who continue to enjoy a pre-eminent role in commerce.

Now, the region's current economic crisis has forced more women to work. Many are starting their own businesses. This may not revolutionize society, but it could be quietly providing a backbone for economic recovery.

West Sumatra at the beginning of the 20th century provides something of an historical parallel to today, says Minangkabau scholar Ranny Emilia of Andalas University in Padang, West Sumatra. In the early 1900s, rebellion against a Dutch colonial tax triggered a disastrous chain of events. The Strait of Malacca closed, the price of cloves dropped and trade was disrupted. Scrambling for cash, many men, who were allowed to manage family property, sold off land without consulting their wives—downright heresy in the Minangkabau tradition.

As a result, women realized that if they didn't help administer their own property, they couldn't control it. They also realized that they needed to create new resources, rather than merely act as repositories for inherited wealth. At the same time, a surge of interest in women's education and trade occurred.

"In the 1910s weaving schools were founded throughout Sumatra, and Minangkabau girls were the homespun experts recruited as faculty," writes Jeffrey Hadler of Cornell University. "Honed in the making of songket, gold-weft cloth, Minangkabau weaving techniques were exported as the market for cotton and silk cloth strengthened."

The women of Minangkabau are once again rising to meet economic challenges. "The men don't have a choice," says Ranny of Andalas University. "They are allowing women to take advantage of any opportunity to stabilize the family income."

But long-term progress requires more than just the financial stability of individual families. As Southeast Asia thinks ahead to the next millennium, a major challenge lies in providing women an equitable role in politics. The arduous task of rebuilding the region's economies depends largely on the political will to root out corruption—a task, many argue, that cannot be left to men alone.

"When it comes to corruption, women are more God-fearing," maintains Hilma Hamid, head of the local parliament in Padang Panjang, West Sumatra. "Women are afraid of hurting their own careers. And they have more shame. Men just don't care."

For the moment, men seem to care most about simply weathering the hard times. And truth be told, Minangkabau men seem more than happy to watch their womenfolk shoulder some of the financial burden. "Women here have a good head for economics. They're strong, dynamic, creative and credible," says Basril Djabar, head of the West Sumatra Chamber of Commerce, with a smile. Like most of his counterparts, Basril hands his earnings over to his wife to manage. "If my wife wasn't Minang, I would run away."

Article 6

Discover, June 1998

Japanese Roots

Just who are the Japanese? Where did they come from and when? The answers are difficult to come by, though not impossible—the real problem is that the Japanese themselves may not want to know.

BY JARED DIAMOND

JARED DIAMOND is a longtime contributing editor of DISCOVER, a professor of physiology at UCLA Medical School, and a member of the National Academy of Sciences. Expanded versions of many of his DISCOVER articles appear in his book The Third Chimpanzee *and in his most recent book,* Guns, Germs, and Steel: The Fates of Human Societies. *For that book, in April, Diamond was awarded this year's Pulitzer Prize for general nonfiction.*

Unearthing the origins of the Japanese is a much harder task than you might guess, among world powers today, the Japanese are the most distinctive in their culture and environment. The origins of their language are one of the most disputed questions of linguistics. These questions are central to the self-image of the Japanese and to how they are viewed by other peoples. Japan's rising dominance and touchy relations with its neighbors make it more important than ever to strip away myths and find answers.

The search for answers is difficult because the evidence is so conflicting. On the one hand, the Japanese people are biologically undistinctive, being very similar in appearance and genes to other East Asians, especially to Koreans. As the Japanese like to stress, they are culturally and biologically rather homogeneous, with the exception of a distinctive people called the Ainu on Japan's northernmost island of Hokkaido. Taken together, these facts seem to suggest that the Japanese reached Japan only recently from the Asian mainland, too recently to have evolved differences from their mainland cousins, and displaced the Ainu, who represent the original inhabitants. But if that were true, you might expect the Japanese language to show close affinities to some mainland language, just as English is obviously closely related to other Germanic languages (because Anglo-Saxons from the continent conquered England as recently as the sixth century A.D.). How can we resolve this contradiction between Japan's presumably ancient language and the evidence for recent origins?

Archeologists have proposed four conflicting theories. Most popular in Japan is the view that the Japanese gradually evolved from ancient Ice Age people who occupied Japan long before 20,000 B.C. Also widespread in Japan is a theory that the Japanese descended from horse-riding Asian nomads who passed through Korea to conquer Japan in the fourth century,

but who were themselves—emphatically—not Koreans. A theory favored by many Western archeologists and Koreans, and unpopular in some circles in Japan, is that the Japanese are descendants of immigrants from Korea who arrived with rice-paddy agriculture around 400 B.C. Finally, the fourth theory holds that the peoples named in the other three theories could have mixed to form the modern Japanese.

When similar questions of origins arise about other peoples, they can be discussed dispassionately. That is not so for the Japanese. Until 1946, Japanese schools taught a myth of history based on the earliest recorded Japanese chronicles, which were written in the eighth century. They describe how the sun goddess Amaterasu, born from the left eye of the creator god

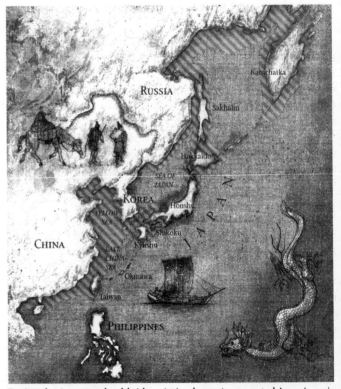

During the ice ages, land bridges (striped areas) connected Japan's main islands to one another and to the mainland, allowing mammals—including humans—to arrive on foot.

Anagi, sent her grandson Ninigi to Earth on the Japanese island of Kyushu to wed an earthly deity. Ninigi's great-grandson Jimmu, aided by a dazzling sacred bird that rendered his enemies helpless, became the first emperor of Japan in 660 B.C. To fill the gap between 660 B.C. and the earliest historically documented Japanese monarchs, the chronicles invented 13 other equally fictitious emperors. Before the end of World War II, when Emperor Hirohito finally announced that he was not of divine descent, Japanese archeologists and historians had to make their interpretations conform to this chronicle account. Unlike American archeologists, who acknowledge that ancient sites in the United States were left by peoples (Native Americans) unrelated to most modern Americans, Japanese archeologists believe all archeological deposits in Japan, no matter how old, were left by ancestors of the modern Japanese. Hence archeology in Japan is supported by astronomical budgets, employs up to 50,000 field-workers each year, and draws public attention to a degree inconceivable anywhere else in the world.

Why do they care so much? Unlike most other non-European countries, Japan preserved its independence and culture while emerging from isolation to create an industrialized society in the late nineteenth century. It was a remarkable achievement. Now the Japanese people are understandably concerned about maintaining their traditions in the face of massive Western cultural influences. They want to believe that their distinctive language and culture required uniquely complex developmental processes. To acknowledge a relationship of the Japanese language to any other language seems to constitute a surrender of cultural identity.

What makes it especially difficult to discuss Japanese archeology dispassionately is that Japanese interpretations of the past affect present behavior. Who among East Asian peoples brought culture to whom? Who has historical claims to whose land? These are not just academic questions. For instance, there is much archeological evidence that people and material objects passed between Japan and Korea in the period A.D. 300 to 700. Japanese interpret this to mean that Japan conquered Korea and brought Korean slaves and artisans to Japan; Koreans believe instead that Korea conquered Japan and that the founders of the Japanese imperial family were Korean.

Thus, when Japan sent troops to Korea and annexed it in 1910, Japanese military leaders celebrated the annexation as "the restoration of the legitimate arrangement of antiquity." For the next 35 years, Japanese occupation forces tried to eradicate Korean culture and to replace the Korean language with Japanese in schools. The effort was a consequence of a centuries-old attitude of disdain. "Nose tombs" in Japan still contain 20,000 noses severed from Koreans and brought home as trophies of a sixteenth-century Japanese invasion. Not surprisingly, many Koreans loathe the Japanese, and their loathing is returned with contempt.

What really was "the legitimate arrangement of antiquity"? Today, Japan and Korea are both economic powerhouses, facing each other across the Korea Strait and viewing each other

through colored lenses of false myths and past atrocities. It bodes ill for the future of East Asia if these two great peoples cannot find common ground. To do so, they will need a correct understanding of who the Japanese people really are.

Japan's unique culture began with its unique geography and environment. It is, for comparison, far more isolated than Britain, which lies only 22 miles from the French coast. Japan lies 110 miles from the closest point of the Asian mainland (South Korea), 190 miles from mainland Russia, and 480 miles from mainland China. Climate, too, sets Japan apart. Its rainfall, up to 120 inches a year, makes it the wettest temperate country in the world. Unlike the winter rains prevailing over much of Europe, Japan's rains are concentrated in the summer growing season, giving it the highest plant productivity of any nation in the temperate zones. While 80 percent of Japan's land consists of mountains unsuitable for agriculture and only 14 percent is farmland, an average square mile of that farmland is so fertile that it supports eight times as many people as does an average square mile of British farmland. Japan's high rainfall also ensures a quickly regenerated forest after logging. Despite thousands of years of dense human occupation, Japan still offers visitors a first impression of greenness because 70 percent of its land is still covered by forest.

Japanese forest composition varies with latitude and altitude: evergreen leafy forest in the south at low altitude, deciduous leafy forest in central Japan, and coniferous forest in the north and high up. For prehistoric humans, the deciduous leafy forest was the most productive, providing abundant edible nuts such as walnuts, chestnuts, horse chestnuts, acorns, and beechnuts. Japanese waters are also outstandingly productive. The lakes, rivers, and surrounding seas teem with salmon, trout, tuna, sardines, mackerel, herring, and cod. Today, Japan is the largest consumer of fish in the world. Japanese waters are also rich in clams, oysters, and other shellfish, crab, shrimp, crayfish, and edible seaweeds. That high productivity was a key to Japan's prehistory.

From southwest to northeast, the four main Japanese islands are Kyushu, Shikoku, Honshu, and Hokkaido. Until the late nineteenth century, Hokkaido and northern Honshu were inhabited mainly by the Ainu, who lived as hunter-gatherers with limited agriculture, while the people we know today as Japanese occupied the rest of the main islands.

In appearance, of course, the Japanese are very similar to other East Asians. As for the Ainu, however, their distinctive appearance has prompted more to be written about their origins and relationships than about any other single people on Earth. Partly because Ainu men have luxuriant beards and the most profuse body hair of any people, they are often classified as Caucasoids (so-called white people) who somehow migrated east through Eurasia to Japan. In their overall genetic makeup, though, the Ainu are related to other East Asians, including the Japanese and Koreans. The distinctive appearance and hunter-gatherer lifestyle of the Ainu, and the undistinctive appearance and the intensive agricultural lifestyle of the Japanese, are frequently taken to suggest the straightforward interpretation that the Ainu are descended from Japan's original hunter-gatherer inhabitants and the Japanese are more recent invaders from the Asian mainland.

But this view is difficult to reconcile with the distinctiveness of the Japanese language. Everyone agrees that Japanese does not bear a close relation to any other language in the world. Most scholars consider it to be an isolated member of Asia's Altaic language family, which consists of Turkic, Mongolian, and Tungusic languages. Korean is also often considered to be an isolated member of this family, and within the family Japanese and Korean may be more closely related to each other than to other Altaic languages. However, the similarities between Japanese and Korean are confined to general grammatical features and about 15 percent of their basic vocabularies, rather than the detailed shared features of grammar and vocabulary that link, say, French to Spanish; they are more different from each other than Russian is from English.

Since languages change over time, the more similar two languages are, the more recently they must have diverged. By counting common words and features, linguists can estimate how long ago languages diverged, and such estimates suggest that Japanese and Korean parted company at least 4,000 years ago. As for the Ainu language, its origins are thoroughly in doubt; it may not have any special relationship to Japanese.

After genes and language, a third type of evidence about Japanese origins comes from ancient portraits. The earliest preserved likeness of Japan's inhabitants are statues called haniwa, erected outside tombs around 1,500 years ago. Those statues unmistakably depict East Asians. They do not resemble the heavily bearded Ainu. If the Japanese did replace the Ainu in Japan south of Hokkaido, the replacement must have occurred before A.D. 500.

Our earliest written information about Japan comes from Chinese chronicles, because China developed literacy long before Korea or Japan. In early Chinese accounts of various peoples referred to as "Eastern Barbarians," Japan is described under the name Wa, whose inhabitants were said to be divided into more than a hundred quarreling states. Only a few Korean or Japanese inscriptions before A.D. 700 have been preserved, but extensive chronicles were written in 712 and 720 in Japan and later in Korea. Those reveal massive transmission of culture to Japan from Korea itself, and from China via Korea. The chronicles are also full of accounts of Koreans in Japan and of Japanese in Korea—interpreted by Japanese or Korean historians, respectively, as evidence of Japanese conquest of Korea or the reverse.

The ancestors of the Japanese, then, seem to have reached Japan before they had writing. Their biology suggests a recent arrival, but their language suggests arrival long ago. To resolve this paradox, we must now turn to archeology.

The seas that surround much of Japan and coastal East Asia are shallow enough to have been dry land during the ice ages, when much of the ocean water was locked up in glaciers and sea level lay at about 500 feet below its present measurement. Land bridges connected Japan's main islands to one another, to the Russian mainland, and to South Korea. The mammals walking out to Japan included not only the ancestors of modern Japan's bears and monkeys but also ancient humans, long before boats had been invented. Stone tools indicate human arrival as early as half a million years ago.

Around 13,000 years ago, as glaciers melted rapidly all over the world, conditions in Japan changed spectacularly for the better, as far as humans were concerned. Temperature, rainfall, and humidity all increased, raising plant productivity to present high levels. Deciduous leafy forests full of nut trees, which had been confined to southern Japan during the ice ages, expanded northward at the expense of coniferous forest, thereby replacing a forest type that had been rather sterile for humans with a much more productive one. The rise in sea level severed the land bridges, converted Japan from a piece of the Asian continent to a big archipelago, turned what had been a plain into rich shallow seas, and created thousands of miles of productive new coastline with innumerable islands, bays, tidal flats, and estuaries, all teeming with seafood.

That end of the Ice Age was accompanied by the first of the two most decisive changes in Japanese history: the invention of pottery. In the usual experience of archeologists, inventions flow from mainlands to islands, and small peripheral societies aren't supposed to contribute revolutionary advances to the rest of the world. It therefore astonished archeologists to discover that the world's oldest known pottery was made in Japan 12,700 years ago. For the first time in human experience, people had watertight containers readily available in any desired shape. With their new ability to boil or steam food, they gained access to abundant resources that had previously been difficult to use: leafy vegetables, which would burn or dry out if cooked on an open fire; shellfish, which could now be opened easily; and toxic foods like acorns, which could now have their toxins boiled out. Soft-boiled foods could be fed to small children, permitting earlier weaning and more closely spaced babies. Toothless old people, the repositories of information in a preliterate society, could now be fed and live longer. All those momentous consequences of pottery triggered a population explosion, causing Japan's population to climb from an estimated few thousand to a quarter of a million.

The prejudice that islanders are supposed to learn from superior continentals wasn't the sole reason that record-breaking Japanese pottery caused such a shock. In addition, those first Japanese potters were clearly hunter-gatherers, which also violated established views. Usually only sedentary societies own pottery: what nomad wants to carry heavy, fragile pots, as well as weapons and the baby, whenever time comes to shift camp? Most sedentary societies elsewhere in the world arose only with the adoption of agriculture. But the Japanese environment is so productive that people could settle down and make pottery while still living by hunting and gathering. Pottery helped those Japanese hunter-gatherers exploit their environment's rich food resources more than 10,000 years before intensive agriculture reached Japan.

Much ancient Japanese pottery was decorated by rolling or pressing a cord on soft clay. Because the Japanese word for cord marking is *jomon,* the term Jomon is applied to the pottery itself, to the ancient Japanese people who made it, and to that whole period in Japanese prehistory beginning with the invention of pottery and ending only 10,000 years later. The earliest Jomon pottery, of 12,700 years ago, comes from Kyushu, the southernmost Japanese island. Thereafter, pottery spread north, reaching the vicinity of modern Tokyo around 9,500 years ago and the northernmost island of Hokkaido by 7,000 years ago. Pottery's northward spread followed that of deciduous forest rich in nuts, suggesting that the climate-related food explosion was what permitted sedentary living.

How did Jomon people make their living? We have abundant evidence from the garbage they left behind at hundreds of thousands of excavated archeological sites all over Japan. They apparently enjoyed a well-balanced diet, one that modern nutritionists would applaud.

One major food category was nuts, especially chestnuts and walnuts, plus horse chestnuts and acorns leached or boiled free of their bitter poisons. Nuts could be harvested in autumn in prodigious quantities, then stored for the winter in underground pits up to six feet deep and six feet wide. Other plant foods included berries, fruits, seeds, leaves, shoots, bulbs, and roots. In all, archeologists sifting through Jomon garbage have identified 64 species of edible plants.

Then as now, Japan's inhabitants were among the world's leading consumers of seafood. They harpooned tuna in the open ocean, killed seals on the beaches, and exploited seasonal runs of salmon in the rivers. They drove dolphins into shallow water and clubbed or speared them, just as Japanese hunters do today. They netted diverse fish, captured them in weirs, and caught them on fishhooks carved from deer antlers. They gathered shellfish, crabs, and seaweed in the intertidal zone or dove for them. (Jomon skeletons show a high incidence of abnormal bone growth in the ears, often observed in divers today.) Among land animals hunted, wild boar and deer were the most common prey. They were caught in pit traps, shot with bows and arrows, and run down with dogs.

The most debated question about Jomon subsistence concerns the possible contribution of agriculture. Many Jomon sites contain remains of edible plants that are native to Japan as wild species but also grown as crops today, including the adzuki bean and green gram bean. The remains from Jomon times do not clearly show features distinguishing the crops from their wild ancestors, so we do not know whether these plants were gathered in the wild or grown intentionally. Sites also have debris of edible or useful plant species not native to Japan, such as hemp, which must have been introduced from the Asian mainland. Around 1000 B.C., toward the end of the Jomon period, a few grains of rice, barley, and millet, the staple cereals of East Asia, began to appear. All these tantalizing clues

make it likely that Jomon people were starting to practice some slash-and-burn agriculture, but evidently in a casual way that made only a minor contribution to their diet.

Archeologists studying Jomon hunter-gatherers have found not only hard-to-carry pottery (including pieces up to three feet tall) but also heavy stone tools, remains of substantial houses that show signs of repair, big village sites of 50 or more dwellings, and cemeteries—all further evidence that the Jomon people were sedentary rather than nomadic. Their stay-at-home lifestyle was made possible by the diversity of resource-rich habitats available within a short distance of one central site: inland forests, rivers, seashores, bays, and open oceans. Jomon people lived at some of the highest population densities ever estimated for hunter-gatherers, especially in central and northern Japan, with their nut-rich forests, salmon runs, and productive seas. The estimate of the total population of Jomon Japan at its peak is 250,000—trivial, of course, compared with today, but impressive for hunter-gatherers.

All through human history, centralized states with metal weapons and armies supported by dense agricultural populations have swept away sparser populations of hunter-gatherers. How did Stone Age Japan survive so long?

With all this stress on what Jomon people did have, we need to be clear as well about what they didn't have. Their lives were very different from those of contemporary societies only a few hundred miles away in mainland China and Korea. Jomon people had no intensive agriculture. Apart from dogs (and perhaps pigs), they had no domestic animals. They had no metal tools, no writing, no weaving, and little social stratification into chiefs and commoners. Regional variation in pottery styles suggests little progress toward political centralization and unification.

Despite its distinctiveness even in East Asia at that time, Jomon Japan was not completely isolated. Pottery, obsidian, and fishhooks testify to some Jomon trade with Korea, Russia, and Okinawa—as does the arrival of Asian mainland crops. Compared with later eras, though, that limited trade with the outside world had little influence on Jomon society. Jomon Japan was a miniature conservative universe that changed surprisingly little over 10,000 years.

To place Jomon Japan in a contemporary perspective, let us remind ourselves of what human societies were like on the Asian mainland in 400 B.C., just as the Jomon lifestyle was about to come to an end. China consisted of kingdoms with rich elites and poor commoners; the people lived in walled towns, and the country was on the verge of political unification and would soon become the world's largest empire. Beginning around 6500 B.C., China had developed intensive agriculture based on millet in the north and rice in the south; it had domestic pigs, chickens, and water buffalo. The Chinese had had writing for at least 900 years, metal tools for at least 1,500 years, and had just invented the world's first cast iron. Those developments were also spreading to Korea, which itself had had agriculture for several thousand years (including rice since at least 2100 B.C.) and metal since 1000 B.C.

With all these developments going on for thousands of years just across the Korea Strait from Japan, it might seem astonishing that in 400 B.C. Japan was still occupied by people who had some trade with Korea but remained preliterate stone-tool-using hunter-gatherers. Throughout human history, centralized states with metal weapons and armies supported by dense agricultural populations have consistently swept away sparser populations of hunter-gatherers. How did Jomon Japan survive so long?

To understand the answer to this paradox, we have to remember that until 400 B.C., the Korea Strait separated not rich farmers from poor hunter-gatherers, but poor farmers from rich hunter-gatherers. China itself and Jomon Japan were probably not in direct contact. Instead Japan's trade contacts, such as they were, involved Korea. But rice had been domesticated in warm southern China and spread only slowly northward to much cooler Korea, because it took a long time to develop cold-resistant strains of rice. Early rice agriculture in Korea used dry-field methods rather than irrigated paddies and was not particularly productive. Hence early Korean agriculture could not compete with Jomon hunting and gathering. Jomon people themselves would have seen no advantage in adopting Korean agriculture, insofar as they were aware of its existence, and poor Korean farmers had no advantages that would let them force their way into Japan. As we shall see, the advantages finally reversed suddenly and dramatically.

More than 10,000 years after the invention of pottery and the subsequent Jomon population explosion, a second decisive event in Japanese history triggered a second population explosion. Around 400 B.C., a new lifestyle arrived from South Korea. This second transition poses in acute form our question about who the Japanese are. Does the transition mark the re-placement of Jomon people with immigrants from Korea, ancestral to the modern Japanese? Or did Japan's original Jomon inhabitants continue to occupy Japan while learning valuable new tricks?

The new mode of living appeared first on the north coast of Japan's southwesternmost island, Kyushu, just across the Korea Strait from South Korea. There we find Japan's first metal tools, of iron, and Japan's first undisputed full-scale agriculture. That agriculture came in the form of irrigated rice fields, complete with canals, dams, banks, paddies, and rice residues revealed by archeological excavations. Archeologists term the new way of living Yayoi, after a district of Tokyo where in 1884 its characteristic pottery was first recognized. Unlike Jomon pottery, Yayoi pottery was very similar to contemporary South Korean pottery in shape. Many other elements of the new Yayoi culture were unmistakably Korean and previously foreign to Japan, including bronze objects, weaving, glass beads, and styles of tools and houses.

While rice was the most important crop, Yayoi farmers introduced 27 new to Japan, as well as unquestionably domesticated pigs. They may have practiced double cropping, with paddies irrigated for rice production in the summer, then drained for dry-land cultivation of millet, barley, and wheat in the winter. Inevitably, this highly productive system of intensive agriculture triggered an immediate population explosion in Kyushu, where archeologists have identified far more Yayoi sites than Jomon sites, even though the Jomon period lasted 14 times longer.

In virtually no time, Yayoi farming jumped from Kyushu to the adjacent main islands of Shikoku and Honshu, reaching the Tokyo area within 200 years, and the cold northern tip of Honshu (1,000 miles from the first Yayoi settlements on Kyushu) in another century. After briefly occupying northern Honshu, Yayoi farmers abandoned that area, presumably because rice farming could not compete with the Jomon hunter-gatherer life. For the next 2,000 years, northern Honshu remained a frontier zone, beyond which the northernmost Japanese island of Hokkaido and its Ainu hunter-gatherers were not even considered part of the Japanese state until their annexation in the nineteenth century.

It took several centuries for Yayoi Japan to show the first signs of social stratification, as reflected especially in cemeteries. After about 100 B.C., separate parts of cemeteries were set aside for the graves of what was evidently an emerging elite class, marked by luxury goods imported from China, such as beautiful jade objects and bronze mirrors. As the Yayoi population explosion continued, and as all the best swamps or irrigable plains suitable for wet rice agriculture began to fill up, the archeological evidence suggests that war became more and more frequent: that evidence includes mass production of arrowheads, defensive moats surrounding villages, and buried skeletons pierced by projectile points. These hallmarks of war in Yayoi Japan corroborate the earliest accounts of Japan in

Chinese chronicles, which describe the land of Wa and its hundred little political units fighting one another.

In the period from A.D. 300 to 700, both archeological excavations and frustratingly ambiguous accounts in later chronicles let us glimpse dimly the emergence of a politically unified Japan. Before A.D. 300, elite tombs were small and exhibited a regional diversity of styles. Beginning around A.D. 300, increasingly enormous earth-mound tombs called *kofun,* in the shape of keyholes, were constructed throughout the former Yayoi area from Kyushu to North Honshu. *Kofun* are up to 1,500 feet long and more than 100 feet high, making them possibly the largest earth-mound tombs in the world. The prodigious amount of labor required to build them and the uniformity of their style across Japan imply powerful rulers who commanded a huge, politically unified labor force. Those *kofun* that have been excavated contain lavish burial goods, but excavation of the largest ones is still forbidden because they are believed to contain the ancestors of the Japanese imperial line. The visible evidence of political centralization that the *kofun* provide reinforces the accounts of *kofun*-era Japanese emperors written down much later in Japanese and Korean chronicles. Massive Korean influences on Japan during the *kofun* era— whether through the Korean conquest of Japan (the Korean view) or the Japanese conquest of Korea (the Japanese view)— were responsible for transmitting Buddhism, writing, horseback riding, and new ceramic and metallurgical techniques to Japan from the Asian mainland.

Finally, with the completion of Japan's first chronicle in A.D. 712, Japan emerged into the full light of history. As of 712, the people inhabiting Japan were at last unquestionably Japanese, and their language (termed Old Japanese) was unquestionably ancestral to modern Japanese. Emperor Akihito, who reigns today, is the eighty-second direct descendant of the emperor under whom that first chronicle of A.D. 712 was written. He is traditionally considered the 125th direct descendant of the legendary first emperor, Jimmu, the great-great-great-grandson of the sun goddess Amaterasu.

Japanese culture underwent far more radical change in the 700 years of the Yayoi era than in the ten millennia of Jomon times. The contrast between Jomon stability (or conservatism) and radical Yayoi change is the most striking feature of Japanese history. Obviously, something momentous happened at 400 B.C. What was it? Were the ancestors of the modern Japanese the Jomon people, the Yayoi people, or a combination? Japan's population increased by an astonishing factor of 70 during Yayoi times: What caused that change? A passionate debate has raged around three alternative hypotheses.

One theory is that Jomon hunter-gatherers themselves gradually evolved into the modern Japanese. Because they had already been living a settled existence in villages for thousands of years, they may have been preadapted to accepting agriculture. At the Yayoi transition, perhaps nothing more happened than that Jomon society received cold-resistant rice seeds and information about paddy irrigation from Korea, enabling it to produce more food and increase its numbers. This theory appeals to many modern Japanese because it minimizes the unwelcome contribution of Korean genes to the Japanese gene pool while portraying the Japanese people as uniquely Japanese for at least the past 12,000 years.

A second theory, unappealing to those Japanese who prefer the first theory, argues instead that the Yayoi transition represents a massive influx of immigrants from Korea, carrying Korean farming practices, culture, and genes. Kyushu would have seemed a paradise to Korean rice farmers, because it is warmer and swampier than Korea and hence a better place to grow rice. According to one estimate, Yayoi Japan received several million immigrants from Korea, utterly overwhelming the genetic contribution of Jomon people (thought to have numbered around 75,000 just before the Yayoi transition). If so, modern Japanese are descendants of Korean immigrants who developed a modified culture of their own over the last 2,000 years.

The last theory accepts the evidence for immigration from Korea but denies that it was massive. Instead, highly productive agriculture may have enabled a modest number of immigrant rice farmers to reproduce much faster than Jomon hunter-gatherers and eventually to outnumber them. Like the second theory, this theory considers modern Japanese to be slightly modified Koreans but dispenses with the need for large-scale immigration.

By comparison with similar transitions elsewhere in the world, the second or third theory seems to me more plausible than the first theory. Over the last 12,000 years, agriculture arose at not more than nine places on Earth, including China and the Fertile Crescent. Twelve thousand years ago, everybody alive was a hunter-gatherer; now almost all of us are farmers or fed by farmers. Farming spread from those few sites of origin mainly because farmers outbred hunters, developed more potent technology, and then killed the hunters or drove them off lands suitable for agriculture. In the modern times European farmers thereby replaced native Californian hunters, aboriginal Australians, and the San people of South Africa. Farmers who used stone tools similarly replaced hunters prehistorically throughout Europe, Southeast Asia, and Indonesia. Korean farmers of 400 B.C. would have enjoyed a much larger advantage over Jomon hunters because the Koreans already possessed iron tools and a highly developed form of intensive agriculture.

Which of the three theories is correct for Japan? The only direct way to answer this question is to compare Jomon and Yayoi skeletons and genes with those of modern Japanese and Ainu. Measurements have now been made of many skeletons. In addition, within the last three years molecular geneticists have begun to extract DNA from ancient human skeletons and compare the genes of Japan's ancient and modern populations. Jomon and Yayoi skeletons, researchers find, are on the average readily distinguishable. Jomon people tended to be shorter, with relatively longer forearms and lower legs, more wide-set eyes, shorter and wider faces, and much more pronounced facial topography, with strikingly raised browridges, noses, and nose bridges. Yayoi people averaged an inch or two taller, with close-set eyes, high and narrow faces, and flat browridges and noses. Some skeletons of the Yayoi period were still Jomon-like in appearance, but that is to be expected by almost any theory of the Jomon-Yayoi transition. By the time of the *kofun* period, all Japanese skeletons except those of the Ainu form a homogeneous group, resembling modern Japanese and Koreans.

In all these respects, Jomon skulls differ from those of modern Japanese and are most similar to those of modern Ainu, while Yayoi skulls most resemble those of modern Japanese. Similarly, geneticists attempting to calculate the relative contributions of Korean-like Yayoi genes and Ainu-like Jomon genes to the modern Japanese gene pool have concluded that the Yayoi contribution was generally dominant. Thus, immigrants from Korea really did make a big contribution to the modern Japanese, though we cannot yet say whether that was because of massive immigration or else modest immigration amplified by a high rate of population increase. Genetic studies of the past three years have also at last resolved the controversy about the origins of the Ainu: they are the descendants of Japan's ancient Jomon inhabitants, mixed with Korean genes of Yayoi colonists and of the modern Japanese.

Given the overwhelming advantage that rice agriculture gave Korean farmers, one has to wonder why the farmers achieved victory over Jomon hunters so suddenly, after making little headway in Japan for thousands of years. What finally tipped the balance and triggered the Yayoi transition was probably a combination of four developments: the farmers began raising rice in irrigated fields instead of in less productive dry fields; they developed rice strains that would grow well in a cool climate; their population expanded in Korea, putting pressure on Koreans to emigrate; and they invented iron tools that allowed them to mass-produce the wooden shovels, hoes, and other tools needed for rice-paddy agriculture. That iron and intensive farming reached Japan simultaneously is unlikely to have been a coincidence.

We have seen that the combined evidence of archeology, physical anthropology, and genetics supports the transparent interpretation for how the distinctive-looking Ainu and the undistinctive-looking Japanese came to share Japan: the Ainu are descended from Japan's original inhabitants and the Japanese are descended from more recent arrivals. But that view leaves the problem of language unexplained. If the Japanese really are recent arrivals from Korea, you might expect the Japanese and Korean languages to be very similar. More generally, if the Japanese people arose recently from some mixture, on the island of Kyushu, of original Ainu-like Jomon inhabitants with Yayoi invaders from Korea, the Japanese language might show close affinities to both the Korean and Ainu languages. Instead, Japanese and Ainu have no demonstrable relationship, and the relationship between Japanese and Korean is distant. How could this be so if the mixing occurred a mere 2,400 years ago? I suggest the following resolution of this paradox: the languages of Kyushu's

Jomon residents and Yayoi invaders were quite different from the modern Ainu and Korean languages, respectively.

The Ainu language was spoken in recent times by the Ainu on the northern island of Hokkaido, so Hokkaido's Jomon inhabitants probably also spoke an Ainu-like language. The Jomon inhabitants of Kyushu, however, surely did not. From the southern tip of Kyushu to the northern tip of Hokkaido, the Japanese archipelago is nearly 1,500 miles long. In Jomon times it supported great regional diversity of subsistence techniques and of pottery styles and was never unified politically. During the 10,000 years of Jomon occupation, Jomon people would have evolved correspondingly great linguistic diversity. In fact, many Japanese place-names on Hokkaido and northern Honshu include the Ainu words for river, nai or betsu, and for cape, shiri, but such Ainu-like names do not occur farther south in Japan. This suggests not only that Yayoi and Japanese pioneers adopted many Jomon place-names, just as white Americans did Native American names (think of Massachusetts and Mississippi), but also that Ainu was the Jomon language only of northernmost Japan.

That is, the modern Ainu language of Hokkaido is not a model for the ancient Jomon language of Kyushu. By the same token, modern Korean may be a poor model for the ancient Yayoi language of Korean immigrants in 400 B.C. In the centuries before Korea became unified politically in A.D. 676, it consisted of three kingdoms. Modern Korean is derived from the language of the kingdom of Silla, the kingdom that emerged triumphant and unified Korea, but Silla was not the kingdom that had close contact with Japan in the preceding centuries. Early Korean chronicles tell us that the different kingdoms had different languages. While the languages of the kingdoms defeated by Silla are poorly known, the few preserved words of one of those kingdoms, Koguryo, are much more similar to the corresponding Old Japanese words than are the corresponding modern Korean words. Korean languages may have been even more diverse in 400 B.C., before political unification had reached the stage of three kingdoms. The Korean language that reached Japan in 400 B.C., and that evolved into modern Japanese, I suspect, was quite different from the Silla language that evolved into modern Korean. Hence we should not be surprised that modern Japanese and Korean people resemble each other far more in their appearance and genes than in their languages.

History gives the Japanese and the Koreans ample grounds for mutual distrust and contempt, so any conclusion confirming their close relationship is likely to be unpopular among both peoples. Like Arabs and Jews, Koreans and Japanese are joined by blood yet locked in traditional enmity. But enmity is mutually destructive, in East Asia as in the Middle East. As reluctant as Japanese and Koreans are to admit it, they are like twin brothers who shared their formative years. The political future of East Asia depends in large part on their success in rediscovering those ancient bonds between them.

Article 7

Foreign Policy, Summer 2000

Reinventing Japan . . . Again

by Frank B. Gibney

FRANK B. GIBNEY, *author of* Japan, the Fragile Superpower *(1996) and other works, is professor of politics and president of the Pacific Basin Institute at Pomona College.*

The year is 2050, by accepted international reckoning. In Japan, it is also the fifth year of the Restructure Era, officially named Dai Ni Bunmei Saikaika Jidai (the Second Era of Civilization and Reform) after its illustrious Meiji Era prototype. For traditionalists at the Tokyo University Faculty of Law and Folklore, it is the 2,710th year since the legendary founding of Japan by the redoubtable Emperor Jimmu.

Times are good. Trade and international investment are flourishing, to the point where it is no longer possible—or even important—to pick out which companies are Japanese and which are foreign on the burgeoning stock exchanges. The Nikkei remains well over 30,000, but the Son Nasdaq Exchange is far higher. The lingering depression of the early part of the century is only a memory. A new generation of schoolchildren hardly remembers the waves of rioting and political violence—the so-called Salaryman's Revolt—that swept the country after passage of the National Restructuring Law and the total defeat of the Nationalist Remnant Party (once known as the Liberal Democratic Party). Japanese business rebuilt itself in the chaotic free competition that came with deregulation. New businesses sprouted as venture capital became available, even excessive. Yet thanks to heavily policed share trading and laws regulating expense spending, the "money madness" of the early thirties abated and the gap between the incomes of rich and poor was somewhat diminished.

New national elections are barely a month away. Although the Green/Democratic Party still holds a Diet majority, the Citizens' Reform League has made significant gains, aided by a temporary alliance with the Falun Gong Theosophists. Voter interest is high—and has been ever since the public corruption scandals exposed in the late thirties. (The statutes forbidding the candidacy of Diet members' relatives remain strictly enforced.) The new political alignments also reflect the growing influence of foreign-born voters, who received the franchise as a result of the recent Immigration Reform Act. Chinese immigrants, in fact, have formed their own highly vocal political party. The environmentalist movement remains strong, thanks largely to the leadership of Yoshiko Nakamura, the country's first female prime minister. Her election four years ago, by popular vote, was opposed only by a dwindling band of traditionalists. With 40 percent of the Diet's lower house female, a woman as chief executive is no longer regard ed as a novelty.

Internationally speaking, Japan's prestige has never been higher. For some years the country has been the prime mover in a worldwide campaign to save the environment by reducing global warming. And since Japan became a permanent "great power" member of the U.N. Security Council, the electorate has given wide support to the dispatch of Japanese troops for peacekeeping operations. When the Self Defense Forces Second Division, fresh from its successful campaign against African fascists, paraded through the Ginza under the U.N. and Japanese flags, the enthusiasm of the welcoming crowds made its own statement.

As a key member of the newly formed Northeast Asian Nations' Association—known familiarly by its acronym NANA—Japan plays a leading role in regional security. Organized just 10 years earlier, after the religious wars in China threatened regional peace, this federation of Japan, Korea, Taiwan, and Russia's maritime provinces has proved a strong stabilizing force—as the recent joint naval maneuvers in the Gold Sea (formerly referred to as the Sea of Japan or East Sea) indicated. The revised Japanese-U.S. Security Treaty has been extended to cover NANA (with joint use of bases), while Washington has agreed that under the circumstances there is no longer any need for American forces to be stationed in the NANA area. (Since the departure of the last U.S. troops, Okinawa's new gambling casinos have transformed that island's economy.)

With Korea now unified and Nanjing the capital of a democratic Chinese Republic, few security problems remain. Thus the thrust of NANA is largely economic. While competition still exists among its leading members, the prevalence of multinational Internet groups and, above all, the dramatic globalization enforced by the knowledge industry have done away with much of the old nationalisms. The widened scope of the Asian Initiative Venture Bank, in which Japan plays a leading role, offers a steady safeguard against the kind of debilitating economic crises that devastated some Asian countries near the end of the last century. The East Asian free-trade area, now expanded to include the renewed economic power of a reformed China, has become a force for world stability.

Changes in Japanese society have been a key factor in the new economy's success. Since women became equal partners in the marketplace, the old Confucian relationship between the sexes has given way to a healthy equality, enforced by a network of antidiscrimination laws. The willingness of more young couples to marry and have babies has finally put a stop to the alarming decline in the birthrate. According to the most recent census figures (December 2049), Japan's total population has stabilized at 110 million. True, this represents a decline of 17 million over the past half century, but it still stands as a distinct improvement over the shocking 100 million figure projected for 2050 at the beginning of the century.

Over the last few decades, as people's subservience to government bureaucracy has diminished, nongovernmental organizations have proliferated. Thanks to the downsizing of the Education Ministry after the school breakdowns of the early twenties, a new generation of Japanese children—more creative and individualistic than their elders—has shown itself capable of scientific and technological innovation on a broad scale.

A truly international mind-set has developed. Once hidebound Japanese universities have opened their gates to foreign students and professors. Thousands of Asian students now flock to Japan as much as the United States or Europe for higher education. Improvements in Japanese language instruction have made it possible for American and European students as well to take advantage of Japanese educational progress. Since the national universities were internationalized, an increasing minority of faculty members are non-Japanese. Japanese novelists are justly world famous. Even local poetry competitions have increased their popularity. Conversely, *manga* publishing has languished, ever since the scandal following the notorious TV *anime* series, *The Marquis de Sade in Love.*

The Japan of 2050 shows us one more very basic change from its late 20th-century past: a realization by Japan's people of the terrible harm done by the ancient Imperial Army in the slaughter of millions of their fellow Asians during the so-called China Incident, and the pillage and rapine of the Pacific War.

This shift was less the result of official government apologies—late and niggardly as they were— than the work of journalists and young writers in recollecting the shameful deeds of their elders.

One must add that the courageous visit of the young emperor to Nanjing, early in the century, to make his personal apology, did much to dissipate the widespread resentment of those deeds. It was, after all, the emperor's army that had done the damage. The resignation of the emperor from his official office later in the century, with his move to Kyoto to act as a kind of guardian of Japan's cultural heritage, formalized the end of his house as a national political symbol. . . .

A TALE OF TWO TRANSFORMATIONS

The reader will forgive me for the fanciful but I hope harmless picture that I have here painted. This depiction is not a serious piece of futurology. But it is much more than personal daydreaming. The situation here projected 50 years into the future includes many developments that, I think, should happen to Japan over a long period—indeed, that could. For in the 150-odd years now ending, Japan has undergone historical transformations so radical as to be unbelievable even to those who lived through them.

In 1860, with the assassination of Naosuke Ii, the superbureaucrat of the Tokugawa Shogunate, the long-isolated nation, already torn by fratricidal strife and threats of foreign invasion, seemed on the brink of anarchy. Who would have thought that the next 50 years would bring into being a tight new nation-state, an unprecedented cultural revolution, and a military power strong enough to defeat China and Russia in two wars? Such military strength was only part of the Meiji success story. Western commentators marveled at the success of Japan's political leaders in creating a new nation-state with laws, a constitution, and what by the 1920s looked like a successful Asian remake of the Western democratic process, including political parties, a popular press, and widening political suffrage.

Now fast forward to the 1960s. Transformation again—and even more amazing! Who could have imagined that a nation devastated in a total war, starved and bombed almost into extinction, an international pariah, its faiths and ethic hopelessly shattered, could in a few decades transform itself into a global economic superpower? With the same intensity that an earlier generation had devoted to military conquest, the managers, workers, and bureaucrats of the 1970s and 1980s pursued unprecedented economic expansion. Stunned competitors in the United States and Europe made a living legend of "Japan, Inc.," while a new generation of bureaucrats and business executives throughout East Asia scurried to learn from the "Japanese model."

How did the same people whose progress excited the world's envy for most of the past half century fall into a virtual "slough of despond"?

Innumerable commentators (myself included) have written books and papers about Japan's indisputable economic successes—an extraordinary combination of technological innovation, marketing skill, and production savvy. Not the least element in that success story was the guidance and help of a government that lavished the same single-minded concentration on peacetime economic advancement that political leaders normally reserve for war and survival. The world marketplace was helped and stimulated by Japan's success—despite the casualties of economic competitors outflanked and outsmarted in a classic demonstration of what economist Joseph Schumpeter called capitalism's inevitable "creative destruction."

THE PRICE OF DISLOCATION

All of which made the world wonder when the collapse of the "bubble economy" of the late eighties began almost a decade of national economic stagnation. Whatever happened? How did the same people whose progress excited the world's envy for most of the past half century fall into a virtual "slough of despond" in its final decade?

The paralysis that overtook Japan in this period was not simply economic. A whole society seemed frozen into a passivity that bordered on anomie. Stories proliferated about the venality and corruption of a bureaucracy once renowned for probity and efficiency. Party leaders famous for their political agility resorted to stopgap measures based on a combination of deficit spending and wishful thinking. Business executives watched sales sag and jobs go, but continued to express satisfaction with their lot in public opinion polls. Schools, once the pride of the country, reported increasing cases of bad behavior, as apathy gave way to student violence. Emergencies such as the Kobe earthquake in 1995 found government authorities weak and wanting. And the poison gas attack by the cult Aum Shinrikyo in Tokyo a few months later brought to light the simmering discontent of young, bright people with their society. Almost everywhere in the country one found laxity instead of industry, complaints instead of solidarity, in difference instead of enthusiasm. Japan's society seemed like a spool of thread, once too tightly wound, now spinning haphazardly off its center.

How could a people change so drastically in such a short time? Had it ever happened before? On reflection, I can recall a similarly sharp change earlier in the century—and a decline of far worse enormity. Rereading history books and the testimony of those who experienced the transition from the Taisho democracy of the 1920s to the militarism of the 1930s, I marvel at how quickly a society with give-and-take politics, a strong labor movement, active and contentious writers and intellectuals, and an urban populace with a jazzy pop culture could be transformed into a nation in armor.

I think the explanation for the apathy and anomie of the 1990s is much the same as that for the militarization of the 1930s: Both typified the reaction of a talented, disciplined, capable yet intensely emotional people to sudden, sweeping change. "Psychohistorical dislocation" is the term used by Robert Jay Lifton, a scholar and psychologist who has spent much time in Japan. In a recent book highlighting the Aum Shinrikyo phenomenon, Lifton argued that, thanks to the dramatic transformations of the Meiji era and the aftermath of World War II, Japanese society in the last 150 years has experienced more wrenching historical and psychological upheaval than any other:

> When exposed to such extremes of social, political, and cultural transformation and collapse within such a short period of time, people experience what I call psychohistorical dislocation, a breakdown of the social and institutional arrangements that ordinarily anchor human lives.
>
> What are impaired are the symbol systems having to do with family, religion, social and political authority, sexuality, birth and death, and the overall ordering of the life cycle. Symbols and rituals by no means disappear, but, because less effectively internalized, come to feel less natural and more coercive. People experience a profound gap between what they feel themselves to be and what a society or culture expects them to be.

Lifton's "dislocation" manifests itself in a variety of ways. It may seem superficially effective in situations such as the "catch-up" modernization of the Meiji Restoration or the drive behind the post-1945 "economic miracle," when the Japanese people exhibited an almost superhuman energy in pursuing very tangible national objectives. Both of these great national efforts, it should be noted, were provoked by foreign pressures.

The bad effects of dislocation are more obvious in the Japanese people's almost total acceptance of what amounted to emperor worship after the flawed constitutionalism of Meiji; the popular militarism of the 1930s, when—even granting the sinister coercions of the authorities—most political dissenters readily recanted and switched sides. Then there was the inability of Japan's elite to recognize the need for peace overtures during World War II, despite the widespread realization among decision makers after the 1944 defeats in the Marianas that the war was lost. Finally, we have the present era of virtual inaction of politicians and voters alike in the midst of economic disasters that would have provoked popular demonstrations and swift government changes in any other modern democracy.

Such behavior goes beyond mere conformity. Rather it betrays a national cult of mutual dependence, one first explained by the psychiatrist Takeo Doi in his memorable book Amae no Kozo (The Structure of Dependence), which struck me forcibly when I read and wrote about it almost 30 years ago. As Doi describes it, this craving for dependence comes from both top and bottom in Japanese society. The junior person in a relationship—be he or she a child, worker, or minor bureaucrat—has strong feelings of *amae,* a special word that expresses far more than the English translation of "dependence," toward someone of higher position. The higher-up, in return, feels a strong sense of obligation toward his juniors. Historically, this craving for dependence has rumbled like a huge geological fault beneath the admirably structured landscape of Japanese society. It is the dark side of such widely praised traits as the sturdy work ethic, the extraordinary capability of consensus, the Confucian harmony in neighborhoods and the workplace, even the storied traditions of *giri-ninjo* (duty and human feelings).

The *amae* syndrome has shown itself in many ways real or fancied—from the indulgence shown to badly behaved small children to the loyalties of striking trade unionists to their companies, to the popular trust in the wisdom and goodness of bureaucrats and the solicitude of permissive governments for voters' feelings. It is a mark of this collectivist society that most of its members expect to be taken care of in some way. This built-in feeling of dependency is not, however, based on the conscious notion of giving one's due in return. There is not much sense of the Western give-and-take in Japan. Instead, there is a vertical sense of hierarchical obligation, running equally strong whether up or down. The dependent individual is obliged to work hard and do his job, but he looks to the one above him for his stability in society. Conversely, it is the job of the person in authority over him to sense his needs and attend to them—almost irrespective of the services performed.

Over the century and a half of Japan's self-modernization, the dependency tradition has hindered Japanese society from realizing the democratic realities promised by politicians and social reformers. (In the process, it has also totally confused foreign observers who keep stubbornly—in the face of all evidence to the contrary—expecting the Japanese to behave like them.) The role of the individual citizen remains a largely foreign concept. The very words "independence," "individual," and "citizenship"—however ably translated—assume in Japan a meaning quite different from Western political originals. The contrast between traditional Japanese concepts and political forms borrowed from others is at the root of what psychiatrists such as Lifton term "psychohistorical dislocation."

This split social personality becomes most evident in unexpected emergencies. True, in earlier times of crisis Japan reacted well, often magnificently, to emergencies. At the time of Meiji—the storied "first opening" of the country—a galaxy of great people emerged to guide a veritable cultural revolution. And at least for a time they ignored old relationship networks

in so doing. Men such as Toshimichi Okubo, the architect of the modernized nation-state, the constitution maker Hirobumi Ito, and the brilliant educator Yukichi Fukuzawa were truly larger than life. The political, economic, and cultural figures of Japan's restoration days were comparable, in American terms, to the Founding Fathers, such as Thomas Jefferson, James Madison, Benjamin Franklin, Alexander Hamilton, John Adams, and George Washington.

Far more than their American forerunners, however, the Japanese "founders" had to bridge extraordinary cultural gaps to attain their mission. The businessmen and bureaucrats of the post-World War II period—widely heralded as Japan's "second opening"—responded with equally amazing alacrity to the challenges they faced. Former Prime Ministers Shigeru Yoshida and Hayato Ikeda and entrepreneurs such as Akio Morita and Shoichiro Honda picked up the pieces of their country and rebuilt it spectacularly, with a speed no one thought possible.

In both these situations, however, the Japanese were responding to a tangible stimulus, a provocation from outside the country—a threat, in other words. The emergencies they dealt with were clear and the need for "opening the country" obvious. Not so with the current depression, even though the threats of change that it embodies are just as serious as the transformations of past decades. Admittedly, the threat was at first not obvious. Yet the reaction of the Japanese people to the economic crisis of the 1990s was sluggish and complacent. However much foreign and Japanese journalists and economic experts warned the public about the crumbling of Japan's cozy postwar financial system, the dangers were by no means clear, for the government did its best to paper over the difficulties with vast amounts of deficit financing. However much some younger bureaucrats expressed their concern, their bosses in the ministries were confident that just a little more "business as usual" could fend off any crisis. The challenge posed to Japan's postwar economic system by new knowledge industries and very real globalization was hard for populace or politicians to grasp.

Despite repeated steroid injections of public-works spending that would make even a hardened Keynesian blush, Japan's economy has remained stagnant.

The current three-party coalition, made by party bosses without popular consent or understanding, is obviously a device to keep power among the partners first, with reform or restructuring priorities running a distinct second. One searches for adequate words to describe the chain of interlocking commitments and factional deal making. During the late 19th century, a similarly faction-ridden French government was contemptuously called *la republique des copains* ("the republic of buddies"). The term could fit Japan's government today. No more obvious indicator of politicians' nonchalance could be found than the selection of Keizo Obuchi, a veteran party hack, as prime minister. At a time when political leaders should have been rousing the nation to face new challenges, the Obuchi cabinet, like the parties it represented, tried to paper over every problem, rather than attempt to solve it.

Despite repeated steroid injections of public-works spending that would make even the most hardened Keynesian blush—the public debt is a record 130 percent of Japan's GDP—Japan's economy has remained stagnant. Unemployment hovers near a record 5 percent (double that if measured by U.S. standards) and no steps have been taken to cope with the problem of a rapidly graying society. The impending early summer election, made inevitable by Obuchi's illness and resignation, has only intensified the ruling coalition's efforts to halt desperately needed economic deregulation out of deference to long-favored voting blocs, such as farmers and inefficient local industry.

Why have Japanese voters not risen up and forced a new government into being and demanded real reforms? True, it is difficult to mobilize a wave of popular indignation in Japan that would seem normal in other democracies. The threshold of political indignation in Japan is unbelievably high. Still, Japan is a democracy, with freedoms and electoral rights guaranteed, at least on paper. Why do people not use these rights available to them? Where are the civic organizations demanding justice and equal rights? For the answer we must go back to the cult of dependence, the *amae* syndrome, which is strengthened by a complacency that is itself the product of Japan's long-continued economic successes and rising living standards.

AN END TO COMPLACENCY?

The Japanese generation that was born in the 1940s and 1950s—the counterpart to America's baby boomers—has been called the *dankai no sedai* (roughly translated as "the generation of clods from the housing developments") by Taichi Sakaiya, the incisive essayist and former Ministry of International Trade and Industry (MITI) official who is now in charge of the Economic Planning Agency. By this appellation, he means the generation of "salarymen" and their families who grew up in an age of affluence and simply took steady progress for granted—as something that was given to them. As such,

they lost whatever capacity they possessed to strike out and think independently.

The truth is that for two generations Japan has been living off the social and political, as well as economic, capital accumulated in past years. In a dynamic world situation, a leadership that shrinks from international diplomacy and does little to reform a badly worn education system and nothing to retrain its workers can no longer rely on the old cop-out: *seikei bunri* (the time-hallowed fictive principle that economy and politics can be separated into different water-tight compartments). Unfortunately, it is all of a piece. A society is more than an economy.

Yasuhiro Nakasone, one of the few Japanese politicians who qualifies as a world-class statesman, recently wrote in the magazine *Chuo Koron*: "Politics, economics, and society—we have three bubbles that burst on us. Yet blueprints for dealing with the 21st century—middle- and long-range both—are sadly lacking. The Meiji Restoration established our nationhood. Douglas MacArthur's postwar reforms gave us democracy. At the moment we need statesmanship in no way inferior to the Satsuma/Choshu reformers or MacArthur." It was Naohiro Amaya, another former MITI bureaucrat but also an independent economic thinker, who first developed the idea of a "third opening" as vital to Japan's advancement. Many others have echoed the phrase, Nakasone among them. All the experts agree on the need for a restructuring, both of the society and the economy, to compete in a global world. How to do this is the question.

Various foreign analysts keep warning us that Japan is "at a crossroads" and must soon make the most drastic changes. But for how many years have they been saying that? The redoubtable Peter Drucker, a scholar-journalist with a marvelously broad perspective of history, differs. Drucker was one of the first international authorities to identify, explain, and chronicle the rise of the economic miracle in Japan. Asked about the need for drastic restructuring in a third opening of Japan during a recent conversation, he expressed his wariness over taking extreme measures. According to Drucker, Japanese leaders have consistently displayed a talent for "muddling through" and finally letting problems solve themselves.

I myself do not think that the Japanese public, press, or politicians are equal to drawing up the sort of medium- and long-range blueprints that Nakasone envisioned—not to mention making the huge national effort necessary to turn a corner of some real or fancied crossroad. Nor will muddling through work anymore, as Obuchi's successor Yoshiro Mori is already discovering.

I firmly believe, however, that the third opening has already begun, that it will continue, and that it will ultimately transform Japanese society, as well as its economy and political leadership. The third opening is nothing more nor less than the changes now occurring through the "computerization" of the Japanese people. The spread of the Internet, the rise of e-mail and e-commerce, the pervasive use of networking—the impact of the Knowledge Era is at last sweeping Japan. It is impossible to prevent, given the globalization of modern industrialized society. Changes in Japanese business already reflect this. The inability of small companies to compete with big ones, for example, is being rectified. The knowledge industry goes beyond size. Individual talent and initiative are essential to success. A salaryman with a personal computer is a different person from the well-mannered cog in a production machine that he has been in the past. As the great Russian Jewish author Isaak Babel once wrote during Russia's revolutionary wars, in quite another context: "A Jew mounted on a horse is no longer just a Jew."

Why, then, are these gathering changes apparently so hard to face, despite the abundance of warnings from press and punditry? Or has it become a negative attribute of Japan's nascent "knowledge society" to understand a problem without doing anything about it? Part of the difficulty lies in accepting changes that often represent a denial of what the Japanese have achieved and how they have achieved it. To cite a few examples: The very idea of mass production that powered Japan's "economic miracle" is now being drastically modified, if not in some cases reversed. Capital for ventures is becoming available as the dominion of the old conservative banking system shatters in a global economy ruled by the price of stock shares, not the extent of bank loans. The entrepreneur, always present in Japanese business, is now seeing his role vastly magnified. But the docile and loyal "company man" is disappearing in an era of corporate downsizing and waning security guarantees. So is the bureaucrat of the old protectionist Japan.

These changes will soon engulf the entire society, with no possibility of a return to old standards. Already the expanding Internet is giving Japanese patients the ability to question doctors (an act once regarded as a form of secular sacrilege), look for the lowest prices when shopping, and form a new kind of grass-roots political and social groups. An alert government, seeing the handwriting of globalization writ large on the wall and trumpeted over the Internet, would bend every effort to smooth the path of change by restructuring the economy in drastic ways, as well as by moving the society to meet the approaching changes in education, child rearing, and cultural attitudes. This shift is taking place in Korea, which has improved its economy greatly while making way for social and political change. But in Japan the opposite has happened. A government ruled by old-fashioned cronyism has done its best, Micawber-like, to assure the people that things are getting better, thanks to its tepid and incremental pa sses at "reform." Complacent to begin with, Japan's people have been lulled into a false sense of security by their government's optimistic statements.

Modern Japan through the first two openings has continued to be a "vertical" society. The writ of Confucian governance has run strong. The national cult of dependency ran upwards and downwards in differently marked out areas, but lateral instant communication and exchange are replacing the ordered guidelines of the past. In this new "lateral society" of the Internet world, Japan's younger generation must become individually assertive in a way that was never expected of its elders.

Will the new generation measure up? Will the current generation prove flexible enough to engineer a new tradition? There are glimmers of hope. In January, the voters of Yoshinogawa, in Fukushima Prefecture, rose up to denounce a new dam proposed by the Construction Ministry—by a 10-to-1 majority. Ministry officials at first announced that they would have the dam built anyway—part of the Liberal Democratic coalition's stubborn plan to raise Japan's economy by continual injections of spending by the party's construction business cronies. Merely a local incident, one may say. Yet the opposition of the Yoshinogawa people may hold a great deal of meaning for the nation's future.

If more voters take advantage of their own democracy, the Japan of 2050 may indeed be a revivified country, fit to play the part in a globalized world that its past hard work deserves. Who knows? Some portions of my rather sweeping vision of the future may even turn out to be true.

WANT TO KNOW MORE?

For good current appraisals of Japan's economic troubles, see Gavan McCormack's *The Emptiness of Japanese Affluence* (Armonk: M.E. Sharpe, 1996) and Richard Katz's *Japan, The System That Soured* (Armonk: M.E. Sharpe, 1998). *The Age of Hirohito: In Search of Modern Japan* (New York: Free Press, 1995) by Daikichi Irokawa and W.G. Beasley's *The Rise of Modern Japan* (New York: St. Martin's Press, 1990) provide useful general background. Frank Gibney's *Japan, the Fragile Super-Power,* 3rd rev. ed. (Rutland: Tuttle, 1996) surveys the history of post—World War II Japan, with special attention given to the economy and cultural phenomena such as amae. For a detailed exposition of *amae*, consult Takeo Doi's *The Anatomy of Dependence* (New York: Kodansha International, 1973), a translation of the original amae *no Kozo*. Naohiro Amaya discussed his argument for a "third opening" of Japan in *Eichi Kokkaron* [An Argument for a Smart Country] (Tokyo: PHP, 1994). Robert Jay Lifton's Destroying the World to Save It (New York: Henry Holt, 1999), quoted in the article, examines the Aum Shinrikyo terrorist cult and its background in Japanese society. Michio Morishima's *Naze Nihon Wa Botsuraku Suru Ka* [Why Will Japan Collapse?] (Tokyo: Iwanami Shoten, 1999) provides an up-to-the-minute rundown of the country's problems, along with a thoroughly pessimistic view of Japan in 2050. Readers can find a review of the book in the Spring 2000 issue of FOREIGN POLICY.

For links to relevant Web sites, as well as a comprehensive index of related FOREIGN POLICY articles, access **www.foreign-policy.com**.

Article 8

Foreign Affairs, July/August 2000

Japan: A Rising Sun?

M. Diana Helweg

M. Diana Helweg is Project Director of the Council on Foreign Relations' Task Force on the Japanese Economy and Research Fellow at the John Tower Center for Political Studies, Southern Methodist University. The opinions expressed herein are her personal views.

THE THIRD WAY AND PROSPERITY

AS JAPAN enters the twenty-first century, it sits on the brink of the biggest transformation in its history. Some observers want to write off Japan as stuck in a cycle of debt and deflation, but today's structural reforms in the Japanese financial system are quietly setting the stage for an economic revolution. Although rebuilding the Japanese economy will be no easier now than it was during the Meiji Restoration or after World War II, the present changes are more fundamental than anything the country has ever seen. Japan's revolutionary path will utterly transform it from the state-run industrial powerhouse of the twentieth century toward an innovation-driven, globalized economy of the twenty-first.

This transformation, however, is not necessarily synonymous with economic recovery. In fact, an overhaul of this magnitude is likely to shrink the economy as Japan initially encounters higher unemployment, lower capital investment, and other deflationary problems. But these downturns are temporary. Japan's financial sector is being reformed—forcibly changing the way companies do business. As it develops, this new financial system is sowing the seeds for an entirely new Japanese economy—one driven by innovation and competition among small and medium-sized enterprises and high-tech companies. The old economic guard of Japan still receives most of the world's attention. But the real action is taking place in the smaller, newer, more creative enterprises that will bring twenty-first-century prosperity to Japan.

THE EMPEROR'S LAST STAND

THE SPARK that ignited Japan's economic revolution was cast by the collapse of the nation's financial system in the late 1990s. Until that collapse, Japan preferred that the government have nearly total control of the economy. While other modern democracies relied on free-market competition to distribute capital, Japan spent the last century perfecting a financial system that allocated capital according to government criteria. In fact, the recent affinity for state-directed economic policies began much earlier. When Japan isolated itself from the rest of the world during the Tokugawa era (1603–1867), the shogunate tightly controlled all domestic activity, from transportation to commerce.

Unsurprisingly, when a few outlying samurai families decided to replace the Tokugawa shogunate with Emperor Meiji and begin crafting Japan's first modern economic system in 1868, they continued to emphasize the state's control of capital. The self-appointed reformers who shaped the Meiji government wanted a modern economy that could support industrial growth at a pace necessary to catch up with Western countries. To achieve such rapid-fire industrialization, the Japanese government—rather than individuals—took the risks and footed the bills. Some of the wealthy merchants who had made their fortunes during the Tokugawa period led and partially funded the new enterprises, but they worked in tandem with the state, not as independent entrepreneurs. Most merchants became political actors collaborating with the central government, and many started the government-supported, family-run companies and banking houses (*zaibatsu*) that fueled much of Japan's growth in the first half of the twentieth century.

Japan steadfastly resisted a short-lived effort by the postwar American occupation forces to introduce competition-based capital markets in the late 1940s. As a result, its economy in the 1950s looked very much like it had in the 1930s, with the state, appointed banks, and large companies working together. The government soon established several financial institutions to apportion capital to companies and sectors it deemed important for economic recovery. The government-run banks included a long-term credit bank, an export-import bank, a development bank, a bank for small-business financing, a foreign-exchange bank, and even a bank for long-term lending to agriculture, forestry, and fishery ventures. The government used this network of banks to make Japan globally competitive by promoting specific industries such as steel, auto manufacturing, and consumer electronics for export. It also used the banks to subsidize small rice farmers and domestic food processors, protecting them from international competition. Eventually, economic growth became managed almost entirely by bureaucrats.

As a result, private funding sources did not develop, in turn making it nearly impossible for new companies (whether foreign or domestic) to enter the marketplace. Being listed on the Tokyo Stock Exchange (TSE) was practically impossible for any company that did not have 25 to 35 years of proven profits. And initial public offerings rarely occurred. Individuals who took the risk of raising capital but failed never got a second chance.

State-controlled competition was further complemented by another postwar development in the Japanese industrial structure. The *keiretsu* or "enterprise group" system was a new concept that resembled the family-run *zaibatsu* of prewar days. The *keiretsu* linked several companies from each major industrial sector, all funded by a main bank. The relationships between these companies were cemented through a shareholding system that allowed *keiretsu* members, primarily main banks, to hold up to 60 percent of each other's shares to prevent outside takeovers. The *keiretsu* groups often locked suppliers and distributors into exclusive relationships to ensure a smooth and dependable distribution chain. Over time, these relationships overshadowed other factors such as efficiency, price, and product. Thanks to the *keiretsu* structure, large, family-owned companies like Mitsubishi, Sumitomo, and Mitsui became as prominent in the postwar economy as they had been before.

The key to the *keiretsu* structure was the main bank. A main bank allocated capital among its *keiretsu*'s companies based on both government industrial policy and the priorities of the group. The main banks had the duty of distributing cash, policing investments, and overseeing the industrial development of their groups by deciding who would get capital and when. The lack of true equity and capital markets only enhanced the status of the main banks and further entwined Japan's industrial and financial sectors.

The system of market manipulation—rather than market competi-

CORBIS-BETTMANN

On the road to reform? Business students march, Tokyo, December 1998

tion—worked after the war just as it had in the Meiji (1868–1912) and Taisho (1912–26) eras to propel Japan into prosperity and global standing. Once Japan achieved its postwar goals, however, the *keiretsu* method of indirect financing began to unravel. Banks and businesses committed themselves to unnecessary expansion and unsustainable projects. Companies began to invest in ventures outside their core business areas, and their main banks granted unwise loans based on relationships rather than economic viability or detailed business plans. Within a decade, excess capital became excess capacity.

In the late 1980s, Japan's bubble economy, inflated by exorbitant real estate values, collapsed. Soon after, companies found they could not repay their loans due to failed investments and a deteriorated economy. Nor could banks demand that their loans—many of which had been secured with insufficient collateral, primarily overvalued real estate—be repaid. At first, many banks attempted to hide these bad loans. But after a few years, some of Japan's biggest financial institutions found them-

selves on the verge of collapse. In November 1997, Sanyo Securities, Hokkaido Takushoku Bank, and Yamaichi Securities all closed their doors, sending the country into shock.

THE BIG BANG

THE GOVERNMENT'S inability to rescue Sanyo, Hokkaido, and Yamaichi brought Japan to the stark realization that its financial system had failed. Fortunately, the failure came right after Prime Minister Ryutaro Hashimoto and his cabinet announced the 1996 "Big Bang" initiative to liberalize the Japanese financial system. The coincidence breathed new life into what would otherwise have been token financial-sector reform. Somewhat inadvertently, and certainly contrary to the expectations of the government officials who crafted the Big Bang, these reforms provided the makings of a revolution.

Hashimoto's initiative got its name from the Big Bang launched under Prime Minister Margaret Thatcher in the 1980s to reform the British economy. Implemented in 1998, Japan's Big Bang was intended to make the financial system more transparent and accessible by loosening the insurance and securities sectors, providing tax cuts for corporations, and restricting the regulatory power of the Ministry of Finance by establishing the Financial Reconstruction Commission, the Financial Supervisory Agency, and an independent Bank of Japan. Despite some backsliding, the improved market access and reconfigured financial regulatory bodies created by the Big Bang have allowed new entrants into the marketplace to challenge the former giants.

Ironically, the Big Bang also gained strength from the economy's refusal to recover. As the Japanese economy steadily tightened, it created additional pressure on the banking sector to open up to competition and consolidation. For example, the bad loans on the books of many banking and securities institutions resulted in a profound scarcity of capital. In response, the Japanese government stepped in to prop up a few banks with large infusions of cash and began letting banks write off bad loans in the spring of 1998. Other financial institutions, however, such as the Long-Term Credit Bank of Japan and Yamaichi Securities, were less lucky and had no choice but to seek foreign capital to survive. And with main banks reluctant to lend to their related companies within the *keiretsu* system for fear that the loans would not be repaid, some Japanese companies, too, had to ask foreign investors to come to their rescue. The resulting flow of foreign direct investment (FDI) has yet to slow.

Today's climate for doing business in Japan is the best ever.

Although still low in comparison with FDI in other countries, the amount of foreign investment in Japan has increased more than 100 percent since 1997. In 1998, Japan had 1,542 cases of

FDI, with a total value of $10.5 billion. The 727 foreign investments made in the first half of 1999 alone—totaling $11.3 billion—exceeded the entire total for 1998. American companies accounted for 60 percent of Japan's FDI in the 1998 fiscal year; European companies accounted for most of the FDI into Japan in the first half of the following fiscal year. Foreign investment shows no signs of abating, especially since many in the foreign community believe the current climate for doing business in Japan is the best it has ever been.

Accompanying the new influx of FDI are other opportunities for creative new Japanese companies to access capital. To support small and medium-sized non-*keiretsu* enterprises that had always faced limited access to capital, Japan created a new market last year. It plans to inaugurate another one this summer, this time springing from the Japanese entrepreneur Masayoshi Son's plan to create a NASDAQ Japan for relatively young, unproven companies to raise capital through public offerings—threatening the dominance of the TSE. In response, the TSE rushed to preempt Son's efforts by opening MOTHERS—Market of the High-Growth and Emerging Stocks—in December 1999. In less than a year, the time it took a company to "go public" in Japan dropped from 30 years to a month. Boasting 7 listings, with 20 more in the wings, the new mothers emphasizes transparency over past performance and eases the demanding criteria that the TSE uses for listing stocks on its main board. NASDAQ Japan, meanwhile, has three times as many applicants as it expects to list once it opens. Although the price of each of the newly listed stocks has declined since the recent technology stock crash, one by as much as 90 percent, initial public offerings are still flooding the marketplace. And thanks to Japan's new financial liberalizations, investment banks from abroad can underwrite many of these projects.

Under the Big Bang reforms, foreign investment banks can also participate in the mutual fund and retail banking sectors by taking advantage of new possibilities in personal savings. Last year, the Japanese government passed a law establishing U.S.-style 401(k) retirement

plans for employees. This defined-contribution plan differs from the defined-benefit plan that Japanese employers have always provided. For the first time employees will help invest their own retirement funds. Moreover, these changes come at the same time that many individuals' ten-year deposits in the Postal Savings System—a quasi bank run by the government—are maturing. The Postal Savings System has begun paying out some of its approximately $980 billion in matured deposits to individual investors. Although initial statistics show that as much as 96 percent of the payouts has been reinvested in the Postal Savings System, almost half of the money has been put into short-term accounts until better long-term investments are found. Even if most of the payouts remain in the system, investment analysts expect that Japan's low interest rates, along with the government's eventual reduction of insurance limits on bank deposits, will still encourage some to put their savings in equities.

Many large U.S. investment firms, such as Goldman Sachs, Merrill Lynch, and Fidelity, are hoping to take advantage of these new opportunities by persuading the average Japanese family to deposit its bundle of savings with their money managers. In addition to using advertising and free training seminars to educate investors, most foreign firms have teamed up with Japanese partners to help make investors more comfortable doing business. For example, Merrill Lynch hopes to make the troubled retail facilities it gained in buying part of the failed Yamaichi Securities profitable by offering American-style investment services to Japanese consumers. Putnam has formed alliances with Japanese entities, and Salomon Smith Barney has created a wholesale banking venture with Nikko Securities. Fidelity had initially planned to sell its products alone, but it recently partnered with the Bank of Tokyo-Mitsubishi and the Nomura-Sumitomo brokerage consortia to market its funds to Japanese consumers. If these consumers achieve greater returns on their retirement investments, the increased dividends will relieve the pressure on Japanese families to save so much. Consumer spending could go up, spurring a demand-led growth that would help pull Japan out of the economic doldrums.

EYES ON THE BOTTOM LINE

THE CHANGES in Japan's financial sector are all the more revolutionary because they have begun to spill over into the industrial structure of the economy. The influx of FDI has done more than just give capital to needy Japanese banks and companies. Foreign investors are bringing technical expertise and know-how to current restructuring efforts all over Japan. Japanese companies have begun seeking out their own consolidations and mergers in order to receive the same opportunities that foreign investors have seized. They are voluntarily restructuring themselves and sloughing off extraneous investments, focusing instead on the most profitable areas of their business. Motivated by the rising cost of capital, universally low profits, growing global competition, and the information revolution, many Japanese companies have now set their eyes on the bottom line.

By putting profits above personal and business loyalties, companies are even crossing *keiretsu* lines. The most publicized and prominent such deals are the Sumitomo-Sakura bank merger and the newly christened Mizuho Bank, made up of Dai-ichi Kangyo Bank, Fuji Bank, and the Industrial Bank of Japan. The Asahi-Tokai-Sanwa bank merger has also received a lot of publicity for consolidating three major banks—even though each maintains weak *keiretsu* connections. But these merged entities cannot boast about their efficiency or streamlined integration. Mizuho, for instance, expects to take three years to complete the consolidation it began in 1999. (Similar mergers in Europe and the United States usually take between three and six months.) Nonetheless, these consolidations are the first steps in the right direction—particularly since they would have been unthinkable five years ago.

Japanese companies have set their eyes on the bottom line.

As companies interact across traditional *keiretsu* lines, they are also unwinding their linked shareholding practices. Japanese companies trying to make themselves attractive for consolidation no longer want to have 40–60 percent of their shares held by related companies. Indeed, soon after the alliance between Dai-ichi Kangyo Bank, Fuji Bank, and the Industrial Bank of Japan, the latter two announced that they would sell the shares they hold in each other. Nissan—which belongs to the same *keiretsu* as Fuji Bank—announced the sale of its shares as part of a major restructuring plan launched soon after it merged with the French auto company Renault. As these and other shares free up, public companies can sell shares on the TSE. Any company, public or private, can use stocks to purchase assets under the newly legalized stock-swap transaction mechanism, which allows companies to purchase other companies using shares instead of cash. Nippon Paper Industries is doing just that with Daishowa Paper. Or shares can be used to compensate or reward employees, as Nissan-Renault began doing this year.

With company shares becoming valued commodities, they have taken on new importance in the eyes of shareholders, brokers, and management. The newly diversified pool of shareholders—including employees, unrelated companies, and foreign purchasers—can now pressure a company's management to obey international standards for corporate governance. Although few shareholders exercise these powers, over the past couple of years business terms like "corporate governance," "shareholder value," "return on assets," and "return on equity" have become commonplace in business conversations in Tokyo. The practice of such concepts has admittedly not been as pervasive as the discussion

of them. Still, Sony and Toyota have reduced the size of their corporate boards, and Sony and Fuji Xerox have brought in outside directors. Gradually, companies are shifting their priorities from growth and long-term investment to more immediate concerns like restructuring and profitability.

As part of this transformation, the management-employee relationship is evolving as well. Employment guarantees by big companies and the tradition of seniority-based pay have begun to deteriorate. Companies on the verge of bankruptcy can no longer guarantee that they will keep their employees for life. So most firms pay employees to retire early and have restructuring plans that rely on attrition and hiring freezes, rather than on U.S.-style mass firings. In addition, companies have begun to switch to merit-based pay systems that include stock options and other incentives. This is especially true in the financial, high-tech, and other young growth sectors, but even old-guard Japanese companies are following suit.

Companies trying to enhance their profitability are finally being forced to relinquish their traditional preference for relationship-based business practices—effectively weakening the *keiretsu* system. Although the interconnected relationships run deep in Japanese society, large Japanese companies are beginning to examine business deals on the basis of profit, not corporate connections. These considerations have trickled down to distribution channels as well. Supply chains for distribution in Japan have begun to respond more to pricing and competitive delivery than to company obligations. Nissan-Renault, for instance, caused quite a stir when it announced its plan to trim its supplier relationships by 50 percent over the next three years. But this move is a sign of the times: if Nissan-Renault's plans succeed in saving Nissan from collapse, other traditional Japanese companies will follow suit.

REFORM.COM

ALTHOUGH CHANGE is spreading throughout Japan, revolutionary reforms are most prominent in two new sectors of the economy: communications and technology, the same two sectors driving growth in the United States. In the final quarter of 1999, Japanese telecommunications and personal computer companies saw triple-digit increases in their gross profits. In the information-technology sector, productivity has risen 7 percent and output is growing at an annual rate of 12 percent. These changes herald a transformation of Japan's entire economic structure.

Why are these sectors more instrumental to Japan's economic revolution? First, unlike old economic sectors such as steel and agriculture, the government does not overregulate the high-tech sectors. Second, individuals in communications and technology are intoxicated with the prospect of catching up to Western technology and profits. And third, those not in the field feel the pressure to develop new models so that they, too, can take advantage of the profits overwhelming Internet ventures everywhere. The ideas and successes of many Japanese start-ups and technology ventures have attracted unprecedented levels of FDI.

As in the United States, this new economic growth has spawned an entire culture. Japan has its own version of Silicon Valley, named Bit Valley, in Tokyo's trendy Shibuya district. The name plays on the word *shibuya,* which means "bitter valley." In just a few years, a lonely high-tech field populated by a few Japanese entrepreneurs has blossomed into a full-fledged industrial sector. The risk-averse label often applied by outsiders to Japanese business no longer fits the multitudes of Japanese students and others who gladly pass up the safety of long careers with traditional corporations to make it big on the Internet.

The telecommunications industry, too, is undergoing revolutionary changes. Domestic and foreign phone companies are threatening the Nippon Telegraph and Telephone Corporation's monopoly by laying their own cables, thereby providing services without paying NTT connection charges. Foreign companies are breaking into the Japanese telecommunications sector in other ways as well. Last sum-

mer, Cable and Wireless, a British company, outbid NTT to obtain a controlling interest in the Japanese long-distance carrier IDC. Despite NTT's supposed domestic advantage, Cable and Wireless won approval for its bid from Japanese shareholders.

NTT's exorbitant rates have kept Japan's household use of the Internet at less than half the U.S. rate—a sticking point in U.S.-Japanese deregulation talks. But NTT is now facing competition from its own cellular subsidiary, Do-CoMo. Because of NTT's monopoly, cellular phones are less expensive to use than regular phones. As a result, practically all Japanese adults and most teenagers own mobile phones, and DoCoMo has the mobile market cornered. Unlike its parent company, DoCoMo is competitive in pricing, service, and innovation. Recently, it began offering a service that links telephones to the Internet with fun graphics and easy-to-use technology. This service, currently enjoyed by one out of twenty Japanese, lets the personal Internet user escape high connection costs and shows how quickly Japanese technology is moving in leading sectors.

THE CHRYSANTHEMUM AND THE SWORD

OPTIMISM, yes. But Japan's economic revolution will not succeed without a fight. A great deal of resistance remains in some Japanese circles—justifiably so. The revolutionary changes in the Japanese economy will not bring growth right away. Nor will they end the economic ups and downs any time soon. But eventually, all these fears will dissipate because the economic revolution will bring jobs. It will undoubtedly take a few years for the reforms to become entrenched and for new businesses based on a competitive model to establish themselves. But once they do, jobs will increase and consumer demand will grow.

The speed and efficacy of these reforms will largely depend on how companies and consumers implement them. Unfortunately, the current reforms are

neither comprehensive nor universal, and many of the changes have been implemented piecemeal—offering too little, too late. The transition to consolidated tax standards, which would encourage corporations to purchase and revitalize companies with large liabilities, has been delayed for another year, reducing investors' confidence in businesses' commitment to reform. The restructuring plans announced by big companies too often rely on attrition and no new hiring, rather than on layoffs, to eliminate redundancies. And although new financial agencies like the Financial Supervisory Agency and the Financial Reconstruction Commission have engaged in significant reforms, they are still limited because of lingering ties to the conservative Ministry of Finance, despite nominal efforts to establish their independence.

Many Japanese companies have backed themselves so far into the corner in terms of overcapacity, unpaid loans, and redundant work forces that they have virtually no good choices left. Without restructuring, they will collapse. But waiting until the last moment to change has only narrowed their avenues of escape. The understandable fears caused by this predicament have translated into high personal savings coupled with low consumer spending, leaving Japan in an interminable deflationary spiral. The enormous fiscal stimulus packages of the 1990s that the government implemented to break this deflation produced nightmarish budget deficits. The combination of all these factors causes some observers to throw up their hands in frustration.

But there is still cause for hope. The revolution underway in Japan will recreate the nation from the inside out. Slowly but surely, Japan is shifting from state direction to a free market. Indeed, Japan's social and political structures already reflect some of these changes. For example, the bureaucracy in various ministries now holds less control over the prime minister's office and the parliament than ever before. Newly elected Diet members are drafting reform legislation with the input of foreign businesses and government officials rather than of bureaucrats.

True, political reforms lag behind economic changes. During the push for reform in 1994, the public showed new interest in eliminating closed-door politics and one-party dominance in favor of greater accountability. But complacency has set in as the ensuing reforms proved superficial. As shown by the lack of public objection to the Liberal Democratic Party's backroom power shift from the dying Keizo Obuchi to the new prime minister, Yoshiro Mori, the public does not expect much disclosure or transparency from its politicians. Fortunately, today's economic reform does not hinge on political reform. The government is not driving the current revolution; business is.

HERE COMES THE SUN

AS JAPAN comes into its own, it could challenge the United States on several fronts. Economically, if Japanese high-tech companies ride the next wave of technical developments—as they are poised to do—they could cut the value of U.S. high-tech stocks and cause the American stock market to drop. If the U.S. stock market falls prey to a high-

tech correction, Japanese technology companies could move into the lead.

With regard to national security, a strong Japan that has redefined itself may feel less inclined to follow U.S. policies in the Asia-Pacific region or around the world. If Japan adopts new alliances or policies that harm U.S. interests or undermine U.S. troops in Japan or South Korea, the United States will have to rethink its military strategy for the region—especially if Japan amends its constitution to permit a standing military force. An economically strengthened Japan might also create power and prestige issues for the United States if other Asian countries look to Japan, not the United States. Despite its recent economic weakness, Japan secured its position as a leader in the region by consistently giving money to the other Asian nations throughout the financial crisis of 1997–98 while the United States held back.

Despite these risks, a vibrant, prosperous Japan is in America's best interests. Therefore the United States should commit to three courses of action. First, American businesses should not underestimate Japanese businesses' potential. As they continue to wrestle with formidable competitors in the global economy, U.S. firms should plan on a reinvigorated Japan in five years or so.

Second, the U.S. government should continue to work with and support Japan as it struggles to resuscitate its economy. Japan needs to know that America is standing by, and the United States needs to provide quiet and steady support that will not humiliate Japan. Third, the U.S. government and the public need to recognize that America has little leverage over Japan's economic or strategic policies. So it must revise its expectations of the U.S.-Japan bilateral relationship. America's best leverage in this era of globalization is its economic strength. Doors around the world open to U.S. corporations because of the promise of prosperity. The door to doing business in Japan is now open wider than ever. The next U.S. administration should take advantage of this access to support domestic Japanese forces already at work for change. These domestic forces are the engines of the Japanese economic revolution and the key to its inevitable recovery.

The Japanese moved further and faster than anyone ever expected during both the Meiji Restoration and the recovery after World War II. There is every reason to expect that Japan will do so again—winning the battle between status quo and reform. But more fundamentally than even the Meiji and postwar periods, this revolution will rebuild Japan's economic foundation from the ground up.

Article 9

The Asian Wall Street Journal, May 15–21, 2000

Keeping Investors Coming

The Makeup of Foreign Direct Investment in Asia Changes Dramatically

By JAMES T. AREDDY

Dow Jones Newswires

Asia started the 1990s facing a gold rush of foreign investors and ended the decade hosting a fire sale for them. Now, if Asia doesn't pursue more privatization and liberalization, the level of foreign direct investment just might come tumbling down.

FDI in Asia took off in the 1990s and stayed high. That heft and consistency despite the region's financial crisis gave FDI an image as globalization's crown jewel. For the most part, it doesn't flee at the first sign of trouble, always looking for a better return, says Dennis Tachiki, an FDI analyst at Fujitsu Research Institute in Tokyo.

The desire to keep investors in place—and keep new ones coming—proved a strong motivation for Asia to remain plugged into the world economy in the past few years. "FDI is the one flow that did not leave," Mr. Tachiki says. Recall how Malaysia carefully crafted its 1998 currency controls so as not to hurt foreign manufacturers and other direct investors.

'Greenfield' Projects

During the crisis, caution about building new "greenfield" projects was offset by fresh opportunities to buy existing businesses. Some companies responded to lower Asian currencies and deflating prices by expanding their production lines and increasing output.

Yet, while FDI proved less fleeting than other kinds of investment, some signs suggest that positive flows aren't a given.

Anglo-Dutch consumer-products giant **Unilever** NV "put a foot on the brakes" and shelved Asian expansion plans when markets plunged in 1997, according to its regional president, Andre Van Heemstra. With economies growing again, Unilever says it is merely "resuming" those plans, not rushing to play catch up. Importantly, expansion will also involve better integration of its existing Asian facilities, Mr. Van Heemstra says.

And the FDI competition from other regions is increasing. "Even prior to the crisis, investors were looking at other regions," according to Richard Newfarmer, an FDI specialist at the World Bank in Washington.

Challenge From Latin America

Annual FDI in East Asia's developing countries jumped by a factor of eight between 1980 and 1990, by another five times by the end of 1996, but by less than 3% since then, according to World Bank numbers. It has remained near historic highs, but last year slipped for the first time in more than a decade, by 4.1%. Meanwhile, in each of the last three years, the sale of state-owned companies in Latin America allowed that region to challenge Asia in FDI for the first time in decades.

Mergers and acquisitions were an important part of Asian FDI growth in recent years, and because the crisis had roots in the financial economy. M&A activity has been concentrated in banks and securities firms. These tend to sell for the most eye-catching prices, but they are fundamentally distress-related transactions and don't tell much about prospects for future FDI flows. "I would expect this to be a short-lived phenomenon," says Mr. Newfarmer at the World Bank.

More important, few economists predict the kind of stunning consumption growth that would fuel another round of heavy investment in manufacturing plants. "The problem is there is too much capacity," says Greg Fager, Asia director at the Institute for International Finance, a Washington-based bank-advisory group.

Consider China. It had no crisis like South Korea and Southeast Asia, but saw FDI slide 31% from 1997 to 1998 and probably more last year. In that market, both banks and companies are skittish about debt-financed expansion, meaning that they will set a higher threshold for taking on risk.

Backdrop of Political Risk

A new backdrop is political risk—and not only in Indonesia. That can dampen investor sentiment faster than anything else. "Almost every investor sees risks galore," says Christine Wallich a senior director at the Asian Development Bank in Manila.

The relatively constant levels of FDI in Asia have masked a dramatic change in its composition. Through the 1980s and 1990s, multinational corporations staked claims to cheap Asian labor with huge new factories that cranked out goods for export, in hopes of eventually tapping the growing local markets. "Those were the frontier days," says Mr. Fager.

Now, government policy adjustments have supplanted heady economic growth and local demographics as the FDI driver in Asia. Some rule changes are real, such as South Korea's suddenly opened regime, and others are anticipated, such as China's impending entry to the World Trade Organization.

Profit Is King

"Before the crisis, everyone just thought of growth, growth, growth. The speed was sometimes more important than cost," says Victor Lo, chairman and chief executive officer of Hong Kong electrical-components maker Gold Peak Industries (Holdings) Ltd. The new king is profitability, and "it has made us a lot more cautious and conservative," he adds.

Overall, the size of individual deals is also shrinking. Analysts figure that money is flowing into technology startups and service businesses, such as distribution networks and supermarkets, which often don't require the capital outlay of factories and often don't employ as many people.

Tellingly, the value of planned FDI in South Korea jumped an impressive 35% to $2.7 billion in the first quarter from the year before. But individual project numbers tripled from the year before, to 926 from 308.

Smaller-sized deals are also the strategy of investors form Japan and elsewhere in Asia, but not the Western multinationals.

The deluge of M&A is another key difference in the kind of FDI Asia is getting today. Up from practically nothing, fully half of the transactions being done in South Korea these days involve the purchase of an existing operation or the buyout of a partner, according to Paik Chang-gohn, chief of the Korea Investment Service Center in Seoul. "Even hostile M&A is allowed," he says.

Bright Side for Foreign Companies

A bright side for multinationals has been the chance to buy out debt-strapped local partners and gain control of ventures for the first time. No more "one bed, two dreams," according to Unilever's Mr. Van Heemstra.

Despite the sudden plateauing of FDI levels in the late 1990s, Mr. Newfarmer says a number of "pull factors" should ultimately keep investment in Asia high during the coming years.

The crisis didn't erase the factors that drew foreign companies to Asia in the first place, including high savings rates, large populations, growing disposable income and young demographics. "A number of fundamentals still stand upright," says Mr. Van Heemstra.

China's entry to the WTO should draw investment to that country and is likely to fuel policy liberalization elsewhere in Asia, raising the region's overall attractiveness, analysts say.

Where analysts see the most potential growth for FDI is in Asia's services industries, which foreign companies continue to find difficult to enter in many countries. Compared with allowing foreign investment in for-export manufacturing and other big job creators, tapping this newer investment vein will carry the political consequences of directly challenging the often-lumbering domestic players.

"The head and the heart of the governments may be in different places," says the ADB's Ms. Wallich.

Privatization will be another important way for governments to draw more FDI. Government-led bank bailouts in several countries have put bigger slices of the economies into public hands, and tight budgets should prompt them to make huge-sized asset sales in the next few years.

The unloading of government-owned companies in Brazil and Argentina in recent years goes a long way toward explaining why South America drew $33 billion more FDI in the last three years than the developing countries of East Asia.

Article 10
World Policy Journal, Fall 1998

The Privilege of Choosing
The Fallout from Japan's Economic Crisis

Masaru Tamamoto

Masaru Tamamoto is a senior fellow at the World Policy Institute and visiting professor in the Faculty of Law at Ritsumeikan University, Kyoto. He divides his time between Japan and the United States.

Not long ago, during Japan's bubble years of 1985–90, when, on paper, Tokyo's real estate was worth more than that of the entire United States, there emanated from Japan brave talk of a Japanese way of capitalism superior to America's. Now there is silence, the onslaught of global capitalism having exposed Japanese inefficiencies. Japan's prolonged economic downturn, which began in 1990, marks the end of an era. This is more than a cyclical downturn. The very structure of the Japanese economy, which made for decades of spectacular performance, now inhibits future prosperity. Pundits in Japan sheepishly acknowledge that what was once touted as a unique form of capitalism turns out to have been simply an overregulated version of a single system. The difference between Japan and the United States, it seems, has been a matter of social choice, not a contest of competing capitalisms.

For over seven years, the Japanese government has been trying, without success, to reflate the sagging economy. What accounts for this political paralysis? Beneath the seeming indecision surrounding the formulation of a viable economic policy is a society remaking itself. Deregulation is the obvious path to economic revitalization. But revitalization cannot be accomplished by simply tinkering in the economic sphere. The whole of hitherto heavily regulated Japanese life needs to be liberated.

The protected life accorded the Japanese by myriad regulations and a sense of community as the Japanese know it is the price to be paid for economic revitalization. The earlier deregulations of the British and American economies—heralded as Thatcherism and Reaganomics—were conducted in societies with strong liberal traditions. For Japan without such a tradition, the social implications of economic deregulation go much deeper.

The Making of a Liberal Japan

Japan at the end of the century is in the midst of a grand social transformation. The countless changes already in effect and those yet to come—permitting superior students to skip grades, introducing a more flexible taxi fare schedule, allowing banks to go bust for the first time since the Great Depression, arresting and indicting public officials for corruption and corporate executives for collusion, to name a few—are seemingly discrete. But their cumulative effect over the next generation is likely to result in a fundamental change in the nature of Japan's social relations.

The practice of corporate lifetime employment, the hallmark of the post-1945 social contract, allowed for a system of seniority in promotion and pay. This caused economic distortions, but firms could operate thus because the bureaucracy dampened "excessive" competition through regulations. Now, for firms to remain competitive in the global market, and others to become so, the regulatory fetters must be removed. And that is going to make the practice of lifetime employment an unaffordable social luxury. Corporate employees will increasingly be paid on the basis of merit rather than seniority; indeed, the unproductive have already begun to lose their jobs.

In a sense, the country has embarked on a transition from social democracy toward liberal democracy. Japan is not commonly thought of as a social democracy. It has been long ruled by a conservative party, and social democracy as an ideology has not been deeply embedded. Still, post-1945 Japan has arguably bee more consistently social democratic in its policies than many of the social democracies of Western Europe. The problems associated with an aging society and with needed administrative reform and economic and social liberalization that Japan faces are shared by Germany, France, and Italy. All four countries made similar beginnings after the devastation of the Second World War, although Japan went further than any of the others in its welfare programs. Similar, too, have been their postwar relationships with the United States, from which now comes the primary impetus for economic globalization.

Japan's post-1945 social contract has had as its goal equality of results. The difference in wealth between the top and bottom income groups is the narrowest among the G-7 countries. And nearly 90 percent of the Japanese people identify themselves as belonging to the middle class. This egalitarian redistribution of wealth and privilege has been effectuated by government regulation. Social democracy tends to be paternalistic and communitarian, whereas liberal democracy tends to be individualistic. In liberal democratic America, the primary task of government is procedural—to create a level playing field, the emphasis being on equality of opportunity—and consequences are left open. America's creed of individual autonomy tolerates inequality to a degree unacceptable in a Japan whose creed is equality of result—"sameness." But Japan cannot continue to afford the luxury of ensuring sameness much longer.

A freer and more open society that is harsher and less deferential is in the making. For now, the vast majority of the Japanese public remains oblivious of the direction in which the many, seemingly discrete, changes that are taking place in their society point. If a plebiscite were to be held today, the Japanese would be unlikely to approve the kind of society that is being forged. But there is evidence of a growing recognition and even resignation among the public that the proven and comfortable ways of the post-1945 social contract cannot hold.

Many among the political and business classes sense the urgency for reform, though their words and deeds are less than coherent. But then, such is the characteristic pattern of change in Japan. The sum of seemingly disparate changes over a period of time makes for a grand transformation whose logic becomes apparent only after the fact. Blueprints are rarely found, for a clear articulation of goals hinders change by providing targets to those who oppose them. In consensual Japan, all must agree; majority rule over minority disaffection is the mark of a more competitive society. Major change in Japan thus tends to come anonymously and amorphously. (Depending on the degree to which Japanese society liberalizes, the amorphous nature of change may be coming to an end. An infusion of genuine debate should come hand in hand with liberalization.)

The Great Watersheds

The importance of the changes under way in Japan may well turn out to be equal in magnitude to that of the upheavals in 1868 and 1945, the two great watersheds in Japan's modern history. The Meiji Restoration of 1868 saw a feudal society recreate itself as a modern society. With defeat in 1945 came the collapse of the authoritarian imperial order, which gave way to a more democratic polity. In both instances, the basic rules and manners by which the Japanese related to each other were altered. By law and social norm, the public personality of the individual was remolded. The Meiji Restoration (the word in Japanese, *ishin,* means to make new) abolished the hereditary samurai ruling class and forged a more meritocratic society. The American-written constitution of 1946 redefined the Japanese personality from imperial subject to democratic citizen. In contrast to these earlier watersheds, the current transformation lacks a formal turning point. Later, with hindsight, we may recognize a Heisei Restoration, named after the reign of the present emperor. Be that as it may, the present transformation promises an equally radical alteration of social relations and the individual's public personality.

In short, Japan's communalism is giving way to a newfound recognition of individual liberty. As society becomes increasingly competitive, the assumption of consensus and sameness as public goods will be altered. The Japanese person now defined by community will become more individualistic as the community becomes less a source of protection, welfare, and an ordered life. As the individual is endowed with greater autonomy, he will have to take greater personal risks. Wealth and privilege will become more the function of personal choice and less the result of overall social gain.

Inequality and envy are emerging social traits. The notion of individual rights, alien to the Japanese tradition, will have to be articulated and practiced. How much longer will this notion continue to be confused with selfishness, as has been the case in Japan since the introduction of the concept with the country's opening to the West? We shall see over the next generation to what extent regulation and attendant protection, the basis of Japan's social democracy, were excessive as well as contingent on the Japanese ability to define community.

With the Meiji Restoration, following more than two centuries of seclusion, Japan actively and fully embraced international capitalism—and paid the price in 1945. Since 1945, Japan, positioning itself as a subservient member of Pax Americana, has adopted a passive attitude. This Japan endured the onslaught of international capitalism, giving way when necessary, often feeling violated. But the memory of 1945 continued to offer assurance of the wisdom of passivity. Despite the prowess of its export industries, Japan's much larger domestic manufacturing sector and almost its entire service sector, including the realm of capital, remained shy of active engagement with international capitalism. This passivity served the country well for decades, but the recent economic downturn has revealed the true degree to which Japan has become enmeshed with international capitalism, and how untenable the pretension to passivity is.

The 2020 Worldview

Pessimism clouds the horizon as Japan enters the twenty-first century. The reform-minded of the political and business classes are driven by the 2020 worldview. The year 2020, a generation away, looms as a symbol of great hardship to come. In that year, one in four Japanese will be over 65 years old. Japan then will be the most aged of the advanced economies, burdened with hitherto unparalleled welfare obligations. Given present trends, population growth will become negative from 2007 onward; gross domestic product will decline after 2025; and the tax, welfare, and medical payments of the average wage earner will triple by 2020. The country is already heavily indebted. Public debt per capita is the highest among the advanced economies and exceeds GDP.

Immigration offers a way out of this predicament, but crowded and culturally insular Japan does not see an infusion of foreign blood as a viable solution. The Japanese aversion to immigration is even greater than in similarly geriatric and parochial Western Europe. The increasing number of women among producers and wage earners will help alleviate the burden, but not to a sufficient degree. Simply put, the welfare state, the equalizer of capitalistic competition and democratic egalitarianism, faces a major crisis. The 2020 worldview demands that economic productivity and efficiency be raised through liberalization.

Japanese policymakers have been acutely aware of the 2020 predicament for some time. A move from direct to indirect taxation as the primary source of government revenue and administrative reforms to curtail the bureaucracy's regulatory powers in order to enhance competition and efficiency have been put forward as key policies to brace the country for 2020.

Herein lies Japan's difficulty in formulating a viable reflationary policy for the current economic downturn: the obvious options will be counterproductive over the long term. Reflation by increased public spending will increase public debt. Moreover, it empowers the bureaucracy. The government's decision to raise the sales tax from 3 percent to 5 percent in 1997,

adding depressionary pressure when reflationary logic called for a tax cut, is telling. The shift from direct to indirect taxation—instituting a sales tax and incrementally raising it, while reducing the income tax—has been in the works for nearly 20 years, which is how long it takes to form a consensus in Japan. Just as the effort comes to fruition, however, the timing proves to be wrong. The problem is that once a consensus is formed, it is rigid.

The Politics of Denial

In consensual and rigid Japan there are no crises. There are only catastrophes. When a devastating earthquake hit Kobe in 1995, the Swiss government offered to send in a canine search-and-rescue team. The Japanese authorities wasted precious time while they considered what to do about the lengthy quarantine requirement for bringing dogs into the country. To Americans, whose politics thrive on the routinization of crisis, for whom crisis management is a hallmark of strong leadership, the Japanese ineptitude in a crisis is striking. The hallmark of Japanese leadership, in contrast, is the ability to forge consensus and avert crisis, to eliminate contingency. So, when contingencies occur, as is natural in the affairs of man, Japan tends toward paralysis. This reflects the Japanese aversion to bestowing concentrated powers upon any one individual. Legally, the Japanese prime minister has tremendous powers; in practice, he is one among equals. Here, too, the idea of sameness is operative.

The initial response in Japan is always to deny a crisis exists. Only when a situation becomes unbearable, only when catastrophe is at hand—as now, after seven years of economic stagnation with the possibility of a depression looming—can the old consensus be broken and the course of events redirected. It was not until the spring of 1998 that the administrative vice minister of the economic planning agency publicly acknowledged the pending financial calamity and the government announced a package of emergency tax cuts and public spending (thereby straying from the consensus of 2020).

The worldview of 2020 reveals another striking feature of Japanese politics: the ability of the Japanese to think and plan for the long term. (In American politics, in contrast, the four-year presidential election cycle is often as long as the future can be considered.) This, in turn, rests on the stability and workings of the national civil service. Because the governing Liberal Democratic Party (LDP) has been continuously in power since its founding in 1955, save for a few short years in the mid-1990s, and because it has relegated much policymaking and legislating to civil servants, planning has been unaffected, on the whole, by the vicissitudes of the electoral process. (The American gift of democracy, in this sense, has remained nominal.)

Whatever the pros and cons of long-term planning, Japan has reached the limits of bureaucratic control. At the core of the 2020 worldview is administrative reform: to break the immense hold of bureaucratic power, to release society's energies hitherto dampened by regulation, to create a liberal Japan that is economically more efficient and socially more imaginative and diverse. Yet, Japan's current crisis is aggravated by the fact that the very bureaucracy that must deal with this crisis is under attack.

The Direction of the Tide

March 1998. Another official of Japan's Ministry of Finance takes his life. He had been accused of taking bribes in exchange for leaking privileged information to banks. More arrests of government officials for corruption follow.

Finance Ministry officials, the best and brightest, mostly graduates of Tokyo University's Faculty of Law, are a proud lot. They know that it is they who govern, and they are sure of their wisdom and power. At the age of 22, after passing the higher civil service examination, finance officials are made forever. They enter a world that is secretive and compartmentalized, where accountability and transparency are not demanded. Given the ministry's immense regulatory powers, the temptation to become corrupt has been great. And many have succumbed to the temptation. For lavish meals and golf outings, officials have performed favors and leaked information.

That was the cozy organic world of Japanese government and business. That was the arrogant world the select few joined at age 22. But suddenly there is accountability; the prosecutor's office moves in to expose the illegality of the organic relationship. For those of heightened pride, the fall is far. Once powerful officials are stripped of honor, as the ethic of a larger world is imposed upon them. Their secretive world falls apart, and for some suicide offers the only expression of shame and atonement.

The politicians seek to enfeeble the bureaucrats; the battle cry of the government is administrative reform. But the government is at war with itself, because the politicians and bureaucrats depend on each other. The politicians are poised to regain leadership in government. The bureaucrats are naturally resistant. But the prosecutors' ability to shatter the cozy world of officials indicates the direction of the tide.

In the post-1945 order of things, Japan's bureaucracy has done more than administer. It has functioned as the nation's legislature and has assumed the role of the judiciary. Customarily, almost all the bills submitted to parliament have been written by the bureaucracy. Laws have been worded so as to give the bureaucracy great leeway in interpretation and enforcement. Over the decades, this habit has deprived parliamentarians of the ability to write law. Moreover, the bureaucracy has, in a sense, stood above the law; no court has the power to make the bureaucracy divulge any information it deems detrimental to the public good—woe to the complainant against any government ministry.

Unchecked and unbalanced, the everyday sources of bureaucratic power derive from the bureaucracy's monopolistic control of information and its role in issuing authorizations and permits for activities covering the minutest details of Japa-

nese life. The bureaucracy constitutes the core of Japan's government: one petitions government, one does not make demands of right as a citizen.

Some call this Asian authoritarianism. The Japanese bureaucracy possesses the Confucian "mandate of heaven" to rule for the people. It is Walter Lippmann's prescription for distributive rather than participatory democracy (in America) come to life. Japan's government by bureaucracy is the natural outcome of a belief in the public's incapacity for critical judgment and self-government; it is government of the administrative elite par excellence. The question to ask about government therefore is, to borrow Lippmann's words, "whether it is producing a certain minimum of health, of decent housing, of material necessities, of education, of freedom, of pleasures, of beauty, not simply whether at the sacrifice of all these things it vibrates to the self-centered opinions that happen to be floating around in men's minds." The achievements of post-1945 Japan bear the administrative elite's mark. But the mandate of heaven is being withdrawn.

To Be Accountable

The word "transparency" has lately entered the Japanese political lexicon. There is a concerted effort under way to introduce independent political think tanks into the Japanese scene to break the bureaucracy's monopolistic hold over information and knowledge. This is part of the liberalizing process in preparation for 2020. Another novel word making an appearance is "accountability." Japanese use the English word; there is no precise Japanese equivalent. Pundits define it as the responsibility to explain. Naoto Kan, now leader of the opposition Democratic Party, was instrumental in introducing the word, and he did so in a manner that has made him probably Japan's most popular politician.

In 1996, when Kan was appointed minister of health and welfare in a coalition government, he showed that government ministers do not have to remain mere figureheads of the state bureaucracies. Until Kan came along, the public generally understood that ministerial appointments, lasting on average no more than a year, were parliamentarians' awards for seniority—cynics would say that political appointees did not even need to know the location of their offices. When Kan took office, the Ministry of Health and Welfare was being sued by Japanese hemophiliacs, the majority of whom had been infected with HIV through contaminated blood products licensed by the ministry. They accused the ministry of criminal negligence; some accused its officials of murder. For nearly a decade, the plaintiffs had demanded disclosure of ministry records, without which the courts could not proceed. The ministry stonewalled, saying that the pertinent records could not be found. Kan ordered that the records be found, and they were produced within a few days. Resignations and indictments followed.

The HIV affair caused the public to wonder whether its trust in the administrative elite was misplaced. After all, regulations were supposed to prevent such accidents. The HIV affair, which cast doubt not only on the bureaucrats' technical competence but on their morality as well, accelerated the demand for transparency and accountability. The consensus as to the need for administrative reform may have been slow in coming, but it is surely forming. Japan's liberal transformation is under way.

At Your Own Risk

On Miami Beach, there is a sign warning: "Swim at your own risk. No lifeguard on duty." According to the Japanese order of things, the authorities are not supposed to erect such notices. If one were to appear on a beach in Japan, there would be public outrage over what would be seen as government irresponsibility. The public would be confounded and demand to be told whether or not it was safe to swim. For the government to tell its people to act at their own risk would simply not make sense. "The Japanese take it for granted that we are always under government protection—that even as adults, choosing our own pursuits in the freedom of our leisure hours, we are protected by some government office responsible for our recreation," notes Ichiro Ozawa, another reform-minded politician.

Nonetheless, the Japanese are being ripped from their protective womb. Heretofore, for example, the government has guaranteed all bank deposits in full; now there is a plan to institute an FDIC-type limit on government insurance. When that limitation and warning is in place, the government will begin to allow more uncompetitive banks to fail. A more liberal Japan, a Japan in which the people will be told to swim at their own risk, is in the making.

That said, the government will have to oversee an uneven distribution of risks, favoring the aged over the young in the society of 2020, and this will transform the political atmosphere. Whereas all political parties today claim to be national parties, in a more competitive Japan which cannot afford sameness political parties are likely to tend to partisanship. The recently created opposition Democratic Party already looks to the interests of urban dwellers and consumers, while the long-ruling LDP is focusing on farmers and big business, for example. It remains to be seen how the hitherto risk-averse Japanese will respond to the new burden and privilege of choice that such a change entails. One thing is certain: what emerges will violate the Japanese notion of harmony.

Note

This essay was written under the auspices of the Program on Asian Studies at Florida International University.

Article 11   *Look Japan*, August 2000

Who Wants To Be a Volunteer?

Japan International Volunteer Center founder Hoshino Masako reviews the history of volunteerism in Japan and considers the reasons for the surge in interest in nonprofit activities today

By HOSHINO Masako

The author is president of the Japan NPO Center and a professor at the Faculty of International Studies, Keiai University.

I founded the Japan International Volunteer Center (JVC) in 1980 with the aim of helping refugees in Indochina. As its representative, I went to Brighton, England in 1982 to attend a subcommittee meeting on the theme of refugees at the International Conference on Social Welfare. At that time I had the good fortune to meet Alec Dickson (1914–1994). Known as the "father of volunteerism" for founding such organizations as the Community Services Volunteers and Volunteer Service Overseas in the 1950s, Dickson was held in very high esteem by his British countrymen.

Dickson talked to me about the activities of his Community Services Volunteers program, which involved giving young prison inmates the opportunity to work in facilities that housed people with serious disabilities. His philosophy was that even the most hardened criminal was made that way by his environment and always had some element of goodness inside that could be drawn out through volunteer activities.

To illustrate his point, Dickson told me about one particular case he had seen. The young man in question never knew his father and was not loved by his family, leaving him a rough life in which he got in the habit of stealing at an early age. Finding himself in jail, he was given the opportunity to work at a care facility for people with disabilities as part of Dickson's volunteer program. He was assigned to an elderly woman who could not speak and was confined to a wheelchair. After completing his day's duties and preparing to return to the jail, he felt the steady gaze of the woman and knew that she was sorry to see him go and wanted him to visit her again. As someone whose own mother had never missed him, he found this experience deeply moving. In fact, Dickson said, he was transformed in that moment. Even after all the bad things he had done, here was someone who looked forward to his visits. The young man faithfully finished out his prison term and was fully rehabilitated.

After telling me this story, Dickson asked me who was being helped by the volunteer organization I had founded. Having established it for the purpose of assisting refugees in Indochina, I answered "Indochinese refugees." To this he said no, the true beneficiaries are the volunteers themselves. This conversation with Dickson contributed greatly to my understanding of the true nature of volunteerism.

VOLUNTEER ROOTS

According to Dickson, the word "volunteer" first came into use in Great Britain around the seventeenth century. As far as he could determine, the oldest confirmed use of it is on the gate of an Anglican church built in the seventeenth century, where it says, "Volunteers from the citizenry were recruited to ward off foreign invaders." Dickson also contends that it was around this time that the concept of "civil society" emerged, through which citizens began to take an active role in resolving the problems faced by the community. According to Dickson, the respective developments of volunteerism and civil society are closely linked.

Viewed in this way, we can say that volunteerism got its start in Japan after World War II. However, there were volunteer-like, nonprofit organizations operating throughout Japan in the Edo period (1603–1868) and even earlier.

One typical example was the form of cooperative labor known as yui (literally, "tying"), through which community members would help each other plant and harvest fields, rebuild homes, thatch roofs, and engage in other activities that could best be done by a group. (The groups themselves were also known as *yui*.) Similar examples of nonprofit groups included *kō* (associations or fraternities associated with particular religions) and *za* (guilds of merchants or artisans engaged in the same line of work).

Strictly speaking, however, these mutual-aid organizations, which supported territorial society, were different from modern volunteer organizations because people had no choice but to participate. For true volunteerism to occur, there must be no coercion and no seeking of reward, and all participants must be treated equally.

Similarly, the private schools of the Edo period and the various charitable organizations of the Meiji period (1868–1912) were not run by ordinary citizens, but by a small coterie of religious benefactors or special philanthropists who sought to change society and help the downtrodden in the spirit of service, sometimes going to extremes of self-sacrifice.

As one might expect, the rise of modern volunteerism in Japan after World War II is related to postwar economic development. Today, citizen–led volunteer activities are gearing up in other Asian countries that are developing economically. According to the publication, "Emerging Civil Society in the Asia Pacific Community," which was published under the supervision of the Japan Center for International Exchange in 1998, surveys conducted in Singapore, Indonesia, Thailand, and other countries show that volunteerism has increased as the negative effects of more affluent lifestyles have become more pronounced, such as environmental degradation and the widening gap between rich and poor.

The same thing happened in Japan in the 1960s. Against the backdrop of accelerated economic growth rooted in postwar recovery, the nationwide Japan Youth Volunteers Promotion Council was founded in 1963, and the Osaka Voluntary Action Center, whose citizen-based volunteer activities had a tremen-

dous impact on subsequent Japanese volunteerism, was established in 1965. Meanwhile, also in 1965, the Japanese government established the Japan Overseas Cooperation Volunteers (JOCV).

In the 1970s, other national and local government agencies got behind the volunteer movement. The Ministry of Education, for example, recognized the importance of volunteerism as a means of vitalizing social studies courses, and the Ministry of Health and Welfare began providing support to social welfare groups.

Then, in response to the flood of refugees coming out of Indochinese nations such as Cambodia, Vietnam, and Laos, various overseas volunteer groups were founded in Japan in the 1980s. Many of these groups are still active today, including JVC and the Japan Sotoshu Relief Committee (now called the Shanti Volunteer Association). This development can be viewed as the internationalization of Japanese volunteerism. Establishing links with such organizations as Oxfam in Great Britain, Doctors Without Borders in France, and CARE in the United States, these groups quickly introduced new ideas to Japan.

NPOs: AFTER THE QUAKE

Perhaps the most important milestone in the history of Japanese volunteerism came with the Great Hanshin Earthquake that struck in January 1995. It's estimated that 1.5 million volunteers participated in rescue and recovery activities in the aftermath of that disaster. As a result, general participation in volunteer activities broadened throughout Japan, and the term NPO (nonprofit organization) came into common use. Another impetus to the development of volunteer groups and NPOs in the 1990s was the fact that it became clear that governmental agencies could only go so far in solving the various problems confronted by society.

Riding this wave of support, the Law to Promote Specified Nonprofit Activities (the NPO Law) went into effect in December 1998, with the stated aim of: "contributing to the public interest by promoting the sound development of volunteer activities and other specified nonprofit activities freely performed by citizens as a contribution to society."

Specifically, the law opened the way for NPOs to incorporate. Although the content of the law is still inadequate (no provision is made for special tax breaks, for example), it has nevertheless served to greatly raise public recognition of NPOs that have acquired corporate status. Also, college graduates have become increasingly aware of NPOs as possible employers.

Although Japanese understanding of NPOs is deepening, some people believe that such activities should be 100 percent volunteer, and that no one should receive salaries. In view of the high cost of transportation and food, however, potential volunteers sometimes find it difficult to foot the bill themselves. In Great Britain, which is highly advanced when it comes to volunteerism, it was long believed that participants should not be compensated. According to Jennifer Hargreaves, professor of the Sociology of Sport at Roehampton Institute (London), however, this attitude has changed, becoming more in line with Dickson's philosophy that any person who wishes to volunteer should be able to do so regardless of his or her circumstances. For example, if people of meager economic means are always on the receiving end of volunteer activities, they will never regain their personal well-being. Since volunteer activities help the volunteers themselves, more and more British have come to believe that poor people who wish to volunteer should be compensated to cover transportation and lunch expenses, and be given access to projects that do not require them to pay anything out of their own pockets. As a result of this shift in attitude, slum residents are now participating in volunteer activities while receiving stipends for necessary expenses from appropriate institutions or churches.

As this example shows, the question of whether or not a person should be compensated for volunteer activities is not a substantial problem. Although volunteers play a crucial role in NPO operations, all NPOs themselves are not volunteer organizations. In cases where donations and subsidies are inadequate, many NPOs are compelled to engage in fund-raising activities, which is why there's a column on the application form for corporate status that asks what kind of activities the organization engages in

for profit. Such activities are permitted as a means of self-support when they are consistent with the purpose of the organization. For example, an organization that receives donations in the form of real estate are allowed to use rental fees on the property to offset their operating expenses.

The relationship between volunteers and paid staff can be likened to the relationship between an apple's flesh and its core. Through the vitalization of NPO activities, one hopes that the number of volunteers will increase. Without a healthy and solid core, however, it's impossible to grow a large, sound apple. In other words, if the number of people who make their living working for NPOs increases, we can expect to see a corresponding increase in the number of volunteers. Unfortunately, however, Japan is still at the stage where projects that should be properly funded by government agencies are being implemented by volunteers who serve as "cheap labor" due to lack of funds. When participation is inadequate in major sports events sponsored by local governments, for example, "volunteers" will often be "mobilized" to pick up the slack. Purely voluntary activities conducted by citizens are still in short supply in all sectors, which is why government agencies must sometimes manage things through top-down measures.

The Japanese are still concerned with such things as which agency has jurisdiction over a given volunteer activity, and I think it will take some time before true volunteerism takes root. Given the current political and economic situation, however, I don't think that government agencies have the power necessary to regain control of the NPOs and volunteer organizations. With the government in a weak position, citizens have an opportunity to act.

A SECULAR PURSUIT

In the West, religion-based volunteerism is extremely common, but such activities are not as common in Japan. Churches do of course engage in various service activities, but secular volunteerism is generally more active. Some Japanese NPOs are organized around a religious leader or other charismatic individual

who provides the motivating force and pulls other people along. The most active organizations, however, are not dominated by any individual and share information among all members, who have access to the decision-making process. These organizations are popular among young people and, because they operate in a field where there is greater equality between the sexes, tend to enjoy the active involvement of women. In fact, women have recently emerged as the leaders of many NPOs.

Because legal support and human resources are only now emerging, we can expect the number of NPOs to grow in the months and years to come. At the same time, the strength of the citizenry and their work as volunteers will be put to the test.

I teach at a university, and when I tell my students that they can choose any form of volunteer activity as long as it helps other people, they often find themselves at a loss. In the course of their student lives, many of them have never undertaken a project on their own initiative, and think it's strange that they can earn college credit by doing something of their own choice. In the same way, I'm sure that there are many "volunteers" who respect the harmony of the organization to which they belong and hesitate to assert themselves. Even so, I believe that more and more people in Japan are looking at volunteer work as a kind of inner treasure hunt that provides an opportunity for them to discover who they are, what they want to do, and what things make them feel invigorated and most like themselves. I also believe that more and more Japanese agree with Dickson when he says that the humanity of the volunteers themselves is the chief beneficiary of volunteer activities.

With its old system no longer functioning, it sometimes feels as though Japan has entered a tunnel that prevents us from seeing what lies ahead. I believe, however, that we can find light at the end of that tunnel if individuals establish themselves through self-motivated action. Abandoning the pyramidal social organization of the past, Japan can open up a whole new future by building a civil society based on unprecedented lateral networking.

Article 12

Harvard International Review, Fall 2000

The Difficulty of Apology

Japan's Struggle with Memory and Guilt

SHUKO OGAWA

Staff Writer, *Harvard International Review*

The complex relationship between domestic politics and war issues makes war guilt a controversial topic in Japan. Japan's World War II occupation of a substantial part of Asia left indelible scars. The Nanking Massacre and the plight of the Korean "comfort women," stand out among examples of Japanese cruelty during the war era. The "Rape of Nanking" occurred in late 1937 when the Japanese Imperial Army took control of the city of Nanking, China, and slaughtered, mutilated, and raped anywhere between 100,000 to 350,000 Chinese. "Comfort women" were officially provided to the rampaging Imperial troops in the field; as many as 200,000 women, mostly Korean but also Burmese, Chinese, Taiwanese, Filipino, Indonesian, and Dutch, were forced into prostitution.

A Japanese politician's attitude toward the war is a political barometer in the public eye, much as abortion stances or affirmative action are in the United States. The Liberal Democratic Party (LDP), which has largely dominated post-war governments, includes a broad spectrum of factions that receive support from various sectors of Japanese society. Groups on the left, such as teachers' unions and grassroots groups, believe in atonement for war crimes, further compensation to victims, school education imparting the full scope of Japanese atrocities during the war, and caution against a resurgence of Japanese militarism. The revisionist right, especially strong among LDP members, says Japan has financially aided its Asian neighbors enough, and that the bilateral treaties signed with China and South Korea in 1978 and 1965, respectively, exonerate Japan of any further official obligations. Additionally, the social conservatives believe children should be allowed to have pride in their country, and thus not be overexposed to Japanese wartime atrocities.

To this day, despite new evidence pointing to Japanese culpability, various members of the Japanese elite, still largely conservative, continue to downplay or even deny official Japanese complicity in these two appalling episodes. Such stances by men in the highest levels of politics and academia have not assuaged the fears of neighboring Asian nations, or indeed of countries throughout the world, that still harbor deep suspicions about a possible revival of Japanese militarism. However, it has become increasingly difficult for revisionist politicians to express their opinions openly without suffering some political consequences.

AP photo/S. Kambayashi

The Yasukuni Shrine commemorates Japan's war dead, including convicted war criminals.

The Japanese public's perception of World War II is particularly complicated by cultural traits shared by some Japanese, including a deeply ingrained deference toward ancestors, submission to authority and a victim mentality about the atomic bombings of Hiroshima and Nagasaki. After the war, Emperor Hirohito remained on the throne a respected figure—although only as a figurehead—and many bureaucrats who ran the daily government apparatus during World War II retained their positions through the postwar era, denying Japan a clean cultural or structural break with the war. Ideological wars have been fought over the wording of history textbooks, commemoration of the war dead, and personal-compensation demands from foreigners who suffered at the hands of the Japanese during the war.

Developments in the last few years, spurred perhaps by the intense soul-searching conducted in 1995 on the 50th anniversary of the end of World War II, have shown encouraging trends, both public (an official apology) and private (polls showing support for further government action), toward full Japanese acceptance of responsibility. Nonetheless, if Japan wishes to assume a more active international role, it must credibly and completely confront its past. Only then will Japan be able to act freely without inviting distrust about the motives behind its objectives, such as a permanent member seat on the UN Security Council, or a more assertive role in peacekeeping missions. The Japanese government and the LDP, in particular, must realize that in the long run, a clear atonement with full political and social backing best benefit Japan.

The Forgotten Aggressor

The Japanese postwar experience is unique. Unlike Germany, Japan was not burdened by a sense of national guilt for its wartime crimes. Most of its hideous wartime actions took place on the Asian continent, far away from the daily life of the common Japanese, with wartime censors and the silence of returning soldiers contributing to the popular ignorance. Instead, the dropping of the atomic bomb imparted a victim mentality upon Japan; 140,000 perished in Hiroshima and 70,000 in Nagasaki

(including later deaths from radiation) according to Japanese estimates. The victim perspective was prevalent enough that in the immediate post-war period almost no Japanese could naturally conceive of Japan as the aggressor.

The onset of the Cold War and Japan's strategic location as a Pacific rim defense base against the communist Soviet Union and China meant that the US occupying forces needed Japan as a strong ally. General Douglas MacArthur, supreme commander of the occupying forces, pushed for extensive investigations into communist activism in 1949 following large-scale labor strikes that threatened to paralyze the still-fragile Japanese economy, at a time when the United States was attempting to instill democratic values in Japan.

Additionally, at the Tokyo War Crimes Tribunal following the war, MacArthur rejected demands to have Emperor Hirohito brought to trial in order to maintain what little stability remained in Japan. The emperor renounced his divine authority in a radio address to his subjects but refused to acknowledge any wrongdoing on the part of Japan; in any case, he remained on the throne until his death in 1989, personally representing, albeit in innocuous form, a continuation of the prewar mentality into the postwar period.

Significantly, of the seven Class A war criminals (those tried for the crime against peace of waging an aggressive war) sentenced to death at the Tokyo War Crimes Tribunal, only one, wartime Foreign Minister Koki Hirota, was a civilian. Thus, the mindset of civil servants and emerging political figures was changed little as they set about the task of reconstructing the country. Mamoru Shigemitsu, another wartime foreign minister who was sentenced to prison at the Tokyo War Crimes Tribunal, later even re-emerged to serve as foreign minister in 1954. In addition, Article Nine of the new "peace" constitution adopted in 1947, essentially written by MacArthur's staff, obligated "the Japanese people forever [to] renounce war . . . and the threat or use of force." This principle fit perfectly with the postwar Japanese desire to symbolically embrace peace in a world without nuclear weapons as the only nation to have been victimized by military use of the atomic

bomb, not as an aggressive nation that had committed unspeakable atrocities.

The spectacular economic growth in the following decades aided in suppressing much thought about past crimes. The national emphasis was on diligent work and economic development. The Japanese mentality of obedience and individual sacrifice for the economic prosperity of the whole country was one factor in achieving this economic miracle. Government leaders accordingly did not attempt to alter a system of thinking that had led to the populace's blind acceptance of imperial reasoning and World War II in the first place. Nor did the government ever seek to fully inform the people about the extent of Japanese atrocities during the war. Granted little official information, the post-war generation grew up knowing too little about Japanese atrocities to push for a historical reckoning, as opposed to the German students of the 1970s who took to the streets to demand a full confrontation with crimes committed by their parents' generation. Powerful veterans' associations, such as Japan War-Bereaved Families Association (Nihob Izokukai), have always been quick to justify Japan's wartime actions with militant support of the conservative position that Japanese wartime occupation actually freed Asian countries from Western colonial oppression. Supporters of these organizations have a sincere interest in keeping noble the cause for which their loved ones died; thus they zealously protest any government move that "degrades" Japan's position concerning its justifications for the war.

Wall of Silence

Successive LDP governments have habitually taken an ambiguous stance, not fully acknowledging, much less apologizing for, Japanese actions in Asia during the war. The LDP, which came to power soon into the post-war period, saw no reason to re-evaluate Japan's official attitude toward the war as long as the economy was performing well and the party remained comfortably in power.

Progress in confronting the past began only in 1993, after the LDP was ousted from government leadership for the first time in 38 years and Morihiro

Hosokawa of the Japan New Party formed a new centrist coalition government. Hosokawa became the first premier to formally recognize Japanese responsibility, using the term "aggression" instead of "advance" to describe Japan's wartime actions in Asia. Tomiichi Murayama, also a non-LDP Socialist who served as prime minister from 1994 to 1996, was the first to use the word *owabi* (apology) in his 50th anniversary address commemorating the end of World War II in 1995, fulfilling a lifelong personal and Socialist Party goal to atone for Japanese aggression. However, Murayama's drive to have the Japanese Diet (parliament) pass a resolution expressing apology was thwarted by members of the LDP, one of his coalition partners: the final approved resolution expressed only *fukai bansei* (deep self-reflection).

The tortured factional conflicts within the LDP and the fragility of coalition governments in recent years have forced prime ministers to consider party and faction affiliation over personal qualification or opinion for cabinet appointments. LDP members with revisionist views thus gain official positions and sometimes are soon forced to resign whey they fail the media's "test" by uttering revisionist statements during press conferences or interviews.

The Japanese government must realize that in the long run, a clear atonement with full political and social backing would best benefit Japan.

The revisionist view that Japan liberated Asian countries from the stranglehold of Western imperialism seems entrenched in the mentality of even the highest-level cabinet officials, many of whom are LDP members. Murayama, despite his personal conviction, could not control his ministers: in August 1994, Shin Sakurai, Murayama's environment agency director-general and an LDP member, remarked that Japanese war-time occupation of Asia was not intended to be aggressive and was even beneficial by freeing Asia from Western colonial influence: "Because of [Japan's wartime] education, [Asian countries] have a much higher literacy rate than African countries that were ruled by European countries." Murayama subsequently dismissed Sakurai, indicating that such comments could not be tolerated, especially in the face of harsh criticisms from Japan's Asian neighbors, with whom Murayama was attempting to build better relations.

LDP member Ryutaro Hashimoto, who succeeded Murayama as prime minister in 1996, served as president of the Japan War-Bereaved Families Association before attaining the premiership. True to his hawkish background, he commented in 1994 that whether Japan's wartime actions could be considered aggressive was a question of "subtle definition." However, as prime minister even Hashimoto was not able to face down increased criticism from Japan's neighbors as the trend toward better regional relations gathered force. For example, Hashimoto's private visit to the Yasukuni Shrine in 1996 sparked student protests in China and a comment from the Chinese Foreign Ministry that "the feelings of all the people from every country, including China, which suffered under the hands of Japanese militarists," had been hurt. Instrumental in the promotion of emperor worship during World War II, the Yasukuni Shrine is dedicated to the souls of Japanese who died for the emperor, including those of Class A war criminals such as General Hideki Tojo, and is thus seen as symbolic of Japan's imperialist and militaristic past. The annual visits to the Yasukuni Shrine by mostly LDP cabinet ministers and Diet members on the anniversary of the war's end in August, a tradition followed by Hashimoto, has always drawn objections from Asian nations, especially China and South Korea. In the interest of maintaining good relations, Hashimoto had to forgo any further visits to the Shrine during his tenure. He also became more forthcoming in admitting Japanese responsibilities: at a memorial service in 1997 commemorating the end of the war, Hashimoto said that "the war caused much pain and sorrow to people, especially in other countries in Asia."

An opportunity was unlocked in the 1990s, around the 50th anniversary of World War II, to crack the Japanese government's silence on the subject of World War II. Increased pressure from Asian nations and the coming to power of centrist leaders created momentum for a full accounting. In the interest of maintaining friendly relations with Japan's neighbors, even LDP cabinet members have recognized the benefit of maintaining a conciliatory stance. However, a static political elite ruling a system based largely on seniority, loyalty, and backroom deals—typified in the succession of Yoshiro Mori to the prime ministership following Prime Minister Keizo Obuchi's incapacitation—ensure that significant policy changes will meet substantial resistance under the current system, no matter how much the LDP's power has deteriorated.

The Darkest Moments

The Nanking Massacre generates a great range of opinion from politicians, from the extreme conservative stance, which denies that the incident ever occurred, to the radical left position, which argues that Emperor Hirohito should have been tried for complicity in the massacre since his uncle Prince Yasuhiko Asaka was a commander in the Nanking area when the incident occurred. However, reactionary comments by top officials have become increasingly unacceptable: when then-Justice Minister Shigeto Nagano called the Nanking Massacre a "fabrication" in 1994, he was forced to resign by Tsutomu Hata of the Renewal Party, the prime minister at the time.

In the case of comfort women, the Japanese government initially refused to even recognize their existence. It was not until decisive evidence began to emerge in 1992, thanks to the efforts of people such as Chuo University professor Yoshimi Yoshiaki, proving the army's complicity in procuring women for its men, that the government began to reluctantly acknowledge its wartime actions. Reference to comfort women in several school history textbooks beginning in 1996 led to a conservative backlash in which a number of parents sued the textbook companies, claiming that the women had willingly serviced the soldiers.

Some government recognition concerning this matter has emerged in recent years. In 1996 Hashimoto issued a letter to surviving former comfort women expressing "apology and remorse," offering to pay US$18,700 to each survivor from the private Asian Women's Fund (set up by the government). However, the Japanese government still remains cautious about issuing more forthcoming apologies and inviting reparation lawsuits as a result, and some former comfort women still consider Hashimoto's apology inadequate and have refused what they consider is pay-off.

Education and Tradition

Japan's atrocities during World War II remained largely a taboo subject within the country until recently. The Nanking Massacre and the army's use of comfort women are still not very well-known episodes in Japan's history, and there was little pressure from abroad or from authorities within to prod the public into confronting these issues until fairly recently. Instead, public perception was formed by the education of children, especially through school history textbooks. Generally, Japanese history textbooks do mention the horrific details of Japanese atrocities during World War II, but without any discussion of further consequences or reasons behind the actions in question.

The subject of teaching children about Japan's wartime history has developed into a battle between the Education Ministry and teachers' unions. The ministry, which has long remained conservative and sensitive to any negative portrayal of Japanese actions, has continuously clashed with left-leaning textbook writers and teachers who seek to impart the whole, ugly truth of Japanese wartime behavior to children. These teachers who advocate "peace education" believe that only by knowing the whole story of Japanese aggression will children become convinced never to repeat history again. On the other hand,

conservatives believe that such education robs children of pride in their country; as former Justice Minister Seisuke Okuno commented, "Why do Japanese schools teach children things that discourage them from having pride in our country? After all, education will determine the future of our country."

An official apology has been tendered, but that should not foreclose further overtures.

Most textbooks do provide reference to the Nanking Massacre, but without any discussion. This lack of detail in textbooks may be due largely to the Japanese education system, which emphasizes rote memorization to prepare students for the rigorous entrance exams into prestigious high schools or universities, the outcome of which will determine much of the course of the rest of their lives. Material presented in history textbooks, which must cover everything from Japan's ancient roots up to present history, therefore consists mostly of concise presentation of facts; unlike the hefty history textbooks used in the United States, some of which analyze the significance of historical events, most Japanese textbooks are the size of a medium-sized novel and merely mention historical dates and occurrences. Japanese students are thus discouraged from forming opinions on Japan's historical actions even if textbooks do mention them. In such an environment, when getting into a prestigious high school, university, or company is a measure of success, the Japanese public in general has become very detached from its own country's wartime experiences. The public's ignorance of Japan's actions as the aggressor can be said to stem from these institutional and cultural factors.

However, popular attitudes are slowly evolving as well as a younger, more open-minded generation comes to age. In the 50th anniversary ceremonies

of the Hiroshima and Nagasaki bombings, special mention was made of victims of Japanese aggression, perhaps a sign that Japan, at least on the local level, is now more willing to acknowledge the country's actions that started the war and ended in the dropping of the atomic bombs. Peace museums documenting Japan's aggression in Asia have sprung up over the country. After Hosokowa issued his apology in 1993, 76 percent of the people surveyed agreed with Hosokawa's decision to recognize Japanese "aggression," according to a poll by *Asahi Shinbun*, an influential Japanese daily. An *Asian Wall Street Journal* poll found that 61 percent of Japanese felt that "Japan hasn't done enough to take responsibility for its role in World War II." Thus, official attitudes by the LDP elite not withstanding, the view that the Japanese suffer from "historical amnesia" concerning Japan's actions in Asia is clearly not as true as it might have been a decade ago.

More Than Sorry

The reactions of Asian nations to Japan's halting progress toward and beyond a clear apology have understandably been tense and cautionary. Because of the extensive political maneuvering behind official apologetic overtures on wartime actions, deep doubts cannot be erased about Japan's sincerity in its commitment to face its actions. Most Asian countries are also in the uncomfortable position of having been both the victims of past Japanese aggression and the benefactors of postwar Japanese aid and trade. Dependence on Japanese assistance may explain why some Asian nations are reluctant to demand an outright apology from Japan or to support local groups that are demanding compensation from the Japanese government. The passage of the 50th anniversary of the war's end and the progress already made by former governments has led to a loss of momentum for further actions, as recent governments have seen the issue of World War II come up less often.

Japan must overcome its concern with saving face and recognize that truly "saving face" lies in seeking full reconciliation with its wartime victims. The current official government position, limiting its options to building better relations with its neighbors through increased trade and investment, is insufficient. True, government-to-government compensation claims were addressed by law long ago, but individual victims of Japanese aggression often did not receive anything. An official apology has been tendered, but that should not foreclose further overtures. Increased activities by concerned citizen groups and opinion polls should tell the government that the public does wish for a historical reckoning with past crimes. In order to earn the unreserved respect and trust of all nations, especially of its Asian neighbors, Japan must confront its past in its entirety. Acknowledgment of guilt will enable Japan to assume without suspicion the greater role in regional and international arenas that it has sought in recent years. A Security Council Permanent Member seat, for which Japan has expressed an avid desire, cannot come to pass while some nations—especially those that suffered Japanese conquest—still fear the return of Japanese militarism. And this shadow of doubt will follow Japan as long as any aspect of responsibility for its wartime conduct is left in the dark.

Article 13

Far Eastern Economic Review, July 16, 1998

JAPAN

Arthritic Nation

The greying of Japan isn't a worry for the next millennium: It's already a problem. Conservative habits of an ageing population have a lot to do with why Japan can't seem to restructure its economy.

By Peter Landers in Kiryu, Japan

No doubt about it, says Shigeo Shimizu, this recession is bad. The neighbourhood near the Kiryu River where he makes black sashes for funeral kimonos used to be lined with textile factories; now they're mostly gone, and people who are still hanging on say orders are sharply down since last year.

Yet when parliamentary elections roll around on July 12, the 66-year-old Shimizu will be voting, as always, for a candidate from the ruling Liberal Democratic Party. He doesn't suppose the LDP will do much to make the bad times better, but the opposition candidates make him uneasy. "I can't vote for someone I don't know. At least I know these guys," he says.

As Japan ages, older voters like Shimizu—wary of risk, averse to changing old habits—have come to make all the difference. In the voting for half of parliament's upper house, the LDP is certain to win more seats than any other single party. It may even capture a majority of the 126 at stake. If past trends hold up, fewer than one in four Japanese in their 20s will vote. But more than two-thirds of those over 65 will—and overwhelmingly they'll vote for the conservative ruling party. In the 1995 upper-house election, those aged 55 and over accounted for an estimated 48% of all voters.

Elections show how the ageing of Japan, once seen as a trend whose impact was still a few decades away, is already making itself felt. With a solid base of conservative supporters, politicians have no impetus to force through the painful restructuring that's needed to revive Japan's economy—seen as vital to the recovery of Asia. And the frugal spending habits of an ageing population do little to stimulate demand.

Japan isn't an old country, yet. Only 16% of its people are over age 65, a proportion comparable to most European nations. But that's changing. The average Japanese woman lives 83.6 years, and the average man 77. Meanwhile, women are having fewer babies than ever. So by 2035 or so, the government estimates that more than 30% of Japanese will be 65 years or older.

But forecasts like those have prompted even middle-aged Japanese to turn conservative politically, and to try to hang on to their financial assets. Concerned that state-supported pensions will dry up, consumers are clinging to their savings like never before and as a result deepening Japan's recession. And young people feel increas-

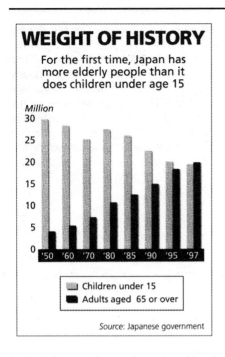

WEIGHT OF HISTORY

For the first time, Japan has more elderly people than it does children under age 15

Million

Legend:
- Children under 15
- Adults aged 65 or over

Source: Japanese government

ingly left out of a society that doesn't seem to involve them.

Places like Kiryu, a textile manufacturing town of 120,000 people 100 kilometres north of Tokyo, give a sense of things to come. About one in five people in town is over 65, partly because competition from South Korea and China has hurt the textile industry and sent young people seeking jobs elsewhere. The town's commercial district is filled with clinics, dentists' offices, and companies offering services for the home-bound elderly. Night-time is quiet, with only a handful of teenagers hanging out in the main shopping area.

In elections, it's elderly activists who ensure victory. At the Kiryu office of LDP candidate Hirofumi Nakasone—a son of former Prime Minister Yasuhiro Nakasone—35 activists gather one night, all but one or two over 60. They're organized block by block, and a sign exhorts them to "steamroller" their neighbourhoods and trade guilds with get-out-the-vote efforts.

That kind of organization virtually assures Nakasone and the other LDP candidate of capturing the two seats open in Gumma Prefecture, which includes Kiryu. The economy is barely discussed; instead, the activists in Kiryu respond to ties of obligation. Many of the prefecture's legislators are on Nakasone's side, says Katsumi Okabe, who owns a dyeing factory. As a result, he tells the group, "We should

try to preserve their face by making absolutely sure to get more votes than the other camp."

The scene in Kiryu is repeated in LDP candidates' campaigns nationwide—and that's significant long after the election.

Take Hiroshi Kobayashi, 65, who runs a tiny sash factory with his wife in Kiryu. "We've never had a recession as bad as this," says Kobayashi. Yet he smiles through the bad times, saying he never expected to be rich, anyway. "As long as we have enough to eat, why be gloomy when you can be cheerful?" he asks.

Radical steps such as deregulation or encouraging corporate restructuring don't appeal to elderly voters since the benefits, as with U.S. deregulation in the late 1970s and 1980s, may take a decade or more to appear. "Older people don't think about what youth needs when voting," says Yumiko Ehara, an associate professor of sociology at Tokyo Metropolitan University. "They vote to protect their own interests and try to keep the status quo."

As Japan's deepening recession over the last year shows, this defensive mentality has spread to millions of middle-aged consumers—who already ranked among the world's biggest savers and have begun preparing for the future by saving more than ever. Household spending, which accounts for 60% of Japan's gross domestic product, was down in May compared with a year earlier and for the seventh month in a row.

On April 1, 1997, the government increased the national sales tax to 5% from 3%. And on September 1 it doubled the percentage of medical costs that

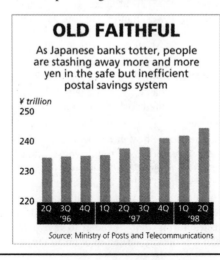

OLD FAITHFUL

As Japanese banks totter, people are stashing away more and more yen in the safe but inefficient postal savings system

¥ trillion

	2Q	3Q	4Q	1Q	2Q	3Q	4Q	1Q	2Q
		'96			'97			'98	

Source: Ministry of Posts and Telecommunications

salaried workers must pay out of their own pockets to 20%.

Even before 1997, some experts had warned that Japan would someday need a sales tax of 15%–20% to support its elderly population, and a government panel had warned in November 1996 that Japan's system of universal health care might collapse in the future. The modest steps of 1997 had the effect of persuading Japanese that these dire predictions were likely to come true.

Isamu Ueda, 39, a member of parliament's lower house from the opposition New Peace Party, says the biggest worry of his constituents in a Tokyo suburb isn't the current economic mess: "The largest concern is for the future . . . A lot of people are pessimistic that they won't get their full pension."

It's not an unfounded fear. Under current projections, 30%–35% of national income would have to go toward pensions, medical care and other welfare costs by 2025, up from around 19%. Low interest rates in recent years have worsened the financial position of the state pension system, which provides basic benefits for all. Many private companies, which supplement government payouts, also suffer from vastly underfunded pension plans. Electronics maker Hitachi said in April it will inject ¥180 billion ($1.3 billion) into its pension plan over the next five years to cover shortfalls, and it is also reducing benefits.

Cash-rich companies such as Hitachi can withstand such a blow, but workers at weaker companies fear they won't get the pension money that they're owed. That gives Japanese all the more reason to save as much as possible now just in case.

For their part, young people tend to tune out of the ageing debate altogether. That's not all bad: Spending by teenagers and people in their 20s represents one of the few supports for the economy. But their ambivalence slows political change.

Shio Yamazaki, the main opposition candidate in the district that includes Kiryu, says young people won't listen to politicians' speeches. "For older people, it's part of their life to vote. They don't ask questions. For the younger generation, voting isn't part of their lifestyle," says Yamazaki after a midday campaign speech outside a supermarket, during which not a single person stops to listen.

But Japan need not be doomed to a permanent recession. Consumer sentiment is notoriously fickle; a series of lucky breaks such as a recent heat wave, which has sent sales of air conditioners surging, might be enough to put Japanese in a buying mood. More deregulatory steps, such as allowing larger housing, could create new demand for durable goods.

Nevertheless, over the longer term, Japan needs leaders who can lay out a persuasive vision of the future, say commentators. With government encouragement, companies could raise their retirement ages and institute more flexible wage scales to make use of older workers' experience without hurting the bottom line. Also, raising consumption taxes and reducing the top income-tax rate from the current 65% would ease the relative burden of salaried workers and perhaps even increase total tax revenue by encouraging people to work more. Pensions could be scaled back for wealthier people without erasing the basic guarantees for a secure old age.

At the moment, however, far-sighted leadership is in short supply. LDP leaders tend instead to pander to older people on fixed incomes by calling for higher interest rates, a move economists say would be disastrous. "The public is ready to take in the real story and choose between the true alternatives," says Ueda, the opposition MP. "But politicians aren't bold enough to take the chance."

Article 14

Look Japan, June 1997

COVER STORY

STATE OF THE STAPLE

CHANGING CONDITIONS IN JAPAN'S RICE MARKETS

JAPANESE RICE CULTIVATION: A KEY TO UNDERSTANDING ASIA'S FUTURE

BY SHŌGENJI SHIN'ICHI

THERE are three points that must be made when considering the future of Japanese agriculture in general and rice production in particular.

The first is Japan's marked lack of self-sufficiency when it comes to food. The country is able to meet only 42% of its requirements (calorie-supply self-sufficiency rate; Ministry of Agriculture, Forestry and Fisheries, 1995).

When the Agricultural Basic Law was passed in 1961, Japan was producing 80% of its food (calorific) requirements. In the intervening years that rate has dropped by half. This is in sharp contrast to other advanced industrial countries, where the self-sufficiency rate has, in the main, been rising. The Ministry of Agriculture, Forestry and Fisheries has calculated the maximum number of calories Japan's agricultural resources could produce assuming minimum nutrition. It says at most 2,100 KCal per person per day (Japanese calorific intake today is an average 2,630 KCal per day). Food is the most necessary of necessities, literally a matter of life and death, and Japan is reaching a precipitous point. For Japan, food is a serious security issue, one that could impinge on its continued existence.

The second point to be considered is the renewed importance of rural, agricultural villages to Japanese society.

After World War II, Japan marshaled all of its resources towards catching up with the more advanced countries of Europe and North America. This was as true of agriculture as any other field, and it remained so from the establishment of the Agricultural Basic Law until very recently. Today, however, Japan has no more examples to mimic. We must create our country and our society from our own concepts. The current debate on agriculture and rural villages must, I think, be seen within this larger context. The "lifestyles" of rural villages must not be preserved merely as relics of the past, but in a living form as a valid choice for members of the society of the future. It is from that perspective that I believe rural villages and agriculture have extremely important roles to play.

The final point is that agriculture in the West is completely different in type and style from that in Japan.

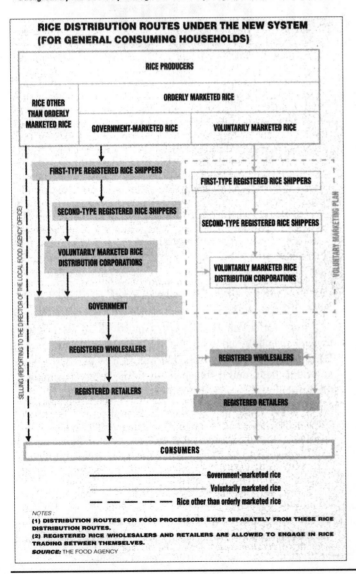

INTERNATIONAL COMPARISON OF MAJOR AGRICULTURAL INDICATORS

	JAPAN	GERMANY	FRANCE	U.K.	U.S.A.	SOUTH KOREA (ROK)	
DAILY PER CAPITA CALORIES (KCAL)	2,899	3,319	3,522	3,174	3,648	3,224	1994
TOTAL LAND AREA (MILLION HA)	37.65	34.93	55.01	24.16	915.9	9.87	1994
AGRICULTURAL LAND AREA (MILLION HA)	5.1	17.29	30.12	17.09	426.9	2.15	1994
FOREST AND WOODLAND AREA (MILLION HA)	25.1	10.7	15.0	2.5	296.0	6.46	1994
NUMBER OF FARMS OR FARM HOUSEHOLDS (THOUSAND)*	3,438	601	501	244	2,088	1,499	1995 (JAPAN, ROK) 1993 (GERMANY, FRANCE, U.K.) 1992 (U.S.A.)
AVERAGE SIZE OF AGRICULTURAL LAND PER FARM OR FARM HOUSEHOLD (HA)*	1.5	28.1	35.1	67.3	198.7	1.3	1992 (U.S.A.)

SOURCES: FOOD AND AGRICULTURE ORGANIZATION
MINISTRY OF AGRICULTURE, FORESTRY AND FISHERIES (MAFF)
EUROSTAT,STATISTICS IN FOCUS.
USDA.
KOREAN MINISTRY OF AGRICULTURE*

Background picture: Rice planting in Shirone City, Niigata Prefecture, early May

PHOTO BY OKI KATSUHITO, JIP PHOTO

RICE DISTRIBUTION ROUTES UNDER THE NEW SYSTEM (FOR GENERAL CONSUMING HOUSEHOLDS)

RICE PRODUCERS

RICE OTHER THAN ORDERLY MARKETED RICE

ORDERLY MARKETED RICE

GOVERNMENT-MARKETED RICE | VOLUNTARILY MARKETED RICE

SELLING (REPORTING TO THE DIRECTOR OF THE LOCAL FOOD AGENCY (OFFICE)

VOLUNTARY MARKETING PLAN

FIRST-TYPE REGISTERED RICE SHIPPERS

SECOND-TYPE REGISTERED RICE SHIPPERS

VOLUNTARILY MARKETED RICE DISTRIBUTION CORPORATIONS

GOVERNMENT

REGISTERED WHOLESALERS

REGISTERED RETAILERS

FIRST-TYPE REGISTERED RICE SHIPPERS

SECOND-TYPE REGISTERED RICE SHIPPERS

VOLUNTARILY MARKETED RICE DISTRIBUTION CORPORATIONS

REGISTERED WHOLESALERS

REGISTERED RETAILERS

CONSUMERS

———————— Government-marketed rice
———————— Voluntarily marketed rice
— — — — Rice other than orderly marketed rice

NOTES:
(1) DISTRIBUTION ROUTES FOR FOOD PROCESSORS EXIST SEPARATELY FROM THESE RICE DISTRIBUTION ROUTES.
(2) REGISTERED RICE WHOLESALERS AND RETAILERS ARE ALLOWED TO ENGAGE IN RICE TRADING BETWEEN THEMSELVES.
SOURCE: THE FOOD AGENCY

Agriculture in Western Europe basically involves fields, with livestock as an appendage. This entails several problems. Unless a certain amount of what is harvested is returned to the earth in the form of fertilizer, European agriculture becomes an exploitative process that robs the land of its nutrients and causes the topsoil itself to run off. The center of Japanese agriculture, however, is the paddy not the field. This is in some senses a recycling-oriented method, a peculiar use of resources that has enabled the same paddies to be used literally for 2,000 years in a row, something inconceivable in the world of fields.

Until South Korea recently joined their ranks, Japan was the only one of the industrialized countries to have a background in rice cultivation. Other rice producers will eventually join them—Thailand, the Philippines, Vietnam—and as that happens Japan will not be so much a peculiarity as merely a holder of a certain share of production on a redefined world stage. Understanding Japan is the key to foreseeing the future of Asia.

Since the 1980s there has been a growing recognition in Europe and North America that, far from being friendly to the environment, agricultural production is actually a burden on the environment. Certainly Japan shares many of these problems. Fertilizers and agrochemicals are made heavy use of in field crops, while paddies give off methane, but Japanese nonetheless see agriculture in general as contributing to the protection of our environment, including our water and soil resources.

On April 18, the Commission on Basic Problems in Food, Agriculture, and Rural Villages began an intensive and wide-ranging debate on the principles that should underlie agricultural policy in today's internationalized world. Certainly it is time for a general reevaluation of the agricultural framework that has been in place since the war. In sum, we must rethink and rework our divisions between those areas best left up to market mechanisms and those that will require a planned, government-led (local governments and even informal organizations at the village level are included in this) approach.

The new Staple Food Law (Law for Stabilization of Supply-Demand and Price of Staple Food), for example, can be seen as a laudable step in this direction—though it does contain provisions both for free production and sales on the one hand, and for supply and demand regulation and price stabilization on the other, systems which seem to be fundamentally in opposition to each other. Rice distribution has been liberalized to the point that rice can be sold when and where one wills, and the barriers to entry, particularly in downstream sectors, are extremely low. This represents a considerable expansion in the scope given to market mechanisms. Over the long term, I think that upstream areas ought to be liberalized in such a way that production too can run without the need for regulation. Our stock piles as well ought to be maintained in a systematic manner so that we are prepared for unseen occurrences, an idea that was never part of the former Food Control Law.

It is my hope and expectation that over the next year or two we will be able to have a frank, open, public debate on food issues.

It is senseless that 800 million people in the world should be starving while the most advanced countries try to hold back their food production. This is one of the things that makes one despair of market economies. They are equally inept at income distribution. As is commonly noted, when total world production is divided by population, there is no food shortage. The problem lies in unequal distribution of purchasing power. Correcting this is the basic solution to food issues. Starvation is therefore not a problem of food so much as income, and the question eventually boils down to how to increase the food production capacities and incomes of developing countries.

What Japanese need to think about are ways to prevent developing countries from becoming any more dependent on food aid (though emergency assistance is obviously excepted from this). Symbolic of this is the idea of "food for work" found in the World Food Program—assistance that improves supply capacity. This is something I think deserves more thought.

The author is a professor at the Graduate School of Agriculture and Life Sciences at the University of Tokyo, his alma mater (Faculty of Agriculture). His publications (joint ed.) include *Nochi no keizai bunseki, Nogyo keizaigaku* (Economic analysis of agricultural land, agricultural economics) and *Kokoro yutaka-nare: Nihon nogyo shinron* (Be rich at heart: A new theory of Japanese agriculture).

Article 15

The Christian Science Monitor, December 8, 1997

A Continental Divide: Who Owns Aboriginal Lands?

Australian leader wants to weaken claims of Aborigines— or call a racially tinged election.

By Lindsey Arkley
Special to The Christian Science Monitor

MELBOURNE, AUSTRALIA
WHAT if the US Supreme Court ruled that native Americans could lay claim to much of the United States?

In Australia, such a real-life drama involving Aborigines and descendants of white settlers has created a national political crisis.

Two high-court rulings since 1992 involving Aboriginal land have led to Prime Minister John Howard threatening to call national elections if the Senate fails to fix the Native Title Act and protect farmers, ranchers, and mining firms from the claims of Aborigines.

Prime Minister Howard says he is only trying to find a middle ground between Aborigines and present landholders. But opponents worry that he is speeding the nation toward an election that will be emotionally charged with racial issues.

With almost two-thirds of Australia's export earnings coming from farming and mining, the issue of who owns rural land—or at least has the right to use it—is important.

The debate has also set many rural Australians, worried about losing their land and livelihoods, against urban dwellers, many of whom would like to see the acrimonious debate over Aboriginal rights end amicably.

The Senate, where Howard's government lacks a majority, refused to pass the amendments last week. And in a rare session over the weekend, the House of Representatives formally rejected proposals by the Senate to change the amendments in favor of Aborigines.

Howard says he would prefer not to see an election fought over racial issues. But he will resubmit his amendments to the Senate in three months. If the Senate again refuses to pass them, the Constitution allows him to call an early election, possibly in mid-1998, instead of when it is required in 1999. Opposition leaders have urged Howard not to do so, saying the campaign would inevitably center on racial issues and prove divisive.

Howard's conservative coalition won a landslide victory at the polls in March 1996 after 13 years of Labor Party rule. But recent polls show the government trailing the Labor Party opposition right now.

Ruling enhances claims

The Native Title Act of 1993 was passed by the previous Labor Party government in response to a historic court ruling that, despite more than 200 years of settlement by Europeans, Aborigines still could claim land rights. The act set out a process for Aborigines to claim ownership if they could prove they had a continuous ancestral link to the land.

Mr. Howard's proposed changes follow another 1996 court ruling in what has become known as the Wik case, brought by Aboriginal people in the northeastern state of Queensland.

In that case, Australia's highest court found that Aborigines could claim title to vast areas of land held by farmers, ranchers, and mining companies under long-term "pastoral leases" granted by

state governments. About 42 percent of Australia's land falls under these pastoral leases.

Although the court ruled that in the event of irreconcilable differences, the current leaseholders would have primacy, the powerful "pastoral" and mining industries were in an uproar.

The Howard government responded by drawing up the amendments. In the words of Howard's deputy, Tim Fischer, the proposed amendments provide for "bucket-loads of extinguishment" to native title rights.

Charges fly

In the ensuing bitter debate, one Aboriginal leader has accused some government leaders of being "racist scum," while a government politician from a rural area urged churchgoers to boycott churches with leaders who sided with Aborigines in opposing the amendments.

Government ministers also raised the possibility that native title claims also could be made on land held under "freehold title," the usual form of land ownership in urban Australia, where two-thirds of the population lives.

This possibility was quickly denied by most legal experts and was not emphasized by the government, but it undoubtedly helped inflame passions.

The opposition Labor Party agrees with the need for some changes to its Native Title Act to create more certainty over land ownership. Labor Party leader Kim Beazley says changes proposed by his party would make the legislation more practical—and in keeping with the Constitution.

"What we in the opposition have been trying to do in the Parliament is to get right all these things that are wrong with the bill—simply to make sure that it's workable and certain in its operation, and strikes a fair balance," Mr. Beazley says.

"If we want reconciliation between black and white Australians, then the best possible thing we can do is get this bill right, so we don't have this divisive bill over and over again."

Aboriginal leaders remain unanimously and vehemently opposed to amending the legislation, saying the existing act provides the means for Aborigines and others to work together to make use of the land.

According to a recent newspaper poll, 52 percent of Australians fear the issue of Aboriginal land claims could divide the nation over race. The current row has occurred just as the prominence of controversial politician Pauline Hanson had begun to fade.

The prime minister was widely criticized for failing to denounce remarks by Ms. Hanson that were seen by many as racist and counter the damage her comments caused Australia's image overseas. Hanson shot to prominence last year when she said Australia was in danger of being swamped by Asians and criticized welfare programs for Aborigines.

Aborigines lag behind

By any number of measures, Aborigines are not faring well relative to other Australians. According to official statistics, the average life-span is 15 to 20 years below the national average of 79 years. Only about one-third of Aborigines complete high school, compared with 77 percent overall. And unemployment among Aborigines is four times higher than the national rate of about 9 percent.

"The indigenous unemployment rate could reach 47 percent by 2006 unless there is an unprecedented expansion in job creation," says John Taylor at the Center for Aboriginal Economic Policy Research at the Australian National University in the national capital, Canberra.

Article 16

Cultural Survival Quarterly, Winter 1998

Burma

Constructive Engagement in Cyberspace?

by Christina Fink

Christina Fink *is the Thailand Coordinator of the Open Society Insitute's Burma Project. She is a cultural anthropologist who has worked with upland and lowland groups in Thailand and Burma.*

As one of the few countries in the world still lacking direct Internet access, Burma is a place where propaganda and rumors abound and hard facts remain elusive. The ruling military junta which renamed the country 'Myanmar,' is bent on silencing democracy activists and subjugating autonomy-minded ethnic minority groups. With non-Burmans comprising almost half the population of Burma, ethnic minority political organizations have demanded a federal state structure or outright independence. Yet they have no chance to express their views publicly in Burma as all forms of media are controlled by the military. Burman and ethnic minority exiles abroad, however, have been able to use the Internet to expose the military's abuses, promote their concerns and demands, and network with other concerned organizations.

Background History of Burma

Some areas of present-day Burma were never under Burman rule and only became a part of Burma when the British drew fixed boundaries around the country a century ago. Since that period, 'Burman' has been used to refer to the ethnic group, and 'Burmese' to all citizens of Burma. During the democratic period, from 1948 to 1962, ethnic armies formed and fought for greater autonomy. Burmese military leaders then used the excuse of instability in the country to justify a military coup and the imposition of martial law.

After General Ne Win overthrew the elected government in 1962, he instituted a repressive system of control and sought to eliminate the ethnic political organizations and their armies, rather than recognize their demands. Gradually removing most members of ethnic minorities from positions of power within the central government, he also sought to 'Burmanize' the population. The military regime indoctrinated the people with the concept that federalism was tantamount to anarchy. Moreover, because of strict censorship laws, ethnic groups were not allowed to publish their political or historical works in Burmese or their own languages. The gap in understanding between the majority Burmans and the ethnic minorities grew, as the Burmans knew little about what the minority groups were really fighting for. While members of ethnic minority groups living in central Burma became absorbed into Burmese culture, in the border areas, nationalist ethnic organizations continued to fight for their territory and the preservation of their cultures.

In 1988, people of all classes, professions, and ethnic groups participated in massive anti-government protests. The military responded by shooting hundreds of protesters, and a new military clique which called itself the State Law and Order Restoration Council (SLORC), took power. The SLORC scheduled elections for May 1990 and numerous new political organizations were formed that included ethnic minority parties. However, even before the election was held, the SLORC began arresting the most active and talented political challengers, including Aung San Suu Kyi, the daughter of Burma's independence hero, General Aung San. When the SLORC's party was overwhelmingly defeated by Aung San Suu Kyi's party, the National League for Democracy, the SLORC refused to hand over power.

The military junta released Aung San Suu Kyi from house arrest in 1995 and through force or persuasion, the SLORC has managed to achieve cease-fires with 15 armed organizations. Still, the SLORC refuses to step down, asserting that only the military can ensure national unity and solidarity.

The SLORC Says 'No' to the Net

The SLORC maintains its power not only through military might, but also by controlling access to information. Although the junta recognizes the importance of communications technology for increasing foreign investment and trade, their fear that the people will rise up again has led them to impose severe restrictions on most forms of communication. Newspapers, TV, and local radio are all controlled by the government. Telephones are tapped, mail is opened, and all fax machines must be registered with the government.

Access to e-mail and the Internet could bring in uncensored information and the chance to network with pro-democracy activists outside the country. To prevent this, in September 1996, the junta decreed that anyone found in possession of an unregistered fax or modem could be imprisoned for up to 15

Situation in Arakan

To: MAYKHA-L@LISTSERV.INDIANA.EDU

Arakan-Burma's back water northwestern province remained as neglected as before, undeveloped and unaffected by the present superficial economic openness witnessed by other parts of Burma. Foreigners are not all allowed to visit Arakan freely except to certain tourist points in guided tours. Northern Arakan with Muslim (Rohingya) majority population remains always tense and volatile. SLORC's policy of genocide and ethnic cleansing against Rohingyas Muslims continues with more ferocity. The Muslims in Arakan have been passing their days in sub-human condition in utter misery, fear and extreme poverty. Ninety five percent of the Muslim population are under-fed and in an environment of continuous fear and having no access to health facilities make them physically crippled. Because of poverty, lack of educational institutions in Muslim villages and a discouraging political and social environment the number of educated Muslims is sharply falling.

Except cultivation and fishing, Muslims have no other employment as the government prohibits to employ Muslims in any governmental service. The Rohingyas have been declared as foreign residents by the SLORC. Whatever the farmers produce from their land, most of it has to be given to the government in the name of tax imposed only on the Muslims. If one fails to surrender the fixed quantity, he is either jailed or his entire land is confiscated. Added to this, the poor cultivators who maintain their living as manual day labourers are often rounded up by the security forces to engage them as slave labourers without payment for a long time making them hand to mouth or to face complete starvation.

years. At the time of writing, the SLORC has not permitted direct Internet service in Burma. There are a few e-mail service providers which call to systems outside the country to upload and download mail once a day, however, only foreign nongovernmental organizations (NGOs), international businesses, and selected members of the military regime have access to these servers. The rest of the people of Burma can only read about e-mail and the Internet in local computer magazines.

Use of the Internet by Exile Groups

Although people inside Burma cannot use the Internet, Burmese in exile have been quick to take advantage of e-mail and the Internet, both to distribute information in a timely fashion and to organize resistance activities. The SLORC has realized that while they can largely control what information comes into Burma, they cannot control what is being said outside the country.

Exiles who immigrated to Western countries and Japan in the late 1980s or early 1990s, found themselves immersed in a computer culture. Those who went back to school quickly picked up how to use e-mail from free classes offered at their universities or informal teaching by peers. In Thailand, e-mail access has been available, although quite expensive, since the early 1990s. But because few Thais were using e-mail, particularly in the border areas, members of the exile community were not exposed to the Internet. Even if they had heard of the Internet, they had only vague ideas of what it was all about.

When the staff of the Burma Project, a nonprofit organization dedicated to increasing international awareness of conditions in Burma, tried to introduce the idea of e-mail and the Internet to ethnic organizations with offices in Thailand, there was little interest at first. However, after seeing how e-mail worked and being urged by colleagues in Japan and the West to get 'on-line,' some young Burmese, Karen, and Mon computer experts asked for help getting e-mail accounts. When

other young exiles saw their peers using the Internet, they asked them what they were doing. Soon, everyone wanted an account.

The Burma Project and other foreign NGOs have provided funding for Burmese political organizations to purchase computers, modems, and e-mail accounts. At first, learning how to use e-mail was done informally by members of Western NGOs and foreigners volunteering or doing research along the border. In the past two years, however, teaching about e-mail and the Internet has shifted to members of the exile groups, who not only instruct other members of their own organizations, but also individuals from other organizations.

In particular, members of Green November 32, an indigenous organization which focuses on human rights and environmental issues, have traveled the border holding month-long computer classes that include an introduction to e-mail and the web. They also do private consultations, helping people upgrade their skills and solve computer-related problems. A self-taught Karen, who learned mostly by studying computer books and magazines, has provided similar assistance for Karens and Burmese living in his border town.

Meanwhile, Mon professionals and students who were raised in Thailand have worked with Mon from Burma to show them how to browse the web and access Mon home-pages. One young Mon from Burma, now living in Thailand, has gone on to develop 17 Mon fonts. While it took him two months to perfect the first one, he can now develop a new font in two days. The fonts are used for Mon language publications produced outside Burma and sent into Mon state. In the future, they will also be used on the Internet.

Ethnic minority groups based in the border areas of Burma and abroad use e-mail and the Internet for three primary purposes: first to distribute news about abuses in their areas, to network with others, and to educate outsiders about their histories, cultures, and political demands. Exile offices of armed resistance groups such as the Karen National Union, the

Karenni National Progressive Party, and the Shan United Revolutionary Army regularly post news on listservs and the Internet about human rights abuses committed by SLORC soldiers in their territories. Written reports are sent from the border areas to offices outside Burma, where they are translated into English and e-mailed around the world. This news is essential because it is virtually impossible for journalists to travel to sensitive areas in Burma, particularly the ethnic minority homelands in the border regions.

A good example of this flow of information is the mass executions of Shan villagers by SLORC soldiers which took place in Shan state in June and July 1997. The SLORC is attempting to wipe out Shan communities in order to cut off supplies and information to the Shan resistance groups. Few people in Burma or outside the country knew about these atrocities until the Shan Human Rights Foundation was able to post reports based on eyewitness accounts on the 'burmanet' listserver and activist Burma web pages.

Established in 1993, the burmanet listserver has become a popular forum for posting news by Burman and ethnic minority groups. With a subscriber list of almost 1,000 that includes foreign embassies, government officials, NGOs, scholars, activists, reporters, and the various political organizations, burmanet provides a place for the often underfinanced and understaffed ethnic organizations to distribute the latest news quickly and cheaply.

Moreover, news from the border areas is often picked up from the Internet, translated into Burmese, and then broadcast back into Burma by radio stations based outside the country such as the Voice of America, British Broadcasting Corporation, Radio Free Asia, and the Democratic Voice of Burma. Thus, the Internet serves as a critical, if indirect link in channeling information from remote areas of Burma back into the rest of the country.

Ethnic minority groups and Burmese pro-democracy groups also use the Internet for networking and international campaigns. Activist groups based in other countries rely on the resistance groups to provide them with documentation and photographs of human rights abuses within Burma to use in

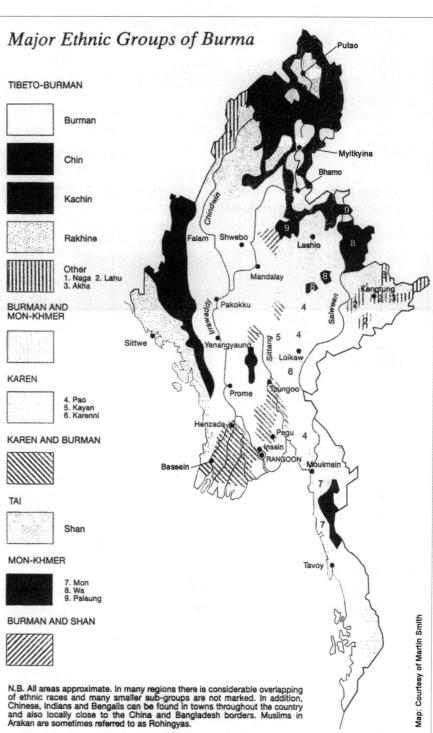

Major Ethnic Groups of Burma

TIBETO-BURMAN

Burman

Chin

Kachin

Rakhine

Other
1. Naga 2. Lahu
3. Akha

BURMAN AND MON-KHMER

KAREN
4. Pao
5. Kayan
6. Karenni

KAREN AND BURMAN

TAI

Shan

MON-KHMER
7. Mon
8. Wa
9. Palaung

BURMAN AND SHAN

N.B. All areas approximate. In many regions there is considerable overlapping of ethnic races and many smaller sub-groups are not marked. In addition, Chinese, Indians and Bengalis can be found in towns throughout the country and also locally close to the China and Bangladesh borders. Muslims in Arakan are sometimes referred to as Rohingyas.

Map: Courtesy of Martin Smith

their campaigns for sanctions, selective purchasing laws, and tourism boycotts. Some exiled Mon and Karen have set up their own homepages with photographs and information about the SLORC's use of forced labor and forced relocation campaigns in their areas, as well as about the plight of refugees in camps along the Thai border. These homepages also contain

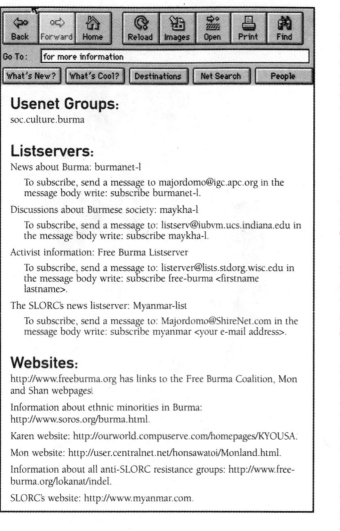

above). Communal tensions between two ethnic groups in Arakan State, the Rohingya Muslims and Arakanese Buddhists, have been exacerbated by the SLORC and thousands of Rohingyas have fled to Bangladesh. By increasing Burmese peoples' knowledge about the two groups, there is a better chance of being able to peacefully reconcile the tensions in the future.

Conclusion

In the mid 1990s, Burmese activists started two listservers: 'free-burma' and 'maykha.' Another listserver, 'burmanet-l,' features newspaper and magazine articles, as well as updates from exile groups based in the border areas around Burma. The SLORC responded in 1997 by creating its own listserver, 'Myanmar-list.' The SLORC posts 'information sheets' and news articles from the state-controlled paper, the *New Light of Myanmar,* on its own listserver, as well as the other listservers which are clearly anti-SLORC. Activists often respond critically to the SLORC postings and SLORC personnel occasionally answer back.

In the past four years, Burmese activists have assembled numerous homepages with news from Burma, information about rallies and boycotts, liberation songs, and speeches by Aung San Suu Kyi. In 1996, the SLORC developed its own webpage, 'Myanmar.com,' which features tourist and cultural information, as well as diatribes against Aung San Suu Kyi and the foreigners who have supported the movement.

The Internet has given members of the ethnic minority groups of Burma an opportunity to explain their cultures and their demands for minority rights, both to Burmans and to the international community. Several of the ethnic minority groups from Burma, have realized the potential of the Internet and have been empowered by the new-found opportunity to voice their concerns publicly and freely.

Although the SLORC has still refused to engage either the pro-democracy, or ethnic minority resistance groups in a genuine political dialogue inside the country, a dialogue of sorts is taking place on the Internet. The SLORC feels compelled to respond to news stories and accusations posted on the listservers, and at least some personnel are regularly reading the statements and arguments of the resistance organizations posted on listservers and websites. The openness and freedom which characterizes the Internet is in direct contrast with the political climate in Burma. Hopefully, as the SLORC and the opposition groups continue to interact indirectly through the Internet, a greater willingness to engage in dialogue within the country will develop.

References

Lintner, Bertil. 1994. Burma in Revolt: Opium and Insurgency Since 1948. *Boulder: Westview Press.*

Smith, Martin. 1995. Censorship Prevails: Political Deadlock and Economic Transition in Burma. *London: Article 19.*

links to Free Burma sites with information about how to participate in international campaigns against the SLORC and where to send letters of appeal for the protection of refugees.

The Free Burma Coalition, which includes ethnic Burmans, ethnic minorities, and foreigners from around the world, has also extensively relied on the Internet to develop a network of college and community groups interested in Burma. Because Internet use is comparatively cheap and fast, it has provided the ideal space for mobilizing people who are geographically separated by thousands of miles.

Members of the ethnic minority groups have also used e-mail and websites to distribute information about their histories, cultures, and political demands. For instance, the Mon homepage (see box for their web address) includes photographs from Mon National Day, Mon dancers, a brief history of the Mon, Mon linguistic information, and press releases from the New Mon State Party, the political wing of the Mon resistance army.

One Rohingya exile regularly posts background information about ethnic groups in Arakan State (also known as Rakhine) on an exile Burmese community listserver, 'maykha-l' (see

Article 17

Parameters, Winter 1999–2000

Living with the Colossus: How Southeast Asian Countries Cope with China

IAN JAMES STOREY

Ian James Storey is a Ph.D. candidate at the City University of Hong Kong. His research interests include international relations and security issues in the Asia-Pacific region. He is currently writing his doctoral thesis on ASEAN threat perceptions of the PRC, and he has recently published articles in Contemporary Southeast Asia and Jane's Intelligence Review.

How the United States manages its future relationship with the People's Republic of China (PRC) seems destined to become Washington's greatest foreign policy challenge in the 21st century. As the PRC's national power grows, it can be perceived as both an economic opportunity and a strategic threat. The United States must fashion a set of policies that takes advantage of the former while protecting its strategic interests in the Asia-Pacific region. China's neighbors are faced with the same dilemma. This article examines how the ten members of the Association of South East Asian Nations (ASEAN),[1] which includes some of America's closest allies in Asia, plan to cope with the rise of China, and how many of them look to the United States to balance the growing military power of the PRC.

The ASEAN states not only reject the more hawkish assessments of "China threat" proponents, but also the arguments of those who seek to paint the PRC as a totally benign power. Instead,

ASEAN takes a more evenhanded approach toward the rise of China, acknowledging that there are both benefits and costs for Southeast Asia. On the plus side, the Association recognizes that the PRC's decision to give priority to economic modernization beginning in 1978 necessitated changes in the conduct of Chinese foreign policy. There was a substantial downgrading of the importance assigned to communist ideology in favor of a more pragmatic modus operandi that emphasized cooperative economic relations with the West and China's neighbors. ASEAN has supported China's opening to the outside world, not only because it enhances the internal stability of the PRC, but also because it creates valuable economic opportunities for ASEAN's members.

However, the ASEAN states are keenly aware that the rise of China is not without actual and potentially negative consequences. Though the word "threat" is seldom voiced, ASEAN officials are apt to highlight a number of "concerns," "problems," or "challenges" when discussing the strategic dimension of ASEAN-China relations. Some of these concerns stem from Southeast Asia's geographic proximity to the PRC, and the huge imbalance between the Association and its northern neighbor in terms of population and territorial size. Moreover, the growing economic and military power of the PRC has served to reinforce historical fears of China as

a hegemonic power. Among these fears are concerns over how China might behave as a great power in the future, where its defense modernization program is leading, and how the PRC intends to resolve its territorial disputes in the South China Sea.

In fact, ASEAN threat perceptions of the PRC differ quite widely among the member states. As a result, the Association has never formulated a corporate policy on China and seems unlikely to do so in the future. However, consensus does exist that engagement is the only realistic policy for ASEAN to pursue. Engagement requires the Association to develop and deepen economic and political linkages with China, thereby weaving the PRC into a complex web of interdependence. ASEAN leaders hope that by engaging China in a security dialogue at both bilateral and multilateral levels, their security concerns vis--vis their elephantine neighbor can be substantially mitigated. However, the ASEAN states also take a realist's approach to the growth of Chinese power, stressing that engagement must have a military-security dimension. Cognizant of their own limited military capabilities to balance China, the five founding members of the Association plus Brunei (the ASEAN-6) recognize the continued need to maintain defense links with external powers—primarily the United States, but also the United Kingdom, Australia, and New Zealand—as a pru-

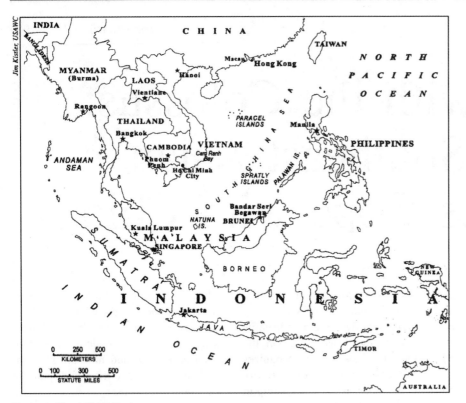

Figure 1. The ASEAN Region.

dent hedge against a more assertive or aggressive China. Though usually played down for fear of antagonizing Beijing, this strategy is referred to by the ASEAN-6 (Indonesia, Malaysia, the Philippines, Singapore, Thailand—and Brunei) as "engagement with insurance."[2]

ASEAN Perceptions of the PRC as a Threat

From among the five major Asia-Pacific powers (the United States, Japan, Russia, China, and India), ASEAN's relations with the PRC are the most problematic. China's sheer size, population, and proximity make it loom large in the geopolitical considerations of all ASEAN states. Stretching back thousands of years, China and the Southeast Asian states have interacted in every conceivable field. Although that history is rich in reciprocal cultural borrowings and mutually beneficial trade, knowledgeable Southeast Asians recall the suzerainty system whereby countries of the region were required to acknowledge China's preeminent position and pay

tribute. The arrival of the Western colonial powers in the mid-19th century terminated this relationship. However, after the establishment of the PRC in 1949, and particularly during the Cultural Revolution of the 1960s, China sought to increase its influence in Southeast Asia by providing both material and moral support to communist insurgency movements whose goal was the violent overthrow of national governments. The leaderships of these insurgency movements were often dominated by ethnic Chinese, reinforcing the notion that Beijing was interfering in the internal politics of the ASEAN states. In the mid-1970s, China's support for the communist insurgents was gradually scaled back as Beijing recognized it needed to win the friendship of ASEAN, both in its efforts to contain the USSR and as a source of investment for furthering economic modernization. Today, no one in Southeast Asia seriously believes that China would again try to promote the export of revolution, largely because communist ideology plays much less of a role in Chinese foreign policy. How-

ever, as will be examined later, concerns still exist that China might try to influence ASEAN affairs through links with the ethnic Chinese.

Southeast Asia is a region of great national diversity, and it is hardly surprising that threat perceptions of the PRC differ widely among the ASEAN states. These threat perceptions result from a combination of factors, including assessments of capabilities and intentions, the geopolitical environment, historical relations, and domestic factors such as leadership perceptions, the role of the bureaucracy and armed forces, and the status of the ethnic Chinese.

At the high end of the threat perception scale lie Vietnam, Indonesia, and the Philippines. Historically Vietnam was subjected to repeated occupations by China, and since 1945 Vietnam has fought hard to maintain its independence from external powers (including Japan, France, the United States, and China). Since 1974 China and Vietnam have clashed militarily on three occasions: in 1974 Chinese troops ejected South Vietnamese forces from the Paracel Islands in the South China Sea; in 1979 the Chinese army launched a punitive attack into northern Vietnam in response to Hanoi's invasion of Cambodia; and in 1988 Chinese and Vietnamese naval forces clashed in the South China Sea over the disputed Spratly Islands. Since then, diplomatic relations have been normalized, but tensions over the South China Sea remain high.

Indonesia's relations with China have been marked by hostility and suspicion ever since the PRC was implicated in supporting the Indonesian Communist Party in an abortive coup in September 1965. Though no evidence was ever produced proving China's complicity in the coup, Sino-Indonesian relations remained ruptured until 1990 when diplomatic relations were finally restored. Poor bilateral relations are partly founded on the ethnic Chinese problem in Indonesia. Indonesian Chinese have not been well assimilated into society and have been the target of repeated persecutions. The government, moreover, questions the loyalty of the Indonesian Chinese and considers them potential fifth columnists.

Philippine threat perceptions of the PRC were moderately high during the first few decades of the Cold War, but dissipated with the Sino-US rapprochement of the early 1970s. However, relations deteriorated rapidly after 1995 when Chinese military facilities were discovered on Philippine-claimed Mischief Reef, close to Palawan Island in the South China Sea.[3] Since 1995 these structures have been upgraded into permanent facilities, and the Philippines now identifies its dispute with China as one of its two most urgent national security problems.[4]

Malaysian and Singaporean threat perceptions today can be characterized as moderate. During the Cold War, Malaysia perceived the PRC to be a threat to the country's fragile postwar independence because of Chinese support for the Malaysian Communist Party (MCP). In 1974 Malaysia established diplomatic relations with China in the hope that Beijing would cut its links with regional insurgents in favor of better state-to-state relations. However, Chinese support for the MCP continued to be a major irritant in bilateral relations until the early 1980s. Malaysia's current prime minister, Mahathir Mohamad, is one of ASEAN's most ardent advocates of engagement with China, and on occasions has sought to rebut any notions of a China threat to the region.[5] However, Malaysia is in direct contention with the PRC over the Spratlys and now occupies five of the islands.

Since Singapore established diplomatic relations with the PRC in 1990, two-way trade has flourished, and the city-state has become a major investor in China. In terms of security, however, Singapore maintains a prudently alert monitorship of the rise of Chinese power. Though not a party to the South China Sea dispute, Singapore's status as a major international transshipment center requires it to place maximum priority on freedom of navigation. Accordingly, Singapore has consistently voiced sup-

port for the continued presence of the US Navy in the region, and has facilitated this presence by offering port facilities to visiting American naval vessels. However, of greater concern to Singapore is the possibility of conflict erupting in the Taiwan Straits. Singapore has significant investments on both sides of the straits, and fears the outbreak of hostilities would severely damage regional stability and hence the prospects of foreign investment in the region, on which its continued prosperity depends.

Thailand and Burma occupy the low end of the threat perception spectrum, though for different reasons. During the first few decades of the Cold War, Thai military governments were fiercely anticommunist (and hence hostile to the PRC) and supported US involvement in the Vietnam War. After the fall of Saigon in 1975, Bangkok's primary external threat perception became focused on Vietnam. Hanoi's 1978 invasion of Cambodia heightened this sense of threat, and Thailand turned to the PRC as a counterpoise. Bangkok was grateful for the military pressure that China applied to Vietnam throughout the Cambodian crisis. Since that time, Thailand's relations with China have been cordial. The two countries have no territorial disputes, and Thailand is one of the largest foreign investors in China. Friendly relations have been facilitated by the absence of an ethnic Chinese problem, as the overseas Chinese have become almost completely assimilated into Thai society.

Since 1988 the interests of the Burmese and Chinese governments have coincided in terms of regime survival. In 1988 Burmese government forces killed hundreds of pro-democracy activists in Rangoon. In June 1989, the Chinese government dealt with protesters in Tiananmen Square using similar methods. Rangoon and Beijing have moved closer together since 1988-89, identifying Western-backed "peaceful evolution" as one of the primary threats to their regime survival. Burma has provided the PRC with diplomatic support, while Beijing has provided Rangoon

with huge amounts of military aid to prop up the regime.

Despite the eclectic nature of bilateral relations, it is possible to identify a number of threats associated with the rise of China that are perceived, to a greater or lesser extent, by ASEAN members. Using Klaus Knorr's classification, we can divide these threats into two main groups, actual and potential.[6]

Actual Threats

The most seriously perceived actual threat from the PRC as seen by ASEAN states is the unresolved territorial disputes in the South China Sea. Although Chinese foreign policy has become much more pragmatic since the introduction of economic reforms in 1978, the PRC shows no inclination to compromise its claims to sovereignty over areas it regards as historically part of Chinese territory. Foremost among these areas are Hong Kong, Macau, and Taiwan. The PRC also claims sovereignty over two major island groups in the South China Sea (the Paracel Islands and the Spratly Islands) and one small group (the Diaoyutai/Senkaku Islands) in the East China Sea (disputed by the PRC and Japan).

The deepest thorn in the side of Sino-ASEAN relations is centered on the Spratly Islands. The Spratly archipelago, located in the South China Sea, is made up of more than 300 rock formations of varying sizes. Sovereignty over the islands is disputed by six countries; the PRC, Taiwan, Vietnam, Malaysia, the Philippines, and Brunei. The PRC, Taiwan, and Vietnam claim the entire chain, while Malaysia, the Philippines, and Brunei claim only part. Sovereignty is contested for two reasons: first, the group of islands occupies an important strategic position, straddling vital commercial sea-lanes linking the Indian and Pacific oceans through which much of the world's trade passes; and second, the area encompasses valuable fishing grounds and is believed to be rich in oil and gas. China, Taiwan, and Vietnam claim sovereignty over the islands on the grounds of discovery and occupation going back several millennia. Malaysia bases its claims on the continental shelf principle, while the Philippines claims sovereignty over part of the group by virtue of proximity. All the claimants except Brunei have occupied islands in the Spratly chain and stationed troops there. The close proximity of the islands and the continued tension between the disputants make the Spratlys one of Asia's most dangerous flashpoints.

Since the early 1990s, the ASEAN states have become concerned at what they see as China's increasingly assertive behavior in the South China Sea. In February 1992 China's National People's Congress passed the Territorial Law of the Sea by which the PRC claimed sovereignty over almost the entire South China Sea. According to this law, Beijing has the right to forbid passage of foreign warships through the area. In response to this law, in July 1992 Association foreign ministers issued the ASEAN Declaration on the South China Sea which urged all claimants to freeze the status quo, shelve the sovereignty dispute, and jointly develop maritime resources. China agreed in principle to the declaration. However, in July 1994 Chinese troops occupied Da Lac Reef near Vietnam. In the same month, the Chinese government published a map of the South China Sea indicating that part of Indonesia's rich Natuna gas field was under Chinese sovereignty. Jakarta asked Beijing to clarify the claims, but it was not until July 1995 that Chinese Foreign Minister Qian Qichen assured Indonesia that the PRC would abide by international law to resolve the overlapping claims.[7]

Although no Sino-Vietnamese military clashes have taken place since 1988, tension between the two countries over the Spratlys remains high. In March 1997 China moved an oil exploration rig into waters claimed by Vietnam, resulting in a tense standoff between the two countries.[8] However, by far the most discomforting incident for ASEAN was the discovery in January 1995 that the Chinese had occupied Philippine-claimed Mischief Reef. ASEAN as a group felt that such actions had negative consequences for regional stability, and indirectly rebuked China, urging the PRC to abide by the 1992 ASEAN Declaration.[9] China and the Philippines subsequently signed a code of conduct that aimed to ease tensions, freeze all further construction activities, and increase maritime cooperation. However, the Philippines has accused the PRC of violating the code on a number of occasions. The most serious infringement of the code occurred in November 1998 when the PRC upgraded its facilities on Mischief Reef to include a permanent brick fortress.

The ASEAN states have been frustrated by what they see as China's duplicitous policy in the South China Sea. On the one hand, China stresses that it wishes to resolve disputes peacefully, abide by the conventions contained in the 1982 United Nations Law of the Sea, and jointly develop any resources with the other claimants. At the same time, China continues to increase its presence in and around the reefs, laying down territorial markers, seizing unoccupied reefs, and building permanent structures. This policy has been variously referred to as "creeping assertiveness," "creeping sovereignty," and more recently by the Philippine Defense Secretary as "talk and take."[10]

Another threat perceived as actual by some members of the Association is the Burmese government's increasingly close relationship with the PRC. As noted earlier, the two governments have moved closer since the crackdowns of 1988-89 against pro-democracy elements. Since then China has bolstered the Burmese regime by becoming its largest supplier of military equipment, valued at more than $1.5 billion (US).[11] There have also been reports that China has established a signals intelligence listening post on Burma's Coco Island in the Andaman Sea near the Indian Ocean,[12] though Burmese foreign ministry officials categorically deny there are any PRC service personnel stationed in the country.[13] Beijing's growing influence in Burma puts Southeast Asia in an uncomfortable strategic position with respect to India, which distrusts China's regional intentions. Some observers have suggested that the ASEAN states' willingness to induct Burma into the organization sprang from a desire to wean Rangoon away from Beijing, though opinion on this issue is split.[14]

Potential Threats

In 1978 the Chinese leadership embarked on a program of economic reform designed to strengthen this vital pillar of national power. The reform program has resulted in phenomenal economic growth, transforming the social fabric of China and boosting the PRC's international stature. ASEAN has been a beneficiary of China's economic modernization, as two-way trade and investment have flourished between the PRC and members of the Association. However, as China's economic base has expanded, more resources have been made available to the People's Liberation Army (PLA), enabling it to embark on an ambitious defense modernization program. Strategic analysts in ASEAN worry about the implications of this arms buildup for regional stability. Moreover, as China's political influence grows, some observers worry that Beijing may seek to reestablish itself as Asia's hegemonic power, exerting pressure on the ASEAN states and interfering in their internal affairs.

Although military modernization was given the lowest priority by the Chinese leadership, since 1989 defense spending has increased annually at a double-digit pace and now stands at $12.6 billion (US).[15] However, some experts have argued that the true level of defense spending is many times this amount, as the official budget excludes associated expenditures such as those for research and development, nuclear weapons, and arms purchases from foreign countries, as well as income generated from the PLA's commercial activities, including weapon sales abroad.[16] Although estimates vary considerably as to how much

China spends on defense, a conservative estimate puts PRC defense spending at two or three times the stated amount, say, $30 billion (US). ASEAN recognizes that defense modernization is the prerogative of every independent state, and since the early 1980s has itself embarked on such programs. However, the pace and nature of the PRC's defense buildup have caused concern. Although China spends much less on defense than the United States, Japan, or Russia, the amount is more than ASEAN as a group spends. If current trends continue, and the Chinese economy grows by 8-10 percent annually, within two or three decades the PRC will have Asia's largest defense budget. ASEAN strategists also have expressed unease at the nature of China's arms buildup. In the early 1990s the PLA began to augment its power-projection capabilities through the purchase of modern fighter aircraft, submarines, transport aircraft, missiles, radars, and in-flight refueling technology, much of it from Russia. The air force and navy have been given priority under the defense modernization program, allowing the PLA to expand its presence in the South China Sea and make its military threats toward Taiwan more credible. ASEAN strategists worry that China may employ its armed forces to settle outstanding territorial disputes in its favor, either directly or by threatening the other disputants.

Knowledgeable Southeast Asians thus harbor concerns about the future relationship between an economically and militarily strong China and the ASEAN states. These concerns center on the notion that a strong China might seek to reestablish itself as Asia's regional hegemon, a position it occupied more than 500 years ago. A hegemonic China would see itself as primus inter pares, and might not treat Southeast Asian countries as equal sovereign states. A hegemonic China might seek to resolve the South China Sea dispute on its terms, denying rich natural resources to the ASEAN states and exercising control over economically vital sea-lanes of communication. ASEAN states with sig-

nificant overseas Chinese minorities also worry about the effect of a strong China on the ethnic balance. A fear exists that the Chinese minorities might somehow be empowered by a strong China, calling on Beijing to intervene on their behalf in times of ethnic tension. Although these are long-term concerns, China's actions today are increasingly being interpreted as precursors for how it might act in the future when the PRC's national power has been enhanced by decades of economic growth. Thus China's assertive actions in the South China Sea and Taiwan Straits are actual threats which tend to reinforce potential threats. Singapore's former Prime Minister, Lee Kwan Yew, has drawn attention to China's power today, and how powerful it might be in the future:

> As Napoleon said, "If China is a sleeping giant, don't wake it up." Now it has awakened, but it [has] not got going yet, it is just, you know, doing morning exercises, a bit of qigong or taijiquan. You reach noontime, when it is completely at the peak of its power, that's a big problem.[17]

Mitigating ASEAN Threat Perceptions

Containment vs. Engagement

A number of strategies have been suggested as to how countries should deal with a rising China. One view holds that the United States and its Asian allies should pursue a policy of containment toward the PRC, restricting China's economic growth by cutting off access to foreign aid and markets, thereby preventing the PRC from achieving the status of a regional superpower.[18] However, among the ASEAN states there is no support whatever for such a policy. Containment is seen as dangerously counterproductive to regional security, as it would push China into a corner, thus reinforcing the position of conservative hard-liners within the Chinese leadership. Hard-line factions would inevitably advocate a more aggressive policy in resolving China's territorial disputes and a more assertive posture

vis-à-vis the United States and Japan. Rather than see this happen, ASEAN governments would like to see China brought out of its corner and made to feel at ease in the international community.

The continued economic development of China is very much in the interests of Southeast Asia, not only because it creates valuable opportunities for the ASEAN states, but also because economic stagnation in the PRC could lead to rising instability or even the territorial fragmentation of China itself. Either scenario could result in a mass exodus from mainland China, with millions of Chinese fleeing by sea to seek refuge in Southeast Asia. Such a large influx of people would impose an enormous strain on the ASEAN states' limited resources, as well as upsetting the delicate ethnic balance. Former Philippines national security adviser José Almonte commented in this regard:

> Among China's neighbors, we in Southeast Asia obviously have the greatest interest in [China's] peaceful transition to economic and political pluralism, if only because the fallout from any political instability in the mainland will be heaviest in this direction.[19]

Though seldom articulated, fears of a weak and disunited China are every bit as great as those of the PRC as a regional hegemon.

The ASEAN states are in agreement that engagement with China is the best policy to pursue. Engagement is seen as beneficial to Southeast Asia in a number of ways. First, the ASEAN states are eager to increase two-way trade and investment with China. China's purchase of goods, services, and raw materials from the ASEAN states, together with foreign investment from the PRC, benefits their domestic economies. ASEAN investment in China, and the consumption of Chinese goods by Southeast Asians, help economic growth to continue in the PRC, thus helping to maintain internal stability. Moreover, by developing and deepening an interdependent relationship with China, the ASEAN states believe the costs to China of any future conflict between it and Southeast Asia, say over the South China Sea, would far outweigh the expected benefits. ASEAN hopes to lock the PRC into a mutually beneficial trade relationship, the fracturing of which would be damaging to China's cherished goal of economic modernization.

Second, engagement enables ASEAN to conduct an ongoing security dialogue with China through which members hope to mitigate and eventually resolve their security concerns. This security dialogue takes place on two levels—the official level, known as Track I, and the unofficial level, known as Track II.

The most important Track I mechanism is the ASEAN Regional Forum (ARF). The ARF, which brings together 21 Asia-Pacific countries, represents a nascent regional security structure. The forum was initiated by ASEAN in July 1993 in response to an uncertain strategic environment brought about by the end of the Cold War. It is designed to tackle security issues through a three-stage process: initiating confidence-building measures among the participating countries, developing a system of preventative diplomacy, and ultimately pursuing conflict resolution.

Track II aims to encourage a continuing dialogue among academics, members of nongovernmental organizations, and government officials in their private capacity to discuss possible solutions to outstanding security problems and generally build confidence between participants. One of the most important Track II forums has been the Indonesian-sponsored workshops on the South China Sea, which have aimed to enhance maritime cooperation among the disputants and protect the region's environment.

Although the ARF process is still at the confidence-building stage and has come in for a good deal of criticism from many Western observers (who see the forum as little more than an ineffectual talk-shop), ASEAN officials are genuinely pleased with the ARF's progress. They argue that getting nearly all the countries of the region to sit down together and discuss security problems

for the first time is a great achievement in itself. As regards China, ASEAN officials note that Chinese representatives to the ARF have become much more comfortable with the discussion process, and have been receptive to the need to build trust among the participants. However, ASEAN recognizes that getting the PRC to move forward on preventive diplomacy and conflict resolution will be difficult, given Beijing's sensitivity toward any issue in which it sees an infringement of Chinese sovereignty, e.g., the Taiwan Straits and the South China Sea. ASEAN is also aware that the whole process could be held hostage to any deterioration in Sino-US relations.

The Military-Security Dimension of Engagement

Economic interdependence and security dialogue are not sufficient strategies in themselves to reduce threat perceptions toward the PRC. The policy of engagement needs a military-security dimension. This dimension is composed of two parts: first, that the ASEAN states maintain modern and credible armed forces to act as a deterrent; and second, that defense links with external powers be maintained.

With regard to the first part, since the early 1980s the founding members of ASEAN have substantially modernized their armed forces (with the exception of the Philippines), with priority to air and naval power. This defense buildup was driven by a combination of factors, including increased financial resources, greater availability and affordability of modern weapon systems following the end of the Cold War, a change in military doctrinal focus from combating internal threats to those from without, and even intra-ASEAN competition. However, limitations have been placed on the military option. It seems highly unlikely that ASEAN will transform itself into a military alliance in the future. The Association has eschewed this option since its establishment for fear of antagonizing other countries that might form countervailing power blocs, and also be-

cause of problems of interoperability among the ASEAN armed forces. ASEAN wants to avoid giving China the impression that the ten countries of Southeast Asia are ganging up on it, which would retard the process of engagement and provide political capital to hard-liners within the Chinese leadership. Moreover, ASEAN is aware that ten small countries cannot hope to balance the PRC by themselves, and that turning the Association into a military alliance would lead to an expensive drain on national resources.

The second part of the military-security dimension is the maintenance of defense links with external powers. Some of these defense links were a product of the Cold War, reflecting a convergence of threat perceptions between the West and the ASEAN-6 toward the USSR, China, and North Vietnam. Such links were understandable given the small size of the ASEAN members and their inability to match the military power of the communist states. In 1991 the Cold War ended, but the ASEAN-6 defense links with external powers have been maintained and in some cases strengthened. New security relationships have also been forged. Defense links with external powers are regarded by the ASEAN-6 as a kind of insurance in case engagement fails, with the Chinese leadership pursuing regional hegemony or a more expansionist policy in the South China Sea. If this occurred, the ASEAN-6 would move closer to their Western allies.

The ASEAN-6 see the continued military presence of the United States as crucial to the maintenance of regional stability. ASEAN-6 officials believe the United States is the only country capable of balancing China, and that Beijing would be dissuaded from seeking regional hegemony so long as Washington stays engaged. Though there are differences over human rights, the ASEAN-6 consider the United States to be a largely benign power, which makes no territorial demands in Asia and has a vested interest in maintaining freedom of navigation. Each of the ASEAN-6 has its own security agreements with the United States. The Philippines and Thailand have had close security links with America since 1951 when both countries signed mutual defense treaties with Washington and allowed the presence of US bases. These bases were removed from Thailand in 1977 and from the Philippines in 1992. However, Thailand continues to hold annual exercises with the United States, and the Philippines recently concluded a Visiting Forces Agreement that will allow US-Philippines exercises to resume. Singapore has long been an advocate of an American military presence in Asia, first as a balance against the USSR, and now as a balance against the PRC.[20] In 1990, Singapore signed an agreement with the United States allowing US naval vessels access to Sembawang Naval Base, and for a US logistics coordinating unit to be based in the city-state. In January 1998 Singapore announced that it would further facilitate the US military presence by allowing US naval vessels, including aircraft carriers, access to the Changi Naval Base upon its completion in 2000.[21]

Malaysia and Indonesia were somewhat ambivalent about the US military during the Cold War, but the rise of China seems to have convinced them of the merits of an American military presence. Beginning in 1997, US warships began making regular port calls to Malaysia for the first time, and the Malaysian Defence Minister said the region needed a continued US military presence.[22] Indonesia also has begun to allow port calls by US vessels in the past few years. Even Vietnam seems to be acknowledging the benefits of a US naval presence in Southeast Asia. In March 1997 the Commander-in-Chief of the US Pacific Fleet, Admiral Joseph Prueher, visited Hanoi. Prueher spoke of a "nascent military relationship" between the United States and Vietnam, and there were suggestions that US naval ships might be allowed to visit Cam Ranh Bay Naval Base.[23] However, despite Vietnam's wariness of the PRC as a potential threat, a security relationship with

the United States is not in the cards in the foreseeable future.

Some ASEAN members are also keen to support military links with the United Kingdom, Australia, and New Zealand. In 1971 these latter three countries formed a military relationship with Singapore and Malaysia known as the Five Power Defense Arrangement (FPDA). Under the terms of the FPDA, the five governments are required to consult with one another if Singapore or Malaysia comes under external attack. The FPDA was originally meant to reassure the two Asian countries that the United Kingdom would not leave a power vacuum behind when it withdrew east of Suez in the early 1970s. In the mid-1990s the FPDA underwent a revival as Malaysia and Singapore became increasingly concerned over Chinese behavior in the South China Sea.[24]

Indonesia is not a member of the FPDA, but it has moved closer to the organization by forging defense links with Australia. In December 1995 Australia and Indonesia signed a security treaty calling for consultations in response to external threats to either country. The agreement was short on details, but was generally reckoned to be aimed at China, having been signed in the wake of the PRC's occupation of Mischief Reef. Brunei, which achieved independence from the UK in 1984, has a defense treaty with London and allows British troops to be stationed in the tiny, oil-rich sultanate. Thus the United States, the United Kingdom, Australia, New Zealand, and the ASEAN-6 are tied to each other militarily through an overlapping web of formal and informal security linkages.[25]

While the ASEAN-6 do want to maintain a defense relationship with the Western powers, there are limitations to this policy. According to ASEAN officials, member governments want the United States to be "on tap, but not on top."[26] This means they do not want to see US bases in Southeast Asia, which would run counter to the 1967 Bangkok Declaration that established ASEAN and the 1971 Malaysian agreement to turn

Southeast Asia into a Zone of Peace, Freedom, and Neutrality. Moreover, they recognize that the policy of engagement with insurance is not without its risks. The ASEAN-6 want defense links with external powers, but they do not want these links to be so strong that the PRC perceives itself to be the object of a policy of containment. The ASEAN states thus have to strike a fine balance between providing for their own security on one hand, and not antagonizing China on the other. Such a strategy is not without its contradictions. ASEAN does not want to see US bases in Southeast Asia, and the United States respects this position by relying on a forward deployment of US naval ships to maintain a military presence. However, ASEAN strategists continually accuse the United States of not being committed to the security of Southeast Asia. They claim the United States is more committed to the defense of Northeast Asia (i.e., South Korea, Taiwan, and Japan) because it has greater national interests there.

Conclusion

China's rise to power has brought both benefits and concerns to the ASEAN members. In the short term, ASEAN hopes to benefit from China's modernization program by taking advantage of economic opportunities. However, as the reform process continues, the PRC is able to augment its political, economic, and military clout in the region. To varying degrees, the ASEAN states worry about how a powerful China will behave, and whether it will use its newfound power to seek regional hegemony, resolving territorial disputes by force and interfering in the internal affairs of its neighbors.

ASEAN supports a policy of engagement with China, hoping that economic interdependence and China's participation in the embryonic regional security architecture will mitigate their security concerns. However, they also take a realistic view of the rise of China, recognizing that the policy of engagement needs a military-security dimension.

The ASEAN-6 need to maintain credible armed forces to act as a deterrent, as well as defense links with external powers, primarily the United States, which is seen as the key balancer. Engagement with insurance is a prudent policy in case the PRC decides to pursue a more aggressive posture in Southeast Asia in the next century.

NOTES

1. In August 1967 ASEAN was established by Indonesia, Malaysia, the Philippines, Singapore, and Thailand. In January 1984 Brunei joined on achieving independence from the United Kingdom. In July 1995 Vietnam became the seventh member of the group, and it was joined two years later by Burma (Myanmar) and Laos. In April 1999 Cambodia became the tenth member of ASEAN.

2. In 1998 the author conducted more than 40 interviews with foreign and defense ministry officials, strategic studies think tank specialists, and academics in the Philippines, Indonesia, Singapore, Malaysia, Thailand, Burma, and Vietnam on ASEAN perceptions towards the PRC. To facilitate academic exchange, all interviewees were granted anonymity.

3. For a full account of the development of Sino-Philippine territorial disputes, see Ian James Storey, "Creeping Assertiveness: China, the Philippines and the South China Sea Dispute," *Contemporary Southeast Asia,* 21 (April 1999), 95–118.

4. In January 1999 the Philippine National Security Council identified the South China Sea dispute and the communist insurgency movement in Mindanao as the country's two "most urgent" national security problems. See "Spratlys, Mindanao Top RP Problems," *Philippine Daily Inquirer* (Manila), 22 January 1999.

5. In 1995 Mahathir said, "It is high time for us to stop seeing China through the lenses of threat and to fully view China as the enormous opportunity that it is.... In my view, to perceive China as a threat and to fashion our security order around this premise would not only be wrong policy, but it would also be a bad and dangerous one." Quoted in "Malaysia Charts China Course," *Far Eastern Economic Review* (Hong Kong), 23 February 1995, p. 32.

6. An actual threat is inferred when the threatener emits definite signals of intent. An actual threat closely approximates the traditional concept of threat as "If you do A, I will do B," where A is detrimental to the threatener and B is detrimental to the target. Such threats are usually declarative in nature. The identity of the threatener is usually clear and the danger posed direct. Actual threats include invasion by foreign powers, territorial seizures, and missile strikes. Potential threats are more ambiguous, as they are not usually accompanied by declarations of intent. Instead, potential threats are perceived after assessing another state's capabilities and intentions. Often such assessments are made when one state increases its military capabilities. Potential threats are seen as threatening the stability of regional systems of order or the international system itself. See Klaus Knorr, "Threat Perception," in *Historical Dimensions of National Security Problems,* ed. Klaus Knorr (Lawrence: Univ. of Kansas Press, 1976), pp. 78–119. David J. Myers adds a useful clarification to this division by arguing that actual threats are more immediate, or closer in time, while potential threats are seen as posing a long-term danger. David J. Myers, *Regional Hegemons: Threat Perception and Strategic Response* (Boulder, Colo.: Westview, 1991), p. 19.

7. "Natunas 'Belong to Indonesia,' " *Straits Times*groun (Singapore), 22 July 1995. Indonesian Foreign Minister Ali Alatas rejected the notion that there was anything to negotiate, declaring China to be "too far away to the north" to have any legitimate claim over the area. See "No Problem with China over Natuna Isles, Says Alatas," *Straits Times* (Singapore) 27 June 1995. In order to reinforce this message the Indonesian armed forces staged massive military maneuvers near the Natunas in September 1996. See "Deep Background," *Far Eastern Economic Review,* 12 September 1996.

8. "Drawn to the Fray," *Far Eastern Economic Review,* 3 April 1997.

9. The ASEAN foreign ministers issued a statement that expressed "serious concern over recent developments which affect the peace and stability in the South China Sea" and urged "all parties to refrain from taking actions that destabilize the region." See "ASEAN Ministers Express Concern over Spratlys," Reuters News Service, 18 March 1995.

10. "Erap Orders Blockade of Mischief Reef," *Philippine Daily Inquirer,* 11 November 1998.

11. Mohan Malik, "Burma Slides under China's Shadow," *Jane's Intelligence Review* (London), 1 July 1997.

12. "Chinese Puzzle over Burma's SIGINT Base," *Jane's Defence Weekly* (London), 29 January 1994.

13. Interviews with Burmese foreign ministry officials, Rangoon, August 1998.

14. Some ASEAN officials opined that the strategic factor was an important consideration in the decision to admit Burma into the organization. Others discounted this factor. Citing ASEAN's long-standing aim of including all ten Southeast Asian countries in the organization, they assert that ASEAN moved to include Burma, Cambodia, and Laos by August 1997, the 30th anniversary of the establishment of the Association.

15. In March 1999 the Chinese government increased defense spending by 12.7 percent to 104.65 billion yuan ($12.6 billion US). "Generals in Vow of Loyalty to President," *South China Morning Post* (Hong Kong), 9 March 1999.

16. See Richard Bernstein and Ross Munro, *The Coming Conflict with China* (New York: Alfred A. Knopf, 1997), p. 24.

17. "China Must Allay Understandable Fears over Its Rise, says SM," *Straits Times* (Singapore), 8 September 1996.

18. For an example of this view, see Charles Krauthammer, "Why We Must Contain China," *Time,* 31 July 1995.
19. José T. Almonte, "The Future of Regional Security," *Kasarinlan* (University of the Philippines journal), 10 (No. 3, 1995), 17.
20. Speaking in 1996, Lee Kuan Yew said, "We have to accept the reality that there is no combination of forces in ASEAN that could stand up to a military confrontation with China. Unless there is an outside force, such as America, there can be no balance in the region." See "Singapore's Senior Minister on Asia's Future," *Business Week* (Singapore), 29 April 1996.
21. "US Aircraft Carriers to Get Access to Changi Base," *Straits Times* (Singapore), 16 January 1998.
22. "US Presence 'Needed in Region,'" *Straits Times* (Singapore), 16 April 1997.
23. "US Pacific Commander Reveals Budding Military Relationship," *South China Morning Post,* 23 March 1997.
24. In 1997 the FPDA held its first combined naval, air, and sea exercise, involving 35 warships, 140 aircraft, and 12,000 service personnel from the five countries. See "First FPDA Joint Naval, Air Exercise," *Straits Times* (Singapore), 1 April 1997. In August 1998 Malaysia announced that it would withdraw from the annual FPDA exercises. The Malaysian government cited budgetary problems caused by the Asian financial crisis, but the real reason was believed to be political tensions with Singapore. Given Malaysia's concerns about the South China Sea it seems unlikely the withdrawal will be permanent.
25. The failure of Indonesia to contain the violence in East Timor in September 1999 could put severe strains on Indonesia's relations with Australia, not to mention its relations with the UN and the United States. See Keith B. Richburg, "East Timor's Capital City Devastated by Fires, Looting," *The Washington Post,* 9 September 1999, pp. A1, A16; and Jim Hoagland, "Out-

Article 18   *Time*, July 17, 2000

Solving the Tibetan Problem

Before it's too late, China and the Dalai Lama must reach a compromise

Tsering Shakya

Tsering Shakya is the author of The Dragon in the Land of Snows: A History of Modern Tibet Since 1947

TIBET IS EVERYWHERE THESE DAYS. ITS IMAGES are used to sell insurance; the dalai lama's face appears on billboards to promote computers; there are countless tibetan festivals and exhibitions. but tibet as an issue in global politics has gone nowhere.

China's leaders lose no sleep over Tibet. They have invested huge amounts of money to improve internal security, making it almost impossible for Tibetans to stage any kind of protest. (Torture and imprisonment inevitably follow any such attempt.) The flight of the 17th Karmapa to India embarrassed China's leaders, providing them with further evidence of what they see as the Dalai Lama's "intrigue" and "insincerity." As a consequence, contact between Beijing and the Dalai Lama has been cut, effectively ending any hope of a negotiated settlement.

Within Tibet the political situation has worsened. The campaigns against the Dalai Lama, his so-called "splittist" followers and religious activities continue. The neighborhood committee meetings are as intense as those during the Cultural Revolution. In March at a meeting of China's National People's Congress, Zhou Yongkang, the Communist Party secretary for Sichuan province (which incorporates large parts of eastern Tibet), announced that the teaching of Tibetan in schools was a drain on government resources.

The flight of the Karmapa and other senior religious leaders and the continual flow of Tibetans over the Himalayas to Nepal and India constitute clear proof of China's failure to win over Tibetans. China has not learned from the Cultural Revolution, the most violent period in Tibetan and Chinese history. The people's religious faith was not even dented, and when the party partially relaxed its control, Tibetan Buddhism rebounded with a vitality that shook the authorities. The party can coerce Tibetans, but it cannot win their hearts and minds. Only a wise and tolerant policy can do that.

The Chinese are confident about their rule in Tibet: they know that however much Tibetans might protest, neither the Dalai Lama nor the Tibetans themselves have the power to dislodge them. Given Beijing's intransigence and the failure of the Tibetan government-in-exile to arouse the people,

China's lack of interest in a negotiated settlement has left the Tibetans hopeless and dejected. Their depression has been further deepened by the Dalai Lama's public statements that he has done his best to find a solution but has so far failed. Among the Tibetan refugee community there is a sense of despair about ever returning to the homeland. An increasing number of Tibetans are looking overseas, especially to the United States, for their salvation. Prospects of a bleak future are driving hundreds of young Tibetans to the West, where they often end up washing dishes in New York, London or Paris.

Hope that international pressure on China could bring change remains unfulfilled. Despite growing popular sup-port for the Tibetan cause in the West, there are no signs that governments are willing to take up the cause in earnest. The Tibetan problem lies at the bottom of the heap. And it is likely to remain there for the simple reason that Tibet has no economic or strategic value for Western governments, and China is not a country that can be bullied. It's not that Beijing is immune to international persuasion—there simply is no concerted pressure on China to relent on Tibet.

The gulf between the Dalai Lama and the Chinese leadership is not insurmountable. A solution can be reached that meets Beijing's security concerns and gives Tibetans a homeland. However, it will require courageous and imaginative decisions on both sides.

Deng Xiaoping's bold "one country, two systems" policy has gone a long way in meeting China's claim to sovereignty while leaving people in other parts of the country to run their own affairs. If Beijing insists that it will talk with the Dalai Lama only in person, refusing to recognize officials appointed by him, he should be prepared to meet with China's leaders. The Dalai Lama has declared he does not want independence for Tibet and is willing to meet Beijing's security concerns by agreeing to relinquish control of foreign and defense policy to China. Beijing in return should recognize that by giving Tibet genuine autonomy its security and status in the world will not be endangered. If anything it will be enhanced.

Article 19

Far Eastern Economic Review, June 10, 1999

Capital Idea

Free-market philosophy, commonly thought of as thoroughly Western, actually has its roots in the ancient annals of China

By Michael Vatikiotis

Michael Vatikiotis is managing editor of the REVIEW.

China may have lagged behind Europe entering the industrial age, but there was much that China contributed to modern European economic thinking. In an earlier age, when Europe was in the throes of the Enlightenment, China was considered more enlightened as it emerged from centuries of centralized bureaucratic control. Modernity often plays tricks with history: China's apparently contradictory embrace of free trade under communist rule today obscures the fact that basic notions of a free-market economy were derived with the help of observations made in China by Jesuit missionaries in the mid-18th century.

The great Scottish economist, Adam Smith, is credited with establishing the principles of modern free-market economics in his *An Inquiry into the Nature and Causes of the Wealth of Nations* published in 1776. Less is known about a French physician-turned-philosopher who inspired Smith when he worked for a time as a tutor to the nobility at the royal court of Versailles. Francois Quesnay first coined the term "laissez-faire" in 1758. He belonged to a group of intellectuals known as the Physiocrats who believed that government policy should not interfere with the operation of natural economic laws.

According to contemporary British philosopher John James Clarke, the source of these ideas was the *Tao Te*

Ching, the basic text of Taoism attributed to the semi-mythical Chinese philosopher Lao-tze. Travellers to the China of the late Ming and early Qing period were impressed by the high degree of organization—exemplified by the exam system—the weak hand of government, and the extensive economic freedom and efficiency this combination produced.

Their Chinese hosts told them that such a state was close to the ideal of *wu wei,* whereby the wise ruler knows that the best way to rule is by doing nothing. "Practise not doing and everything will fall into place," the *Tao* recommends.

For Europeans struggling to escape from the last vestiges of feudalism, the relative sophistication of Chinese abso-

lutism was appealing. Quesnay was so enthralled he was known as the "European Confucius" in his day. The French translated the concept of wu wei into laissez-faire. It had a tremendous influence on Smith, who, historians speculate, would have dedicated the *Wealth of Nations* to Quesnay had his mentor not died shortly before its publication.

Article 20

Far Eastern Economic Review, June 10, 1999

Sticking Point

Acupuncture is adapting to the modern world. But by leaving tradition behind, it risks being reduced to a marginal secondary treatment.

By Adam Brookes

Adam Brookes is a writer based in Beijing.

At the Chongwen District No. 2 Hospital in central Beijing, Mrs. Hao sits serenely on a grubby bed. She has needles sticking out from the corners of her eyes. Three inches long and stainless steel, they protrude like ghastly antennae. But Mrs. Hao would have it no other way. She has been coming every week to the hospital's acupuncture clinic for 25 years. She credits acupuncture with halting chronic eye disease and saving her sight. Each session costs her only 6 renminbi, about 70 U.S. cents. Mrs. Hao snorts that she has "no time for Western medicine. It's expensive, unreliable and has side effects."

After a century of suspicion and decline, acupuncture is making a comeback. In China, it has made a partial return to the place in mainstream medicine that it occupied for the larger part of the millennium. Up-to-date research techniques and clinical trials, imported from the West, have enhanced its effectiveness and broadened its application. Even in the sceptical West, acupuncture is acquiring a measure of the legitimacy

long denied it, although there, its effects on the body's biochemistry are poorly understood, if at all.

As the clinic's presiding doctor, Wang Jingping, flits between bodies stuck like pin-cushions, her deft manipulation of the needle and choice of acupoint—the exact point at which the needle is inserted—would be recognizable to a doctor working 1,000 years ago, in the Song dynasty. But such a time-travelling doctor would find today's acupuncture strangely shorn of the philosophy and cosmology that for millennia have underpinned it. Indeed, some Chinese practitioners are warning that acupuncture risks becoming a victim of its resurgent success. They worry that, as modern Western medicine legitimizes some elements of acupuncture, an ancient, holistic practice risks relegation to the status of a complementary therapy, limited to the treatment of specific ailments.

For, having shed the traditional explanations, we do not know how acupuncture works. In the modern clinic, it has been largely separated from its cultural context. Yet it remains a curiously cultural phenomenon, informed by philosophy, vulnerable to ideological shifts, and a tantalizing mystery to medical science.

Acupuncture's history stretches back into myth. Chinese accounts speak of sharp stones used to stimulate the surface of the body in prehistoric times. The theoretical foundation of Chinese medicine began to emerge in the *Nei Jing Su Wen,* sometimes translated as The Canon of Medicine, which dates from approximately 2,000 years ago.

The Canon describes the grouping of all phenomena within fundamental categories, among them the female and male principles, *yin* and *yang,* and the Five Elements of earth, water, fire, air and wood. These categories engage in a complex system of interaction that animates the cosmos. Traditional medicine in China is built on knowledge of this cosmological system. Health depends on the stable and balanced interaction of key elements and their flow through the body's channels, or meridians. Acupuncture facilitates that balance and flow by clearing and stimulating the meridians. The Canon of Medicine refers to meridians and acupoints, and the corpus of knowledge on their positions and potential grew over the following centuries.

Through the Sui (589–618) and Tang (618–907) dynasties, as printing technology emerged, so did the first manuals on acupuncture. Chinese medicine attracted the attention of the intelligentsia and the patronage of the emperor. Imperial colleges of medicine became repositories of expertise and centres of research. The Song dynasty (960–1279) and the new millennium brought vital developments. In the 11th century the scholar Wang Weiyi collated a vast quantity of information on acupuncture's theory and practice, synthesizing it into a standard model. He made bronze casts of the human body, etched with hundreds of acupoints and the meridians. And he published an exhaustive illustrated manual that still serves as a basis for study 1,000 years on.

The Ming dynasty (1368–1644) saw the compilation of the *Zhenjiu Dacheng,* or Great Compendium of Acupuncture and Moxibustion. Advances in paper and printing technology and communications allowed for voluminous reproduction and wide distribution. But during the Qing dynasty (1644–1912), acupuncture suffered a dramatic fall from grace with the intelligentsia and the Manchu court, which viewed the practice with suspicion. In 1822, it was removed from the curriculum of the Imperial College of Medicine. Although acupuncture continued in widespread practice, without imperial patronage standards fell, and so did respectability.

Ironically, it was communism that rescued acupuncture. Communist guerrillas fighting Nationalists and Japanese in the 1930s and '40s were chronically short of doctors and materiel. Acupuncture's equipment was mobile and its expertise, however low-level, was easily disseminated. After the founding of the People's Republic in 1949, the benefits of traditional Chinese medicine, experienced amid the privations of guerrilla war, remained in the minds of policymakers. They decided that the new socialist health-care system would incorporate both Western and Chinese therapies.

But ideology ensured that acupuncture would not stay the same. Marxist dialectical materialism had little time for yin and yang, or for the holistic methods of diagnosis that Chinese medicine demanded. "Acupuncture, and all of Chinese medicine, became viewed through a Western scientific lens," says Dr. Hu Weiguo of the Chinese Acupuncture Research Institute. "The philosophy underlying acupuncture was cut away in favour of simplistic, mechanistic treatment."

Such a view was encouraged during the Cultural Revolution, when doctors found themselves exiled to impoverished rural communities without drugs or equipment. Once again, they turned to traditional methods. A generation of medical students and rural "barefoot" doctors learned simple, standardized acupuncture treatments. Hu blames the Cultural Revolution for lowering standards and eroding acupuncture's scope.

Yet ideology did generate practical advances, too. In 1972, as U.S. President Richard Nixon oversaw Sino-American rapprochement, the world was treated to breathtaking accounts of acupuncture. Patients in Chinese hospitals sipped orange juice as they were painlessly operated upon, chemical anaesthetic having been largely replaced by electronically charged needles in their ears.

Yet it would take another 25 years for American medical authorities to concede a tiny role for acupuncture in medicine. In 1997, the National Institutes of Health finally said it had found "clear evidence that needle acupuncture is efficacious for adult postoperative and chemotherapy nausea and vomiting and probably for the nausea of pregnancy."

Yet Hu bemoans such a confinement of acupuncture's capabilities. Acupuncture, he asserts, is not just for pain relief, but has the capacity to rehabilitate stroke victims, improve mobility among the paralyzed, cure drug addiction and treat skin conditions and stress-related illness. Perhaps most importantly, he credits acupuncture with the ability, if employed to its full potential, to prevent disease occurring in the first place. "But between them, the Cultural Revolution and Western research have encouraged the belief that the uses of acupuncture are limited," Hu says.

He does, however, take heart from the rapid growth in awareness among consumers in the West of alternative medicine. He is enthusiastic, too, about a growing Western interest in holistic medicine—the belief that doctors should consider the health of the whole person, their psychological state and social circumstances. This, he says, echoes traditional Chinese diagnostic techniques, and could foster a fresh appreciation of acupuncture's potential.

Article 21

The New York Review of Books, May 11, 2000

Indonesia:
Starting Over

Clifford Geertz

1.

"Indonesia has been one of the most remarkable development success stories in the last third of the twentieth century. In the mid-1960s, it was one of the poorest countries in the world, with a per capita income below that of many African and South Asian countries. It had experienced little economic growth for thirty years, it was on the verge of hyperinflation, it was engulfed in political turmoil, and it had begun to disengage from the world community and economy. Living standards were stagnant and about two thirds of the population lived in abject poverty. . . .

"No one at that time would have dared to image—much less to predict—that just thirty years later Indonesia would be regarded as a dynamic "tiger" economy, and a member of that most exclusive club, the World Bank "East Asian Miracle Economies." The notion that Indonesia's economy would expand six-fold over this period, and that according to World Bank projections it could become the world's fifth-largest economy by the year 2020, would have appeared preposterous in the gloom of 1964–66. Yet that is precisely what has occurred in these three decades: economic growth has been among the highest in the world, and it has been accompanied, with a lag, by striking improvements in social indicators."
—*Hal Hill, May 1997*[1]

"[Very] suddenly and unexpectedly, everything collapsed [in Indonesia] in the latter half of 1997 . . . in the onslaught of the Asian financial crisis. The extent of the turnaround is nothing short of astounding. Economic output is expected to contract by about 15 percent, after expanding 8 percent in 1996 and 5 percent in 1997. The single-year collapse in growth is among the largest recorded anywhere in the world in the post-World War period. Millions of Indonesians, many just surviving over the poverty line during the good times, have lost their jobs. Food production has been disrupted. . . . Prices for many export commodities . . . have fallen on world markets. Investors, both foreign and domestic, have fled to safer havens. The banking system is moribund and thousands of firms are facing the prospect of bankruptcy and closure."
—*Steven Radelet, September 1998*[1]

Since Indonesia's sudden reversal of fortune, globalism interrupted, a great deal more has happened there than capital flight, currency collapse, and a tripling of the poverty rate. The regime has changed twice—the regime, not just the government—once abruptly, in a spasm of violence, once glacially, with troubled and unnerving hesitation. The first time, in late 1998, Suharto, the architect, or anyway the godfather, of both the expansion and the collapse, walked away amid wild disorder—race riots, looting, bloody clashes between students and the army, Jakarta on fire, Surakarta ransacked—leaving B.J. Habibie, his just-appointed crony vice president, haplessly behind to sort through the ruins. The second time, a protracted, vastly complicated, ultimately indecisive, but, so it seems, fair and open national election (ninety million voters, forty-eight parties, seven hundred electors) ended last autumn with the midnight designation, by a half-dozen *arriviste* kingmakers, of Abdurrahman Wahid as the new president. An ill, erratic, nearly blind religious intellectual, he had been written off by almost everyone as too frail to serve.

In September, the ex-Portuguese enclave of East Timor, half of a very small island out on the edge of the archipelago, was at last allowed to separate after thirty years of on-again, off-again resistance to annexation, only to be laid waste by Indonesian-armed irregulars, whose savagery brought on a worldwide outcry, an Australian-led UN intervention, the human rights attentions of Mary Robinson, and, just possibly, a revanchist problem for the future. Local violence, some of it ethnic, some of it religious, some of it merely criminal or entrepreneurial, has broken out all over the archipelago, from Aceh and Kalimantan in the west to the Molucas and New Guinea in the east, leaving hundreds dead, thousand in flight, the government at a loss, and the neighbors—Malaysia, Singapore, the Philippines, and Australia, who have minorities (and refugees) of their own who might like to see things generally rearranged—worried.

The army, its leadership divided and threatened with prosecution for war crimes in East Timor and elsewhere, is demoralized, resentful, disprivileged, cherishing enemies, weighing possible strategies. The press has been freed and reenergized: books are no longer banned. Suharto, ill and demonized, is housebound, as incommunicative as ever; and the country's most famous political prisoner, the radical nationalist novelist Pramoedya Ananta Toer, is out and about, giving interviews, accepting tributes, and counseling the youth. Oil looks good again, inflation is down, exports have recovered a bit, bankrupts are regrouping, growth has advanced to zero.

At the same time, militant Islam, NGO environmentalism, populist xenophobia, neoliberal utopianism, Christian apologetics, and human rights activism have all grown markedly in volume, visibility, and the capacity to bring on mass rallies, mobs, and marching in the streets. Factional party politics have returned with a vehemence and complexity not seen since the early Sixties, when Sukarno's "guided democracy," designed to keep him in power for life, collapsed in conspiracy and slaughter. It is a mixed and unsettled, fluctuating picture, without center and without edge—resistant to summary, hard to hold in place. As virtually everything has happened, it seems that virtually anything might; and it is impossible to tell whether all this stir and agitation—what the Indonesians, with their usual gift for verbal camouflage, have come to call *reformasi*—is the end of something or the beginning of something.

What it might be the end of is the political impulse that set the country in motion in the first place. Along with India, Egypt, and perhaps Nigeria, Indonesia has been a prototype of the "emerging," "developing," "post-colonial" country—crowded, splayed, capriciously bordered, and the product of a world-historical shift in the distribution of sovereignty, selfhood, and the power to act. Officially established at the end of 1949, after four years of intermittent warfare against the

Dutch, and nearly forty of agitation before that, the country took shape during the heyday of third world nationalism—Nehru, Chou, Tito, Nkrumah, Mussadegh, Nasser; Dien Bien Phu, the Battle of Algiers, Suez, Katanga, the Emergency, the Mau Mau.

By now, this period—call it "Bandung," after the famous gathering of nonaligned "emerging forces" leaders that Sukarno ("I am inspired. . . . I am absorbed. . . . I am crazed by the Romanticism of Revolution") staged there in 1955—is not even a living memory for most of the population. Its concerns are faded, its personalities simplified; the obsession that obsessed it, and to a fair extent subsidized it, the cold war, has been summarily called off. But the doctrines that were developed then, and the sentiments that accompanied them, continue to shadow the country's politics. A half-century old this coming winter, and just emerging from thirty-two years of business-card autocracy, Indonesia still projects itself as a triumphalist, insurgent, liberationist power.

The question that engages the more reflective Indonesians, and particularly the older ones who have been through it all and seen what it comes to, is of course how far this master idea, with its slogans, stories, and radiant moments, remains a living force among either the country's elites or its population, and how far it has become just so much willed nostalgia—declamatory, a pretense, worn, and seen through, cherished if at all by Western romantics and political scientists. Certainly the history of the country, which has tended to be one of grand promises and grander disappointments in quickening alternation—large plans, large collapses—would seem to militate against the continuing hold of Bandung-size expectations. Neither Sukarno's "old order" populism nor Suharto's "new order" paternalism (the differences between them have been much exaggerated by the partisans of both—their contrasts were mainly in style and presentation, and to some degree in disciplinary reach) was able to impress an identity and a transcending purpose on the society as a whole, to make of it an integral community, real or imagined.

"The Nationalist Project," the construction of an aroused and self-aware

people moving as one toward spiritual and material fulfillment—"An Age in Motion," so the tag says—has become increasingly hard to formulate in believable terms, much less to pursue and carry out. The shaken country that was delivered first, *faute de mieux,* to the unfortunate Habibie in the spring of 1998 (his presidency lasted seventeen months, plagued by confusion and scandal), and then, *in camera,* to the improbable Wahid, had lost more than its bank balance, its equilibrium, and its international reputation. It had lost the power of its history to instruct it.

2.

The man who is expected to correct all this, to right its economy, calm its politics, restore its confidence, reset its course, clear its conscience, and improve its image, as well as, perhaps, to entertain and distract it, is a fifty-nine-year-old veteran politician whom virtually everyone seems to have met (including me: a decade ago, I spent four days closeted with him and a few of his allies in a rest house near Bandung discussing, no less, the future of Islam in Indonesian politics), most seem to have liked, and almost all seem to have underestimated.

Known universally as "Gus Dur," his honorific childhood nickname ("Gus" means "handsome lad"), Wahid was born and grew up at the very center of Javanist Islam and Javanist Islamic politics—his grandfather's famous religious boarding school, or *pesantren,* fifty miles southwest of Surabaya. His grandfather, a personage and a personality, as well as a renowned Koranic scholar, founded the country's largest Muslim organization (it may, in fact, be the largest in the world), Nahdatul Ulama, in 1926—in part, at least, to counter the growth of secularist nationalism, and to strengthen the hold of vernacular piety against modernist innovations flowing out of the Middle East. The tolerant, open, somewhat traditionalist, somewhat inward "Javanese Islam" he represented continues to the present as a major social and religious force. Wahid's father, in the loose, inexplicit sort of way in which power is passed in the *pesantren* tradition, inherited the school, the stature, the program, and the organization;

he was the country's first minister of religion, and a broker of consequence in Sukarno's ideological spoils system, distributing jobs to petitioners in the vast and shambling clerical bureaucracy that to this day regulates mosques, marriages, benefactions, and religious courts.

Wahid, after traveling, studying, and getting himself known in Cairo, Baghdad, and various countries in Europe for awhile in the mid-Seventies, returned to become a widely read columnist at *Tempo,* the country's leading, and later suppressed, news magazine, and to found Forum Demokrasi, an elite ginger group whose criticisms of the establishment drove the government to near-murderous distraction. He also took over the reins of the Nahdatal Ulama organization, which he then promptly separated from the counterfeit political party ("Development and Unity") Suharto had concocted to contain it.

If close-up, been-through-the-mill experience, as well as patience, agility, humor, and a refined sense of timing, is what Indonesia needs, Gus Dur, who is the closest thing to a machine politician the country has, could be the man. Compared at various times to Peter Falk's Columbo, the Javanese shadow-play buffoon Semar, Chaim Potok's lapsed rabbi Asher Lev, Ross Perot, Yoda, and (by his defense minister) a three-wheel Jakarta taxi, Wahid would seem well equipped to weave his way through the densest sort of lunatic traffic.

That much, surely, he demonstrated in his oblique, arduous, and—when we consider what he had to overcome to undertake it—brave and tenacious trek to the presidency. When the electoral process (which was rather more of an enormous, and enormously complicated, straw poll designed to assess the general state of popular opinion than it was a proper selection mechanism) began in late 1998, Wahid was in the hospital, just beginning to recover from the second of two diabetic strokes that had put him in a coma; he had already been damaged in one eye by diabetes, and blinded in the other. Aside from him, there were four leading candidates, thrown up by the convulsions of the previous two years, and off and running: Habibie, the sitting president, an estab-

lishment satrap trying desperately to look like a new broom; Megawati Sukarnoputri, Sukarno's daughter, a reclusive and taciturn, rather standoffish suburban matron, whom an unexpected and uncharacteristic series of strategic blunders on the part of Suharto had transmogrified into a popular hero; and Amien Rais, an ambitious and mercurial Muslim intellectual and sometime college professor who had studied theology and political science at both Notre Dame and the University of Chicago and who had played a leading role in arousing the students in the last stages of Suharto's collapse. Then, in the extra-party, shadow-state style of the Indonesian army, there was General Wiranto, its hesitant and unconfident and soon to be infamous chief of staff.

Other suggestions and possibilities surfaced from time to time. Among them were the Sultan of Yogyakarta; a Berkeley-trained neoliberal economist: the head of Golkar (i.e. "Functional Group"), Suharto's parliamentary party and political arm; and one of Wahid's oldest and, up to that point anyway, closest friends. But for the whole eleven months the lumbering drama took to unfold these five were the dominant players, and they remained so to and through its quite operatic, vertiginous end.

For most of the campaign, indeed until a few half-hours before Wahid squeezed his way in through the narrowest of spaces, the leading candidate, far and away, was Megawati. Despite her heritage and the lingering charisma of her father's name—particularly strong in Java, where power is supposed to pass supernaturally and act thaumaturgically—she was a newcomer to Indonesian high politics, having lived the smooth and upholstered life of a society wife until a spectacular collision with Suharto, which she had neither sought nor wanted, and did not quite know what to do with once it occurred, turned her overnight into the reluctant vehicle of popular outrage.

Seeking, apparently, to test the limits as the general, ill and recently bereaved of his wife, began to stumble a bit, one of his concocted political parties—"Indonesian Democracy," which was designed to contain the nationalist left—installed Megawati as its titular head in December 1993. Suharto, to

whom she must have seemed like the ghost of insurgencies past, wildly overreacted, trying forcibly to replace her with an army-backed puppet. When that failed, leading to a breakaway of the party and the movement of university students into the streets crying for Suharto's head and for his children's fortunes, he sent soldiers and paramilitaries to take over Megawati's Jakarta offices and arrest her supporters. In July 1996, this produced what turned out to be the most consequential "affair" of his regime: thirty or so killed, a hundred-odd arrested, scores of stores, houses, and vehicles burned. This, though no one knew it yet, was the end of the beginning of the end of his rule. Megawati, startled and swept along, was established as the people-power heir apparent, Indonesia's version, culturally reedited, of Cory Aquino.

Despite the divine-right regality which never deserted her and in the end undermined her, Megawati's campaign was an over-the-top, quasi-revivalist, in-your-face affair: frenzied mass rallies, revolutionary symbolism, hypernationalist sloganizing, and a certain amount of putsch-in-the-works and street-tough threats—all of which may have frightened as many people as they attracted, while scholars and journalists talked of civil war and the return of the repressed. Wahid, more or less recovered from the worst of his illness, formed a party of his own and set up an odd, on-again off-again, arms-length alliance with her.

The Islamic right, without a champion of its own or much of a program beyond moralism and xenophobia, attacked Megawati as not really a Muslim but some sort of Javanist Hindu, beholden to Christians and Chinese, possibly a crypto-Communist, and, anyway, a woman. She avoided the press, issued only the vaguest of policy statements (she had been against independence for East Timor and for a pegged rupiah, but she soon glided noiselessly away from these positions). She spoke, she said, with her dead father daily. In the event, after a half-year or so of this, she got a bit more than a third of the vote in the June 1999 elections for the National Assembly, which is convened every five years to designate a president. Habibie, who ostensibly anyway, was Golkar's candidate and particularly strong outside

Java, got a bit more than a fifth; Wahid and his party, whose appeal was localized, a bit more than a tenth; and Rais, who had been expected to do much better, given his popularity with educated Muslims and the Jacobin role he played in the last days of Suharto (whom he called, *inter alia,* "a rabid dog, biting everything," and volunteered to replace immediately), got something less than 8 percent. The stage was set for some serious politicking.

The details of the maneuvering, the alliances, the horse trades, the betrayals, the flatteries, the ear-whisperings, and the pirouettes that took place during the final three days of the "election"—i.e., the opening days of the National Assembly in October—remain, for the most part, both obscure and contested.[2] What is clear is that Megawati was out of her element, and Gus Dur was in the very thick of his. Unwilling, or unable, to cut deals and exchange favors, and apparently convinced that having roused the masses and "won" the election, she could not be denied, she lost every scrimmage at every stage, until in the end only Wahid, who had allied himself with just about everybody else in turn as the process unfolded, was left standing. (Rais, with Wahid's support, became head of the Assembly. Habibie's man in Golkar, a Sumatran named Akbar Tanjung, was induced to desert his boss and, with Wahid's support, became speaker of the Parliament. Wiranto, with Wahid's, as it turned out, retractable support, lobbied vigorously for the vice presidency.) "Wait until next time," Wahid is supposed to have told Megawati, kindly, one imagines: "You need more experience."

When Wahid's selection was announced on October 20, the reaction on the part of Megawati's supporters, who were as convinced as she was of her moral right to the presidency and the illegitimacy, deceitfulness, and corruption of everyone else, was enormous. Violence erupted all over the archipelago. In Bali, where her campaign had begun and her support was perhaps the most passionate (she won 80 percent of the vote there), trees were felled across all the roads, a government office was burned, and youths, attacked a dormitory where Timorese refugees were be-

ing held awaiting their repatriation. Plans were laid to attack the Muslim quarter, the so-called *kampong jawa,* which, had they been carried out, might well have convulsed the entire country. In Jakarta, a large downtown hotel where a huge crowd had gathered throughout the night to hear the outcome was immediately trashed, and angry protesters, weeping and screaming, spilled out into the city. It looked as if the promised civil war, or anyway a sidewalk coup, was at hand.

Wahid instantly changed course, turned away from Wiranto, and appointed Megawati as his vice president. He told her to go on radio and television and calm her followers, which she immediately did, saying, "I am your mother. You are my children. Return to your homes." And, in what, in some ways, was the most startling twist in the whole twisting drama, only slightly more startling than her acquiescence in her own eclipse, they promptly and efficiently obeyed—put away their placards, packed up their revolution, and walked quietly away. Bali was cleaned up in the course of a few hours; it was as though nothing at all had happened. Jakarta remained calm, if shaken. The eruptions elsewhere—in Surabaya, Medan, South Kalimantan, the off-coast islands of West New Guinea—soon fizzled out into scattered clashes. Whether all this was, as some began to call it, a move toward freedom, democracy, maturity, and the free market, or simply another turn in a very old wheel, there clearly was, at last and for the moment, a more or less legitimate, more or less open, more or less consensual government in place.

3.

The great question remaining is, can the government in fact govern? Almost everything about the Wahid presidency, not just the President himself, breathes of the temporary, the ad hoc, the fragile, the jerry-built. Brought into existence by a thrown-together coalition of power brokers who have known one another too long and too well, confronted not by a single crisis but by a flood of them, and lacking very much in the way of either popular backing or a worked-out program, the new regime is reminiscent

of nothing so much as those of the Naguibs, Barzagans, and Kerenskys of the world: distracted, scuttling, more or less well-meaning place-holders in a historical process preparing to run over them. Wahid, or his government, may turn out to be less evanescent than theirs, more consequential, or even more capable of defending its interests. That is at least one prospect insofar as one can speak yet of Wahid's actually having a government, as opposed to just a role (only a few of his cabinet members are his own choices; most are the result of what he himself called "a cattle auction" among the politicos who put him in power). It is clear by now that betting against Gus Dur is a bit of a mug's game. But he has, to put it mildly, a lot to do, and not much beyond his wits and, some say, his supernatural connections to do it with.

The problems facing him are diverse and urgent, each clamoring for immediate attention before they reach, severally or together, a point of no return. They cannot be listed as items for an agenda, because there is no way to put them into a logical sequence of importance and priority. But they fall, more or less, into three broad categories. First (but not, as foreign businessmen anxious to get back to foreign business tend to assume, necessarily foremost) there is the need to reignite the local version of the transnational economy that, between 1977 and 1997, added nearly $400 billion to the GDP, made a few people rich and a fair number middle class, and turned Jakarta, where upward of 70 percent of the activity was concentrated, into a forest of grandiloquent high-rises.

Second, there is the need to rein in and reprofessionalize the army, to halt and reverse the vast expansion of the functions and powers, legitimate, illegitimate, and outright criminal, that it achieved, first under Sukarno, who brought it into the world of commercial management when he dispatched the Dutch and confiscated their properties, then under Suharto, who fashioned it into a furtive, para-government extending the hand of violence into the smallest and most distant corners of civil society. And third, there is the need to respond to an enormous increase in the power of regional, ethnic, racial, and religious forces, most of them not entirely new, entirely unified, or entirely clear in their aims, but all of them newly excited about their development possibilities now that the dominance of Jakarta has weakened and threatening to dismember the country and turn it into a Balkan nightmare.

When we try to sum up the Indonesian crisis in this compound, multiplex way, its most striking characteristic, the one that makes what happens there seem so broadly instructive, is that the immediate and the fundamental are thoroughly tangled up together. The restless surface—the street demonstrations, the regional killings—lies very close to the settled bottom—Muslim/Christian religious division, the political ascendancy of Java and the Javanese. The most pressing issues are at the same time the most far-reaching. Quick fixes, such as reallocation of revenue among regions, and lasting changes are so tightly linked that transient adjustments—a change of provincial boundaries, the dissolution of a government department—have general and enduring resonance. There is no small policy. Tactics are strategy, tinkering is planning, repair is reform, and however ad hoc and pragmatic particular actions may seem to be (and Gus Dur's are both, plus antic, mystifying, offhand, and unpredictable), they are responses to a good deal more than the fragility of the rupiah, the command structure of the army, or the stability of the outer provinces.

Whatever the fate of what some enthusiasts are already calling "the Wahid revolution" and others, less entranced, "what-the-hell-ism" (biarinisme, for the cognoscenti), the path by which it arrives there, and what happens to it en route, should tell us a great deal about what can happen and what cannot, and not just in Indonesia or the post-heroic, post-colonial world, but in the dispersed, borderless, McDonalized, and networked "global civilization" supposedly in the making.

So far as the revival of the neoliberal market economy is concerned—if that, amid the corruption, the waste, and the imaginative misuse of resources, is what it was in the first place—even the quick fixes and the transient adjustments are but scarcely begun. The relatively speedy recovery that the smaller "tigers," Thai-

land, Malaysia, Hong Kong, and South Korea, have experienced has as yet to take place in Indonesia, by far the largest among them. Unemployment is rising, production is flat, and there is little sign of a return of departed capital or departed capitalists, from wherever it is that it, or they, may now be resting, hiding, or beginning a new life.

But the deep-lying issues that any move toward recovery, however hesitant, however slight, immediately uncovers are already, not two hundred days into Gus Dur's term, subjects of heated, what-side-are-you-on political struggle. Every cabinet reshuffle, or rumor of one, every budget recalibration, however modest, every policy proposal, even the most technical or circumstantial—to allow a foreign bid on a state-seized company, to shift ministerial control over a bankrupted bank, to remove a Suharto-era businessman from his Suharto-era business, to renew a standing contract with an American mining company—gives rise to a loud, Aesopian debate which only seems to be about the matter at hand. The real division is over a deep and unresolved, possibly unresolvable, foundational question: How open, how borderless, how transnational an economy do we really want? How "global," how "developed," how "market rational" can we be, should we be, dare we be?

This may seem to be nothing more than the familiar opposition between those who see transparency, trade, and market freedom as the beginning, the middle, and the end of everything, and those who wish to replace what Sukarno fifty years ago called (just before he demolished it, and elections with it) "free fight liberalism" with policies more sensitive to cultural conscience and national feeling. But though free trade and protectionism, comparative advantage and import substitution, foreign capital and domestic ownership remain the poles between which the arguments and accusations move, the experience of the last thirty years has changed the sense of what is at stake in the debate. Having known, now, both the joys and costs of extravagant market expansion and the pains and spin-offs of extravagant market collapse, the Indonesian elites, and a good part of the populace as well, are concerned less with trying to isolate the country from storms of "high," "late,"

"global," "footloose," or "advance" capitalism than they are with enabling it to survive and move forward in the face of them.

There is, as the saying goes, no other option on offer than connecting the national economy to the IMF-WTO-Davos world being put together in the banks and boardrooms of New York, Tokyo, Frankfurt, London, Paris, and Geneva. The trick, if there is one, is somehow to ride out, even perhaps somehow to profit from, what no less a neoliberal than Paul Volcker has called the inevitable train wreck that occurs when grand, unregulated, high-velocity capital flows collide with weak and rickety national economies.[3] Economic nationalism still lives in Indonesia, so do "Asian values," and there are even some relics of the theory formerly known as Marxism lying about. But their promises of empowerment, authenticity, justice, and moral shelter, just yesterday so beguiling, ring increasingly hollow.

The same general picture, the persistence of familiar threats and the inadequacy of familiar remedies, appears in the other matters of immediate concern: the role of the army and the integrity of the state. So far as the army goes, the problem is simple enough on the surface—specifying its function and confining it to it. But after three decades during which local political capacity, the simple ability to manage one's own affairs through one's own institutions, melted away in the face of close-up and pervasive military control, that is not easy to do. The soldiers are dug in, and, in many places at least, removing them would remove as well whatever is left of a national presence and an enforceable order; and they have in any case, as East Timor demonstrated, very little willingness to accept restrictions.

So far as the state's integrity goes, the call to national unity in the name of a shared ideal seems to be a wasting asset. Whatever is going to hold the place together, if, in the face of population movements, regional imbalances, and ethnic suspicions, anything is, it is not going to be settled by an ingrained sense of common identity and historical mission, or by religious, "Islamic State" hegemony. It is going to be something a good deal more patchy, capricious, and decentered—archipelagic. In whatever direction Gus Dur looks with his one good eye there seems nothing to do but hang in there, try something, stay loose, hope for the best, and above all keep moving. Nothing if not mercurial, nimble and ingenious, and, blithely unaware, or unconcerned, that his position is impossible, he seems made for the moment, however long—it could be days or years—the moment lasts.

—April 12, 2000

Notes

1. Hal Hill, Indonesia Industrial Transformation (Singapore: Institute of Southeast Asian Studies, 1997), p. 1, references removed; Steven Radelet, *Indonesia's Implosion* (Harvard Institute for International Development, 1998), p. 1.

2. For the closest thing to a blow-by-blow account of what happened, see Jeremy Wagstaff, "Dark Before Dawn: How Elite Made a Deal Before Indonesia Woke Up," *The Wall Street Journal,* November 2, 1999 (the event itself took place on October 20). For a brief description of the presidential election and its postscript outcome,

see R. William Liddle, "Indonesia 1999: Democracy Restored," *Asian Survey,* XL2, forthcoming (2000).

3. Volcker, the former head of the Federal Reserve, is quoted from a video conference on "international financial architecture" in *The Calendar and Chronicle,* Council on Foreign Relations, March 2000, p. 4:

   I [have] spent my life worrying about [supervision, bank capital standards, disclosure, and risk management] and none of it is going to prevent an international financial crisis.... Large and volatile capital flows are coming up against small, undeveloped financial systems, which is a recipe for a train wreck under the current international structure.

   As an investment banker, Volcker may be suspected of having a bias toward pessimism; but, in the same conference, the former head of Clinton's upbeat Council of Economic Advisors, Laura D'Andrea Tyson, is only marginally more sanguine:

   The goal is not to eliminate financial crises. The question is, can we reduce their intensity and number?. . . I think you can reduce vulnerability by having better regulatory environments, better accounting environments, and greater transparency.

   Since there is no sign on the horizon of "better regulatory environments," or indeed of any serious change in "the current international structure," except perhaps to render it even more unstable by crippling the few institutions that seek to manage it, this is rather cold comfort.

Article 22   *Far Eastern Economic Review,* September 21, 2000

Jakarta's Shame

Indonesia takes the blame as militiamen in West Timor murder three aid workers and drive the UN out of the province's refugee camps

By John McBeth/JAKARTA
and Michael Vatikiotis/WASHINGTON

IT WAS A HUMILIATING moment for Abdurrahman Wahid. At the United Nations' Millennium Summit in New York, the Indonesian president stood with 154 other world leaders as UN Secretary-General Kofi Annan asked for a minute's silence in memory of the Puerto Rican, Ethiopian and Croatian aid workers butchered in a September 6 attack on the UN High Commissioner for Refugees compound in the small West Timor town of Atambua.

With the eyes of the world focused on Indonesia's failure to deal with its side of the Timor problem, Wahid's response has been to blame the international community for not providing enough assistance—or simply to try to redirect attention.

Following the attacks, Wahid was subjected to a litany of outrage from Annan, U.S. President Bill Clinton and other leaders. In a testy meeting with U.S. Secretary of State Madeleine Albright, in which Albright berated Wahid for his

failure to control the militias responsible for the killings, he responded by reminding her he had been swamped with pleas to help resolve international conflicts from the Middle East to Kashmir. He made the same boast in a gathering the next day at Columbia University in New York, where he received an award for his "lifetime contribution to humanity."

A human-rights official in New York says that though Wahid handled the criticism well, "you didn't get the sense he really knows what's going on" in West Timor.

Barely 48 hours after Wahid arrived in New York, machete-wielding militiamen hacked to death the three UN workers, burning their bodies in the street as seemingly outnumbered soldiers and policemen looked on. The next day, eight people were killed in fighting between local villagers and militiamen outside the Betun refugee camp, south of Atambua. As the worst case of violence between locals and militiamen so far,

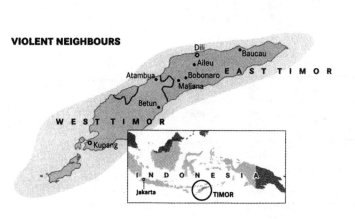

VIOLENT NEIGHBOURS

that incident was yet another sign of rising social tensions across West Timor.

For months now, UN peacekeepers have warned that the Indonesian government's failure to assert its authority has put the province of West Timor in increasing danger of falling under militia control. Annan and U.S. and European leaders have pressed Jakarta for much of this year to rein in the militias; at the summit, the UN Security Council called on Indonesia to immediately disarm and disband them. But a Western military officer who toured the West Timor border region a fortnight before the Atambua attack told the REVIEW: "The Indonesians just haven't provided the resources the problem needs. There doesn't seem to be the will to do anything."

Says a Jakarta-based ambassador: "We just can't understand why the government is allowing one of its own provinces to be subverted." Wahid's weak civilian government, a yawning leadership gap in the Indonesian armed forces, and support for the militias from active and retired military figures are all blamed for Jakarta's failure to impose effective control.

The mayhem was sparked by the September 5 slaying of militia leader Olivio Moruk, who was decapitated and castrated in Betun just a week after Indonesian prosecutors named him as one of 19 people suspected of human-rights abuses in East Timor. Indonesian officials claim he was the victim of a local dispute, but the timing suggested other motives: He was killed exactly one year after his militiamen allegedly slaughtered 200 independence supporters in a church in Suai, on East Timor's southwest coast. Was it revenge or were some of his former military backers enforcing a code of silence?

Only last month, Indonesian Foreign Minister Alwi Shihab said he needed three to six months to close the camps and put a lid on the problem. Since then, little has changed. Two Indonesian infantry battalions are strung out along the 170-kilometre border trying to prevent 200 hard-core militiamen crossing into East Timor. The security forces have done nothing, however, about the militias' control of the refugee camps or their intimidating acts in other parts of the province, including Kupang, the West Timor capital.

Little wonder, perhaps. Eurico Guterres, the leader of the Aitarak militia, which was blamed for some of the worst atrocities in East Timor after the UN-supervised vote on independence last year, now heads the West Timor paramilitary youth wing of the Indonesian Democratic Party for Struggle, headed by Vice-President Megawati Sukarnoputri. Two months ago, Guterres was seen dining with disgraced former special-forces commander Lt.-Gen. Prabowo Subianto in Kupang, suggesting continued military collusion with his militia. Western intelligence agents have seen Prabowo in Kupang three times this year, most recently on August 31.

Megawati, a fervent nationalist, sided with the military over the East Timor issue. She also enjoys good relations with former armed-forces commander Gen. Wiranto, who may yet face trial for failing to stop the militia rampage in East Timor last year that left more than 1,000 people dead.

In a poignant example of just how much Jakarta has lost control in West Timor, regional commander Maj.-Gen. Kiki Syahnakri dispensed with time-consuming clearances and gave the go-ahead for three armed New Zealand helicopters carrying special-forces troops to evacuate 55 UN and other aid workers trapped in Atambua hours after the militia attack. Given the strained relations between Indonesia and the UN authority in East Timor, this was an extraordinary move.

In New York, Wahid asserted the murders were committed to embarrass him, and ordered troop reinforcements into West Timor "to help control the situation." But he expressed no regret over Indonesia's failure to act against the more than 2,000 militiamen in West Timor, and said it would take money from the international community to resettle them in other parts of Indonesia.

New Coordinating Minister for Security and Political Affairs Bambang Yudhoyono, who in a recent published interview did not mention West Timor as being among his priorities, has since promised to restore security and order. He didn't say what he would do about the militiamen, all of whom were originally armed and trained by the Indonesian military. By mid-week, Jakarta was moving at least three army battalions of up to 800 men each into the province.

Now that UN agencies are refusing to return until the militias are removed, aid workers worry about the spectre of famine and the possibility of refugees going on the rampage in search of food. UN officials estimate that 60,000-70,000 refugees would return to East Timor if they were permitted to do so by the militias.

The rest of the refugees include 2,600 former East Timorese soldiers, 8,000 ex-civil servants and their families who would lose their Indonesian pensions if they returned, and others who have been won over by militia propaganda, which teaches camp residents that UN workers will rape female returnees and use the men as forced labour. Senior UN military sources in Dili told the Review that militia recruitment in the camps has in fact accelerated in the past two months.

THE FIGHT in East Timor

An Australian soldier holds his finger tightly on the trigger of his automatic rifle, watching with his unit for movement in the brush across the stream that separates East and West Timor. The threat is real. Two UN soldiers, a New Zealander and a Nepalese, have been killed since late July by pro-Indonesian East Timorese militiamen who have managed to sneak across what has become one of the most heavily defended borders in Southeast Asia.

Lt.-Col. Brynjar Nymo, the Norwegian spokesman for the UN peacekeeping force in Dili, says that as many as 150 militiamen in eight to 10 groups, each of five to 30 men, have managed to cross the 170-kilometre border from West Timor over the past few months.

Maj. David Thomae of the 6 Royal Australian Regiment in Maliana—a town that was almost completely destroyed in violence last year—calls these groups "a completely new type of militia. Last year, they were armed with pipe guns and machetes. Now they carry automatic rifles and hand grenades."

Few people in the border areas doubt the militias are receiving support from the Indonesian military and powerful politicians in Jakarta. Most militia members appear to have had some training in basic guerrilla warfare.

Local villagers are scared. Maria Soares, a young woman in the hilltop border town of Bobonaro, says people "don't dare to go to their fields in the hills, so we are short of food."

Nymo says the numerous refugee camps in West Timor are "the power base of the militias," and their claim to legitimacy is based on the population they control there.

Since last week's murders forced the pullout of UN aid workers from the refugee camps, border security has been tightened, with more Indonesian troops in West Timor, but the remote hills of East Timor's western region are no less tense.

East Timor has been divided into three sectors for peacekeeping purposes: West, with 2,200 men from Australia and New Zealand, with smaller contingents from Fiji, Nepal and Ireland; Central, with 1,026 men from Portugal and a company of Kenyan troops; and East, with 1,636 men from Thailand, the Philippines and South Korea.

The most important sector is, of course, Sector West, nearest the border: Some peacekeepers complain that because the Portuguese in Sector Central do very little patrolling, militias are safe once they have managed to cross Sector West. One 30-man militia group, the largest known, managed to reach Same in Sector Central.

Maj. Thomae says militia activity near the border has increased markedly over the past few months. "We patrol the area constantly," he says, "and our aim is to isolate the militias in their mountain hide-outs, to restrict their movements."

The UN wants to prevent the militias from reaching the local population. It seems to be working. On a recent UN mission on September 1, Australian troops had surrounded rebels hiding out on a mountaintop above Maliana. In order to avoid a raid and possible casualties, troops dropped leaflets by helicopter, urging the group to surrender.

The aim of the militias, UN spokesmen say, appears to be to "wait out the UN," which is supposed to pull out after next year's elections. The people of East Timor see that as an invitation to disaster. As Efren de Guzman, a Filipino Jesuit priest in Maliana, says: "The UN should not leave. When the peacekeepers leave, how can the local people defend themselves?"

East Timor does have its own defence force: the remaining elements of Falintil, the armed wing of pro-independence group Fretilin. And Falintil's commanders may have a unique insight into the tactics of the militiamen. By using the East Timor mountains as a base for ambushes, the militias are copying Falintil's guerrilla tactics during its struggle against the Indonesians.

Domingos Pacheco, a farmer in Bobonaro, says that "we have no security, and the UN's peacekeeping force doesn't know the terrain here. They need help from the Falintil to find the militias' hide-outs."

But under the UN's present mandate, interaction with Falantil is limited. Falintil keeps one liaison officer at the UN peacekeeping-force headquarters in Dili, and three in each of the three sectors. Local commanders in the field also seem to be in favour of more active Falantil participation in tracking down the militias.

Falintil participation is a very delicate issue, especially if Falintil fighters and UN forces start to work together in the border area. Indonesia would consider that a provocation, and could step up support for the militias. Force spokesman Nymo says the first step might be to assign Falintil liaison officers at the company level, not just the sector level.

In fact, the UN presence may have weakened Falantil. When the first international peacekeeping force, Interfet, arrived in September last year, it was under instructions to disarm "all armed factions," including Falintil. After Falintil refused, a compromise was reached, and the force's remaining 1,500 men were put in a cantonment in the small town of Aileu in the hills south of Dili. There they are allowed to retain their guns, but must leave them in the camp when they travel.

Many Falintil troops have been reposted to their local areas or given leave to return to their families, according to a recent study prepared by the Centre for Defence Studies at King's College, London.

A year of cantonment has demoralized the group. "Falintil finds itself marginalized," the King's College report says. The remarkable discipline the fighters showed last year is gone, and members of the group have resorted to smuggling, theft and extortion, according to the report.

It is also difficult today to determine who is Falintil and who is not. Several influential Falantil commanders have left Aileu, guns in hand, and taken up residence in the Baucau area in a group that calls itself the *Sagrada Familia,* or Sacred Family.

Falantil will need support to establish itself as a proper defence force when the UN departs. But the UN cannot provide any training to Falintil, says Nymo. For Falintil to become a proper force, either the mandate will have to change, or Falintil will have to reach bilateral agreements with the defence forces of individual countries.

Even if the UN's mandate is not extended, East Timor would need bilateral defence agreements with countries such as Australia. Defence analysts in Canberra say Australia has to be prepared for a long stay in East Timor.

On September 1, the situation remained tense. UN armoured personnel carriers moved closer to the mountain above Maliana where the militiamen were ensconced, and Black Hawk helicopters ferried supplies. And nervous residents, watching the operation, begged the peacekeepers not to leave East Timor.

Berth Lintner/Maliana, Bobonaro and Memo, East Timor

Article 23

CQ Researcher, May 19, 2000

FUTURE OF KOREA: BACKGROUND

Unified Korea

The Korean people trace their ancestry to members of a branch of the Mongoloid race who migrated from north central Asia and inhabited the mountainous peninsula beginning at least 20,000 years ago.[1] By the fourth century B.C., a tribal kingdom called "Choson" (possible translation: "land of morning calm") emerged near the Chinese border. A period of Chinese rule was followed by the establishment of three separate kingdoms by the 4th century A.D. One of those—the Silla kingdom—overwhelmed the others with Chinese help and unified nearly all the peninsula in 668.

For the next 1,200-plus years, Korea—the Western name is derived from the ruling Koryo dynasty (918–1392)—was a unified country with a single administration, a common language and a proud but troubled history. Korea was a seafaring, trading country with a rich culture derived in part from its larger neighbor China. Koreans perfected printing with movable metal type in 1403—a half-century before Gutenberg—and built the world's first armored warships in the 16th century.

Geographically vulnerable, Korea suffered under repeated invasions. In the wake of Japanese and Manchu invasions in the 16th century, it established a policy of excluding foreigners. Westerners called it "the Hermit Kingdom." The policy gave way as European powers and the United States moved into Asia beginning in the mid-19th century as traders, missionaries and overlords. Korea signed trade treaties with Japan in 1876 and in 1882 with the United States.

Internally weakened, Korea became the prize in a three-power rivalry between China, Japan and Russia. Japan won out after defeating China in one war (1894–95) and Russia in a second (1904–05). As part of a peace agreement brokered by President Theodore Roosevelt, Russia recognized Korea to be in Japan's sphere of influence. Roosevelt won the Nobel Peace Prize as well as Japan's acquiescence to the U.S. occupation of the Philippines. Five years after the Russo-Japanese war's end, Japan formally annexed Korea.

Japan built infrastructure and pushed industrialization but also ruthlessly crushed nationalist sentiment. In one notorious incident, Japanese police in 1919 killed as many as 7,500 pro-independence Koreans by locking them in a church and burning it to the ground. Resistance, driven underground, took two forms: a provisional government in exile in China, headed by the Western-educated Syngman Rhee, and a Communist-led guerrilla movement.

The United States, with its World War II allies Britain and China, endorsed Korean independence in 1943. As the war with Japan neared an end in 1945, Russian forces moved into Manchuria and northern Korea. Belatedly, the United States recognized Korea's strategic importance. On Aug. 10, 1945, two young Army officers—including the future secretary of State, Dean Rusk—proposed that Russia accept the surrender of Japanese troops north of the 38th parallel and the United States occupy to the south. Korea was now divided.

Korea's Civil War

The division of Korea, ostensibly provisional, quickly hardened. When the Soviets reached the 38th parallel, they cut transportation and communications links with the south. Kim Il Sung, a 28-year-old Korean communist, was installed to head the administration. In the south, the United States organized an administration headed by the septuagenarian Rhee, whose fierce anti-communism matched Kim Il Sung's ideological zeal.

Under President Harry S. Truman, the United States won passage of a United Nations resolution in November 1947 calling for peninsula-wide elections for a single government. The Soviets, as expected, refused to admit U.N. electoral commissioners into the north. Meanwhile, the Rhee regime had used a national police force—established with U.S. support—to crush a peasant- and labor-backed communist movement in the south. In the north, the communist regime effected land reform and nationalized basic industry while also suppressing political dissent.

A U.N.-sponsored election in the south on May 10, 1948, produced a national assembly dominated by Rhee's conservative supporters. The assembly wrote a democratic-style constitution and elected Rhee president. The Republic of Korea was officially proclaimed on Aug. 15. In the north, after Soviet-style elections, the Democratic People's Republic of Korea was officially proclaimed on Sept. 9, with Kim Il Sung as premier. Both regimes claimed to govern all of Korea.

Early in 1950, North Korean troops repeatedly engaged in clashes with South Korean forces along the 38th parallel. Nevertheless, the North Korean invasion on June 25 took the United States by surprise. Over the next two days, Truman's secretary of State, Dean Acheson, fashioned the plan to commit U.S. forces to South Korea's defense and to take the issue of North Korea's "aggression" to the United Nations. With the Soviet Union ill-advisedly boycotting the international body, the United States won Security Council approval for a U.N. force to defend South Korea.

North Korea took Seoul on June 28 and pushed South Korean forces back to an enclave along the southeastern coastal city of Pusan. Seoul was "liberated" on Sept. 15 after the famous Inchon landing—planned by Gen. Douglas MacArthur, the American designated as U.N. commander—but retaken on Jan. 15, 1951, after Communist China had joined the war on North Ko-

rea's side. By April, U.N. forces had retaken Seoul and pushed the communist forces back across the 38th parallel. The United States urged peace talks in July, but the war dragged on for two more years.

The war finally ended on July 27, 1953, with an armistice—not a peace treaty—signed only by the United States and North Korea. To the end, Rhee wanted to continue the war to unify Korea under his government. With the end of the fighting, North and South Korea remained as armed camps, each one brooking no internal dissent and planning for Korea's unification on its own terms.

Divergent Paths

The two Koreas took sharply divergent paths over the next four decades. South Korea, backed by U.S. aid, emerged as an economic powerhouse and—by the end of the 1980s—had moved toward more genuine democratization after 30-plus years of autocratic and military rule. North Korea became a closed, determinedly self-reliant society—a new "Hermit Kingdom"—with a cult of personality around "the Great Leader" Kim Il Sung and a reputation as an international pariah.

The rivalry between the two Koreas continued unabated after the war. At first, the North enjoyed significant advantages. Economically, as historian Bruce Cumings points out, the government's command-style industrialization plans pushed the economy forward at "world-beating growth rates" though the 1970s—25 percent per year in the decade after the war, 14 percent per year from 1965–1978.[2] South Korea in the 1950s under Rhee followed free-market policies in theory but government-directed development in fact—backed up by U.S. aid that averaged $600 million per year. Despite the assistance, the economy—in the assessment of a Korean expert—was "a failure" and "an unaccountably expensive one."[3]

Rhee's successor as president, Park Chung Hee (1961–79), is credited with beginning the transformation of the South Korean economy with policies that he called "guided capitalism." While North Korea pursued economic self-sufficiency, Park in the 1970s

pushed the development of export-oriented industries, including steel and electronics.

Meanwhile, North Korea's growth was stalling. South Korea surpassed its communist neighbor in per capita gross domestic product by 1976, according to one Central Intelligence Agency estimate, and by 1986 according to another.

Both countries maintained disproportionately large militaries. North Korea's numerical superiority came to be offset by South Korea's technological advantages and U.S. backing. South Korea remained on edge not just because of the fear of invasion but also because of North Korean terrorism, including an attempted assassination of Park in 1974 and the bombing deaths of four senior South Korean leaders in Rangoon, Burma, in 1983.

Politically, Kim Il Sung held all but absolute power in North Korea as a kind of Confucian-style communist until his death in 1994. He proclaimed a philosophy of "Juche"—roughly, self-reliance—that elevated Korean nationalist identity and opposed "flunkeyism" to foreign powers.

South Korean politics was both unstable and unfree. The autocratic Rhee was toppled by student demonstrations in 1960. Park came to power a year later through a military coup and kept his hold in power thanks to what amounted to a second coup in 1972, when he declared martial law and mandated indirect election of the president. Park was assassinated by his own intelligence chief in 1979.

Another coup thwarted democratization and brought Chun Doo Hwan to power. He designated Roh Tae Woo as his successor and—to quiet student protests—decided to return to direct presidential elections. Roh won a four-way race with 36 percent of the vote in December 1987. Two pro-democracy candidates—Kim Young Sam and Kim Dae Jung—split the opposition majority with 28 percent and 27 percent of the vote, respectively.

Difficult Engagements

North Korea began by the late 1980s to try to emerge from its diplomatic isolation with overtures to South Korea and the United States. The efforts became

Fifty Years After the 'Forgotten' War . . .
Americans Are Starting to Remember

Just two months after enlisting in the Marine Corps Reserve, a raw recruit named Clyde H. Queen Sr. found himself disembarking with thousands of other U.S. servicemen at the South Korean port of Inchon. The September 1950 landing put U.S. troops behind enemy lines and led to one of the first important U.S. victories in the Korean War—the retaking of the South Korean capital of Seoul.

Two months later, Queen found himself in less successful combat: the Battle of Chosin Reservoir deep in North Korean territory near the Chinese border. U.S. forces, virtually surrounded by Chinese troops in weather so cold that rations and blood plasma froze, fought their way south to safety. But some 3,000 Marines were killed or wounded in the action, the United States' first contact with Communist Chinese forces.

Queen, of San Pedro, Calif., recounts his experiences on a Web site maintained by the Defense Department committee commemorating the 50th anniversary of the war's start.[1] His service left him 70 percent disabled and with painful memories, including 150 comrades buried in a mass grave along the course of the Chosin retreat. "That never leaves you," he tells a Navy journalist.

But most Americans seem to have forgotten about the three-year, United Na-tions-authorized "police action." Veterans call it "the forgotten war."

"A lot of Korean War veterans feel bitter," says Army Lt. Col. Jim Fisher, director of operations for the 50th Anniversary of the Korean War Commemoration Committee. "They feel they never received the 'atta-boys' and laudatory comments that a lot of World War II veterans received."

The commemoration officially begins on June 25 with a ceremony at the Korean War Memorial in Washington. President Clinton will be the keynote speaker.

"We want to ensure that our Korean War veterans receive the proper recognition for their service and sacrifice," Fisher says. South Korea will launch its own commemoration the same day. The official U.S. party will include two of the eight Korean War veterans in Congress: Sen. John Warner, R-Va., and Rep. Charles Rangel, D-NY.

The three-year war began with the North Korean invasion of South Korea on June 25, 1950, and ended in an inconclusive armistice signed at the border village of Panmunjom. Technically, the two Koreas are still at war. Many Americans viewed the war as a failure since the communists remained in power in North Korea, and President Harry S. Truman had refused Gen. Douglas MacArthur's request to take the war into Communist China. But David Kaiser, a historian at the Naval War College in Newport, R.I., says the United States, in fact, won the war.

"We preserved an independent, non-communist South Korea," says Kaiser, author of a new book on U.S. policymaking in the Vietnam War. "We got into big trouble" by drawing the Chinese into the war by approaching the Yalu River border, Kaiser continues, "but eventually we were able to win the original objective despite the Chinese intervention."

Despite ample warning, the war came as a surprise to U.S. military and diplomatic officials alike.[2] The Joint Chiefs of Staff viewed South Korea as of minimal strategic value and in 1948 recommended withdrawal of all U.S. troops, even if it led to domination of Korea by the Soviet Union. Two years later, Secretary of State Dean Acheson declared that Korea was outside America's defense perimeter in Asia. Later, Republicans blamed Acheson—but not the military—for inviting the North Korean attack.

North Korea's leader, Kim Il Sung, wanted to unify all of Korea under his communist regime. In the South, Syngman Rhee similarly wanted a unified Korea with himself as president. Kim, a disciplined communist dependent on Moscow's support, sought permission for the attack from Soviet leader Josef Stalin. Stalin initially refused but relented after the communist victory in the Chinese civil war. "I

(continued)

more critical with Pyongyang's loss of ideological patronage from Moscow and Beijing at the end of the Cold War. South Korea, meanwhile, marked its arrival on the world stage by hosting the Summer Olympics in Seoul in 1988. Democratization and economic development proceeded apace over the next decade, but with two significant bumps: a major domestic corruption scandal in 1996 and a significant financial crisis in 1997.

The new decade began with an auspicious agreement between the two Koreas. The "Agreement on Reconciliation, Nonaggression and Exchanges and Cooperation between the South and the North"—initialed by the two governments on Dec. 13, 1991—called for mutual recognition, renunciation of force, peacemaking efforts and economic, social and cultural exchanges.

Three years later, Kim Il Sung agreed to a summit meeting with his South Korean counterpart, Kim Young Sam, who had been elected president in 1992. But the summit was aborted by Kim Il Sung's death in July 1994.

During this period, peacemaking also had been on the agenda for the United States, which began covert diplomatic dialogue with North Korea in October 1988. The dialogue became more urgent against the backdrop of North Korea's nuclear weapons program.[4] U.S. intelligence had detected evidence of a nuclear weapons program—including a possible plutonium-reprocessing facility—by 1986.

The Bush administration shared the information with the Soviet Union, China, Japan and South Korea in 1989, and the story quickly leaked. To try to defuse the controversy, North Korea signed a second agreement with South Korea in 1992 with both countries pledging not to develop nuclear weapons or to possess nuclear-reprocessing facilities.

Pyongyang's resistance to inspections by the International Atomic Energy Agency (IAEA), however, produced a prolonged crisis. North Korea decided in 1993 to withdraw from the Nuclear Non-Proliferation Treaty. A year later, both Seoul and Washington am ready to help him in this matter," Stalin wired the Soviet ambassador in Pyongyang in January 1950.

North Korea began with daunting military advantages. "The North Korean military was larger and better trained, and it had tanks and the benefit of a rather large network of Soviet advisers," says retired Army Gen. Robert Sennewald, who fought in the Korean War and served in the 1980s as commander-in-chief of the South Korean-U.S. military command.

By contrast, South Korea had "a pathetic small force that had not been through any training and had no equipment," according to Katy Oh, a Korean-American and a researcher at the Institute for Defense Analyses. In addition, Sennewald notes, South Korea "had just finished a very divisive war inside their boundaries against communist forces. It wasn't until 1948–49 that the last communist enclave was defeated in southwest South Korea."

Initially, the war went badly for the South Koreans and for the United States. "In the first 120 hours of the war, the South Koreans lost one-third of their officer corps and one-third of their enlisted personnel," Sennewald says. U.S. forces also fared badly because of poor leadership and overconfidence.

Fighting a determined foe on unfamiliar, mostly mountainous terrain, U.S. troops—who made up the bulk of the United Nations command—suffered a series of losses and defeats until finally establishing a defense perimeter in August 1950 outside the port city of Pusan. A month later, though, MacArthur—the World War II hero designated by Truman as the United Nations commander in

chief—brought the Americans their first cause for cheer with an amphibious landing behind enemy lines at Inchon.

After retaking Seoul, MacArthur directed the U.N. forces across the 38th parallel despite warnings that the move would invite Chinese intervention. Pyongyang fell on Oct. 19; the U.N. forces continued northward, closer to the Chinese border. The Chinese responded in massive numbers in November, forcing the U.N. troops back south of Seoul by January 1951. MacArthur responded by calling for aerial bombardment and a naval blockade against China. Defying a presidential directive, he expounded his views in public statements in February and again in March. Truman responded by firing MacArthur on April 10—provoking public outrage. MacArthur returned to Washington and a joint session of Congress as the politically handcuffed advocate of "no substitute for victory."

With MacArthur gone, Truman sought a truce. Cease-fire talks began in July 1951 and dragged on for two years—delayed by North Korea's insistence that its POWs be returned home, willing or not. The war also dragged on. Some memorable battles remained to be fought, including Heartbreak Ridge and Pork Chop Hill. But, as Kaiser notes, there were "no major movements of the front." "The objective of most of the soldiers was to serve 12 months and go home," Sennewald says.

The war provided chastening lessons on the limits of presidential power. Truman's popularity never fully recovered from his dismissal of MacArthur. The Supreme Court's decision barring Truman's seizure of struck steel mills still stands as

a legal barrier to unilateral presidential action in wartime. And Gen. Dwight D. Eisenhower's 1952 victory over the Democratic candidate for president—after pledging to go to Korea if elected—demonstrates a president's difficulty in maintaining political support while waging an inconclusive war.

For Koreans, the war left a bitter legacy: the lasting political and ideological division of their once unified country. "Koreans fought and killed each other," says Oh. "The war provided a foundation for real mistrust and hatred toward each other." Toward the United States, South Koreans have "a deep feeling of gratitude," Oh adds, but tinged with some resentment of the outside assistance.

Americans were ambivalent about the war—and remain so to this day. World War II is remembered in popular culture through scores of epic films. The Korean War is remembered through the movie and later television series "MASH," with its decidedly unglamorous and unheroic depiction of war and military valor.

But Sennewald says we won the war, and that a celebration is warranted. "The legacy is a free country, a country that is economically robust, and a good ally of the United States," he says. "Any veteran of that war should take a lot of pride in that."

1. See http://korea50.army.mil.

2. Background drawn from Stanley Sandler, *The Korean War: No Victors, No Vanquished* (1999). See also John Toland, *In Mortal Combat: Korea, 1950–1953* (1991).

feared war was imminent. But the confrontation ended with a deal brokered by former President Jimmy Carter with Kim Il Sung in June 1994. Kim agreed to freeze the program and allow international inspections in return for broader, official talks with the U.S. Those negotiations produced the October 1994 "Agreed Framework." It called for the U.S. to organize an international consortium to furnish North Korea with light-water nuclear reactors and for North Korea to comply with the IAEA and dismantle existing nuclear facilities once the new reactors were completed.

In the South, Kim Young Sam's presidency ended badly.[5] His government sent former presidents Roh and Chun to prison for maintaining secret slush funds, but Kim caught the backlash instead of gaining credit for dealing with the scandal. The Asian financial crisis of 1997 forced Seoul to accept an International Monetary Fund (IMF) bailout. Kim's popularity plummeted. In December 1997, his pro-democracy rival, Kim Dae Jung, won a narrow victory over a candidate associated with the ruling elite.

The president-elect immediately proposed a summit with North Korean leaders.[6] "Through direct dialogue between North and South Korea, I will look for ways to solve the problems between our peoples," Kim said. North Korea had no immediate official response: An unofficial adviser said the reaction would be "positive, but still a bit reserved."

Notes

1. Background drawn from Bruce Cumings, *Korea's Place in the Sun: A Modern History* (1997) and Oberdorfer, *op. cit.*

2. Cumings, *op. cit.*, p. 423.

3. Jung-en Woo, *Race to the Swift: State and Finance in the Industrialization of Korea* (1991), cited in *ibid.*, pp. 305–06.

4. Background drawn from Oberdorfer, *op. cit.*, pp. 249–336. See also Leon V. Sigal, *Disarming Strangers: Nuclear Diplomacy with North Korea* (1998).

5. Background drawn from *The New York Times*, Dec. 22, 1997, p.A12.

6. See *The New York Times*, Dec. 20, 1997, p. A1.

Vital Speeches of the Day, July 1, 2000

Relations between North and South Korea:
Unification of our homeland

Address by KIM DAE-JUNG, President of the Republic of Korea

Delivered on his arrival from Pyongyang, Seoul, Korea, June 15, 2000

Fellow Koreans, I have just returned home after completing the historic visit to North Korea. I am pleased to be back with my tasks finished as planned. I appreciate very much that the nation supported me, skipping sleep so that I could finish my tasks during my visit to the North.

A new age has dawned for our nation. We have reached a turning point so that we can put an end to the history of territorial division of 55 years. It is my hope that my visit to the North has contributed to peace on the Korean Peninsula, exchanges and cooperation between the South and North as well as to the unification of our homeland.

National Defense Commission Chairman Kim Jong-il's welcome and hospitality was beyond my expectations. He went to the Pyongyang airport to welcome me on arrival and also on my departure. In the course of the talks, there were times when I was desperate, but I tried my level best with faithfulness. Eventually, Chairman Kim extended substantial cooperation, and we were able to reach the agreements we are dedicating to the nation.

When I arrived in the northern capital city, as many as 600,000 citizens came out to greet me. When I was returning, 300,000 to 400,000 Pyongyang citizens turned out. That means that 1 million citizens hit the street on this occasion, and they said that it was the largest turnout in the history of the city. I am very grateful for the welcome which I think is an expression of love as members of one ethnic family. With you, I thank the Pyongyang citizens very much with applause.

The summit was supported by the most encouraging world reaction. All the countries of the world supported it virtually without a single exception. Also I would like to express my gratitude for the support given by the press. In Pyongyang, I watched and read reports by our TV and newspapers. That kind of press support is unprecedented. I wonder if I really deserve that kind of support, but I appreciate it very much. I take it as evidence of the press's fervent interest in reconciliation and cooperation within the nation.

In the summit, I told my counterpart the two of us have responsibilities to the nation and the world. I said that if we fail we will be inflicting tremendous disservice on them. I also said that if we succeed, we will bring about great development and a turning point in the history of the world. I approached the summit with a sense of mission and determination to succeed. At every opportunity, I pledged to myself that I would engage in the summit with utmost faithfulness and wisdom. Our other official and non-official delegates, in their meetings with their counterparts, did their best to further develop the bilateral relationship and offer assistance to my work.

However, more than anything else, the summit meeting itself was the most important. I found that Pyongyang, too, was our land, indeed. The Pyongyang people are the same as us, the same nation sharing the same blood. Regardless of what they have been saying and acting outwardly, they have deep love and a longing for their compatriots in the South. If you talk with them, you notice that right away. That is quite natural because we have been a homogeneous nation for thousands of years. We lived as a unified nation for 1,300 years before we were divided 55 years ago against our will. It is impossible for us to continue to live separated physically and spiritually. I was able to reconfirm this fact first-hand during this visit. I have returned with the conviction that, sooner or later, we will become reconciled with each other, cooperate, and finally get reunified.

I told Chairman Kim, in the waning years of the Choson Kingdom, when the people should have united and hastened modernization, the country was splintered and turned away from modernization. In the end, we earned the sorrow of losing the country, resulting in 35 years of Japanese colonial rule, the division of the country on August 15 (1945), the Korean War and the confrontation across barbed wire. Thus, didn't they give their descendants 100 years of punishment? The world is now entering into an age of the greatest revolutionary change in the history of mankind, which is called the age of knowledge and information.

At a time when the world is also entering into borderless and boundless economic competition, how can we survive if we who are one people waste our energy against each other? On the other hand, even if we cannot unify the country right away, we can open the skies, roads, harbors, we can come and go freely, cooperate with each other, develop the economy together, and have exchanges in culture and sports. Wouldn't the Korean educational tradition and cultural creativity be assets in the age of knowledge in the 21st century? It is not an age of imperialism when the big four powers rule us. On the contrary, the big four powers are our markets, and we can take advantage of them. At this time, if we don't become alert and the South and North don't cooperate with each other but fight, instead, what would be our fate? Therefore, whatever happens, we should not stick to the ideas of communizing the South or absorbing the North. Instead, let us coexist and proceed on the path toward unification. At this time of best opportunities, which is the 21st century, I stressed to the North that we must forge a first-rate nation on the Korean Peninsula. I would like to tell you that they expressed agreement.

My fellow citizens, even though I am telling you all this, it does not at all mean that everything went smoothly, and there is nothing to worry about. This is only the beginning. I am only saying that I came back after looking at the possibilities. It will take time; we need patience. And we need devotion. We also need to look at things from the point of view of the other side. There should not be the slightest wavering in our resolve to maintain national security and the sovereignty of the Republic of Korea. But we must ultimately go on the path toward unification by solving one thing at the time, solving the easiest things first while cooperating with each other and giving consideration to the other side.

I told the North that we should say everything we wanted to say to each other. The gist of what I wanted to say was also written down and handed to them. We talked about nuclear and missile issues. The issue of the U.S. forces stationed in the South also cropped up. The issue of the National Security Law was discussed. Dialogue was very useful, and I was able to confirm that there are things that have a bright prospect for resolution.

Now, I would like to offer a few brief words on the content of the South-North Joint Declaration. First was to resolve the problems of the people independently by ourselves. This was contained in the July 4 Joint Communique. However, I told the people of the North. It is a matter of course that our problems should be solved by ourselves. But it was issued 28 years ago and yet nothing has been achieved. It talked about independence, peace and unity of the people, but nothing has been achieved.

Second, in February 1992, the South and North produced the Basic Agreement in which the two sides promised reconciliation, nonaggression, exchanges and cooperation, and the denuclearization of the Korean Peninsula. But this Basic Agreement, despite all the detailed provisions, has failed to produce any practical result, as did the July 4 Joint Communique that defined the basic principles of reunification. What we need is practical action. The South-North summit meeting this time was a dialogue for practical action. If we repeat saying such words as independence, unification and peace, without any practical moves, the peoples of the world would not trust us any more. Thus, we agreed upon some concrete matters in Item 2 and below, which involve practical tasks. The second paragraph is concerned with a South-North confederation which we have proposed as the formula for reunification. The confederation concept requires maintaining two governments for the two sides as they are now and create a conference of ministers and an assembly with which the two sides can jointly solve problems step by step.

In 1980, North Korea also proposed a confederation system. But this formula would give the central government the functions of diplomacy and defense with the local government exercising only the power of internal administration. That is a totally impracticable idea. In recent days, North Korea has revised this idea and has accepted the notion of allowing the local governments the right to diplomacy and national defense, calling it a "loose form of confederation." This is practically identical with what we have been advocating as the unification formula. Because the two ideas have common elements, we have agreed to hold talks between representatives of the two sides with scholars and other experts joining in the discussions. I believe this will provide a great momentum in searching for practical agreement in the history of the nation's unification movement.

Third, the South and the North will exchange groups of separated family members around August 15 and will solve the question of unconverted Communist prisoners serving long-term sentences (in the South). What I'd like to tell you here is that the focus is the problem of the North Korean refugees and separated families. At the airport today, I talked with Chairman Kim Jong-il about this issue.

I told him to do boldly as they used to say in the North by August 15. Then, I said I would seek to solve the problem of the unconverted prisoners they are talking about and others through consultation with the people.

I told him to do well first, and we agreed to do so.

The Red Cross will start to work right now this month, June. I told him that upon arriving in Seoul I would request the Red Cross to contact the North, and Chairman Kim agreed.

We are unable to know how many people will be reunited with their separated family members, but I report to you here that the first batch of visitors will be considerable as agreed on between the two sides.

Fourth, we agreed to promote the balanced development of the national economy through cooperation and increase cooperation in all fields—social, cultural, sports, public health, environmental and so on. It is true that the North Korean economy is difficult. It is true that our cooperation will be helpful to the North. Both the North and the South will enjoy great benefits when we construct railroads, solve the power shortage, and build roads, port facilities and communication networks in the North, and when we advance into the North by building industrial complexes.

That's not all. Why can't our trains go to London or Paris? They can't because the Kyongui Railroad Line (Seoul-Sinuiju in the North) and the Kyongwon Line (Seoul-Wonsan in the North) are cut. In Manchuria, trains can go there freely, can't they? In case of Kyongui Line, only 25 kilometers of rails are cut. If we connect them, our trains can go. Transportation costs will be reduced by 30 percent and the period of transport will be cut drastically. If we agree with the North to solve this problem, we can drive on without a stop to Europe. If so, there will appear a silk road in the new millennium and the day will come when both the South and the North will be able to enjoy great economic prosperity.

There have been many news reports that North Korea has a very superior labor force. Wages there are very low. Our small and medium enterprises whose competitiveness in the South is low can secure strong competitive power in the North. Both will benefit from this.

What we have to do is to make a hard and fast rule in interKorean relations and that is that only one side, either the North or the South, should not enjoy profits unilaterally. Both should be happy. Then, there will be reconciliation and cooperation. We have to adopt a win-win policy. I reached an agreement clearly with Chairman Kim Jong-il to promote this type

of exchanges and cooperation not only in the economic field, but in all sectors, including sports and cultural.

I do not want to go into detail since I do not have much time. The two sides will designate representatives who will go on discussing the implementation of the issues.

It was difficult to reach an agreement on the issue of Chairman Kim Jong-il's visit to Seoul. But he decided that he would come to Seoul within the agreed time frame. I told Chairman Kim that he has to come to the South to prove to the world the relationship between the two sides will continue to improve. If he does not, then people will think that this was just a one-time event. I told Chairman Kim that he must come to Seoul since an elderly person like me came all the way to Pyongyang.

This basically concludes my briefing.

Let me reiterate once again.

Korea is one country with one ethnic family. Koreans of both South and North have the same behavior and life styles. But it is also true that South and North Koreans lived under different political and social systems. These gaps cannot be narrowed down within a short time. This was the reason that we have not implemented the terms of the July 4 Joint Communique for the last 28 years. We must consider North Koreans as our brothers and sisters. We must believe that they have the same thought. We have to resolve easier and possible issues first, which will benefit both sides. Doing so, we will be able to build up trust and mutual understanding. I will only build the foundation, and the person after me will do better. Most importantly there is no longer going to be any war. The North will no longer attempt unification by force and at the same time we will not do any harm to the North. We have to live together to become a first-rate nation in the world. The four superpowers will no longer be considered imperialists but our markets.

I hope you will treat the North under an assumption that with high intelligence, cultural background, and an information edge, the two of us will explore the world market. Security should be tight, aimed at preventing war and reaching reconciliation and cooperation. When we move this way, God will bless the two nations.

I am confident that we will be able to leave a unified and prosperous peninsula.

Thank you once again. I promise you that I will do my utmost to serve the people of this nation. There is more good news, but I would like to stop here. I wish you all the best.

Thank you.

Article 25

CQ Researcher, May 19, 2000

Future of Korea:
Chronology

Before 1945 *Korea is a unified nation for 1,300 years before Japanese occupation, 1910–1945.*

1945–1970 *Korea is divided with an entrenched communist regime in the North and an autocratic U.S.-backed government in the South.*

1948
North and South Korea are established after U.S. and Soviet Union fail to agree on nationwide elections; anti-Japanese guerrilla leader Kim Il Sung installed as North Korean leader; U.S. picks Western-educated Syngman Rhee as South Korea head.

1950–1953
Korean War ends with inconclusive armistice. More than 54,000 U.S. troops killed.

1960
Student-led revolt topples South Korea's Rhee; moderate government is replaced a year later by military coup led by Park Chung Hee, who begins program of "guided capitalism."

1970s *Abortive effort at dialogue between North and South Korea.*

1972
North and South Korea issue joint statement supporting peaceful unification.

1974
Attempted assassination of Park by North Korea-linked gunman produces tension, recriminations.

1979
Park is assassinated by his intelligence chief; motive undetermined. Return to democratic rule thwarted by Dec. 12 coup led by Chun Doo Hwan, who succeeds to presidency and directs bloody putdown of student uprising in May 1980.

1980s *South Korea surpasses North Korea in economic, military strength; democratization gains in South, no reform movement in North.*

1983
Four ranking South Korean officials killed in Rangoon, Burma, by bomb planted by North Korean army officers.

1987
After pro-democracy protests, Chun's designated successor Roh Tae Woo agrees to direct election of president; Roh wins four-way race on Dec. 16.

1988
Seoul hosts Summer Olympics; South Korea and U.S. seek engagement with North Korea.

1990s *North Korea in economic tailspin after losing Communist patrons; U.S. confronts North Korea over nuclear weapons, missiles.*

1994
Crisis over North Korea's nuclear-weapons program brings war scare in May and June before former President Jimmy Carter's mission to Pyongyang defuses crisis; North Korea's Kim Il Sung dies July 8; U.S. and North Korea complete "Agreed Framework" on Oct. 21 with promise to freeze nuclear program in exchange for civilian nuclear reactors.

1995
Massive floods in North Korea exacerbate food shortages; U.S. and other Western nations begin providing food assistance.

1996
Corruption trials of Roh and Chun bring long prison terms.

1997
Former dissident Kim Dae Jung elected president in South Korea; promises to seek engagement with North.

1998
North Korea fires medium-range ballistic missile Aug. 31 over Japan.

Sept. 17, 1999
U.S. says it will lift some economic sanctions against North Korea, which announces halt in missile testing a week later.

2000s *Plans for first summit of North and South Korean leaders.*

April 10, 2000
Two Koreas announce plan for Kim Dae Jung to meet with Kim Jong Il in Pyongyang June 12–14.

Article 26

The World Today, February 2000

DYING FOR RICE

At the start of the twenty-first century, children in north Korea continue to face death from severe malnutrition. Women do not have rice to feed their babies because the world would rather donate surplus American wheat. US assistance has helped tens of thousands who would otherwise have starved. But food aid now needs to be fine-tuned, more targeted, and directed less by the Department of Agriculture—whose priority is the interests of American farmers—and more by USAID.

Hazel Smith

Dr Hazel Smith is Reader in International Relations at the University of Warwick. She most recently visited north Korea at the end of last year, and is a consultant on the country to UNICEF and the UN World Food Programme.

UNITED STATES FOOD AID FLOWS INTO NORTH KOREA through the UN's organisation, the World Food Programme (WFP), at a cost of $160 million a year. It is the UN's biggest food aid programme. By late last year, although there were signs that wholesale famine had been averted, there was little evidence that the scale of malnutrition differed significantly from that found in a 1998 international survey.

At that time, a staggering thirty-five percent of boys aged twelve to twenty-four months, and twenty-five percent of girls of the same age, were 'wasted'. This technical term accurately evokes the suffering of acute malnutrition where lack of food—combined with disease and illness—threatens life unless there is urgent medical intervention. Survivors may be permanently physically and mentally damaged.

By last year, the humanitarian agencies had a more nuanced understanding of the health and nutritional status of children. This compared with the earlier snapshot of acute and urgent suffering on television, showing skeletal youngsters in nurseries and hospitals. Some severely malnourished and many chronically malnourished children are still visible because of three key problems.

The challenges now are the lack of appropriate foods for young babies; the inability of international aid to reach children not in nurseries, kindergartens and schools; and the specific difficulties of the ten thousand livingin orphanages.

All this is compounded by a lack of reliable information about the health and nutrition of children. Officials jealously guard what they see as their sovereign right to run things in their own country their own way, without foreigners telling them what to do and how to respond to the food crisis.

BABY FOOD

Resident humanitarian agencies—the UN and non-governmental organisations (NGOs)—have found that the very high rate of malnutrition in children under two is caused partly by the lack of appropriate food aid for the weaning child—those who needs to graduate to solid food through a diet of semi-solids. In any country or culture this is difficult, and timeconsuming for mothers and carers who must find suitable food, spend time preparing it and attempt to feed the baby. In north Korea, the problems are enormous. Rice porridge is the traditional and familiar base for weaning food, yet rice is often not available.

Most children of weaning age used to attend nurseries—although no one is quite sure if this is still the case. Nurseries, like the rest of the country, are struggling to cope with a lack of heating in winter temperatures regularly twenty degrees below zero centigrade, electricity shortages and scarce fuel. Workers have to be given time off to look for fuel and food for their families.

It is very difficult to use the unfamiliar commodities provided by international aid, like wheat or wheat flour or corn soya mix. They need extensive preparation and cooking to make them edible—and fuel is scarce. Sugar makes such food palatable but it is often unobtainable. It is not, therefore, easy to persuade children to eat the donated food.

UNREACHABLE

UN agencies and resident non-governmental organisations report healthier children in pre-school or schools than in the previous couple of years, but all are concerned about those non-attending who have no means of receiving UN food, which is distributed only through institutions. NGOs operating in the north of the country say there are obviously malnourished children in cities and urban areas who do not appear to be going to schools or any other institution. Some of this is to do with the floods ofthe mid-1990s. Nurseries, kindergartens and schools were destroyed or severely damaged. Some children are too sick to attend school and others are away from home relatives cope with food shortages. Families in counties where food is not so scarce take care of youngsters from worse-off areas.

LESS CARE

Children up to seventeen in orphanages get priority for governmental and international aid. They are also a cause of concern, despite having had steady access to whatever food and health care the international community has provided.

The lack of weaning food in these institutions contributes to the risk of malnutrition for small children. For older ones, insufficient numbers of care staff pose a problem—they are struggling to cope with a vast increase in numbers in residential care since the mid-1990s.

As with the entire population, the priority of care staff is to find food and basics for their own families' survival. This means regular time off work, leaving even fewer staff to deal with the multiple problems of deprivation where there is absolutely no familial support. The human and physical resources are pitiful compared to the scale of need. This reflects the very serious lack of basic supplies to almost the entire population.

SEEKING HELP

Regular time off work, combined with underemployment because of a lack of materials, fuel and other inputs to keep enterprises functioning, is illustrated by the very common sight of hundreds of women and men purposefully walking, bicycling and hitching rides from county to county seeking self-help solutions to questions of survival.

The formerly extensive state Public Distribution System can no longer guarantee food, fuel and basic supplies, and the government now seems to be permitting, of necessity, some relaxation in controls over internal movement. Another unintended and much less desirable consequence of the food shortages is the shockingly rapid and extensive deforestation in the last five years. People are scavenging for firewood and any possible food source in the once extensive forests and woods.

The chronic food shortages and economic dislocation are a product of both the changed international economic and political landscape and the country's inability to reconfigure its politico-economic relationship to fit into the new global economy.

The abrupt end to cheap oil from the former Soviet Union in the early 1990s, the unavailability of modem technology, the isolation from capitalist markets and the absence of hard currency or capital, combined with a series of natural disasters in the mid-1990s, have all contributed to the tragedy of a nation where sixty-two percent of children are malnourished. Not one child has been unaffected by the lack of food, the absence of basic drugs—like anaesthetics, antibiotics or tetanus vaccine—and medical equipment and the deterioration of water and sanitation systems, contributing to compromised water supplies.

BANDAIDS

The government is responding with the 'second Chollima movement'—a campaign named after a legendary Korean horse that could cover enormous distances in one stride. It is the second such campaign. The first helped rebuild the nation after the devastating Korean war of 1950-1953. The population is being mobilised now to rebuild damaged infrastructure like flood barriers, bridges and roads and to resuscitate production in mines and factories.

The difference today is that the Soviet Union is no longer around to support indigenous efforts through cheap oil, spare parts, technology and barter. The lack of external support, combined with the sheer physical exhaustion of a malnourished population, means that the country can only achieve the crudest bandaid solutions for its struggling economy.

Given that the leadership has no intention of changing its political or economic system to embrace capitalism, in the medium to long term, the main hope is that relations with its oldest adversaries—the United States, the Republic of Korea and Japan—improve to permit capital and technology transfer directly or indirectly through the international financial institutions.

FORGET THE POLITICS . . .

In the short term, the international community can and should do more to assist north Korea's hungry children. Rice should be donated specifically to feed those under two and for nursing mothers. This would encourage longer breast-feeding, which would in turn help protect against malnutrition in babies.

The argument is often made that rice is both expensive and easily divertible to less deserving members of Korean society, such as the military. There are four reasons why this is specious. Firstly, the scale of the international aid effort is such that funds could be found for appropriate food if the political will was there. Secondly, no international agency has ever found evidence of food aid diversion to the military. Thirdly, there would be nothing to prevent the humanitarian agencies negotiating specific protocols with the government to monitor the delivery of rice. Fourthly, the agencies could process the rice so that it was more suitable for feeding babies than adults—thus minimising any likelihood of diversion.

More importantly, both the government and the United States need to rethink food priorities. If the north Korean leadership wants to ensure that future generations are not permanently damaged because of a lack of appropriate food at the crucial growth stage between birth and twenty-four months, it should ensure that all its young children have access to any available rice. Washington could also decide that, instead of prioritising the interests of American farmers and their consequent votes, it prioritises the interests of north Korean babies and their food. These babies need rice—they are literally dying for it.

Article 27

Far Eastern Economic Review, June 10, 1999

Roots of Poverty

The headlong rush into coconut farming in the Philippines early this century has left the country impoverished and underdeveloped

By Rigoberto Tiglao

Rigoberto Tiglao is a REVIEW *correspondent in Manila.*

Folks in a remote village in Lanao del Sur province in Mindanao, at the centre of the Philippines' Islamic rebellion, still tell the story of how things changed forever in their homeland. An American adventurer showed up one day "about the time after the Spanish left" a century ago, the tale begins. He met with the elders of the Muslim village and told them he wanted to turn their cow-grazing land into a coconut plantation. They thought the lands would be used communally, as they always had. The man promised to pay them to plant and maintain the palm trees. For a few years, the Muslims were indeed paid, as a huge coconut plantation took shape.

Later on though, one by one, Christian migrants from Luzon, the Philippines' most populous island, replaced them, and they were asked to leave the plantation. Protesting that they couldn't leave their homes, they were told that the land was not theirs but belonged to an ethnic-Chinese trader, who had bought it from the American. "All over our Moro homeland, this was what happened, how the Christians stole our land," says Amil Alibansa, a fighter in the Moro Islamic Liberation Front, which is waging an armed separatist movement against the Philippine government.

During the first decades of the century, what happened in that Muslim village was repeated not only across Mindanao, but in many other parts of the country. Grazing lands, communal fields, brushlands and forests were converted to coconut plantations. To this day, in rural areas in Mindanao, eastern Visayas and southern Luzon, coconut trees dominate the landscape, covering even steep mountain slopes.

It wasn't always so. When the Spanish claimed "Las Islas Filipinas" in 1521, rainforests blanketed the land. Coconut palms were found mostly in coastal areas. The natives didn't cultivate them, since coconut wasn't a staple food, although it was eaten when rice was scarce and delicacies made from it were common. Perhaps it was most popular for the wine that could be fermented from the water inside the shell.

Indeed, there were so few coconut trees that the Spanish colonizers had to issue a decree in 1642 requiring each native to plant 200 of them. The Spanish prized the coconut's waterproof shell, which they used to caulk their galleons, and its husks, for making ships' rigging. But in 1903, more than two centuries after the decree, only 150,000 hectares, or 5% of arable land, were devoted to coconut trees. Many of them had been planted only in the late 1800s.

Towards the turn of the century, however, the country's landscape began to see a dramatic transformation. Today, coconut plantations cover 3 million hectares, a quarter of the country's agricultural land. During Europe's industrial revolution, events halfway around the globe triggered the change. Two everyday commodities, among the earliest to be mass-produced, were about to radically alter the Philippines.

In Britain, a four-decades-old tax on soap that had been imposed towards the end of the Napoleonic wars was finally lifted in 1853. After the abolition of the duty, soap, which had been a luxury item, suddenly became affordable. It grew so popular in Britain and Europe that its mass production began in the 1860s.

Meanwhile, a second novel commodity whetted Europeans' desire: margarine, patented in 1873. It was invented partly as a result of the two-decade-long Napoleonic wars that made dairy products scarce. The mass migration from rural areas to cities during Europe's industrial revolution had also made butter expensive. Margarine proved a logical and popular substitute: It was cheap yet nutritious enough to feed the legions of new factory workers.

Soap and margarine shared the same raw material: beef fat. Cattle from the American frontier supplied most of it in the 19th century. Beginning in 1887 though, blizzards and drought in the American West almost wiped out the cattle industry; the price of beef fat shot up dramatically. Desperate British and European soap and margarine manufacturers turned to vegetable oils as less-expensive replacements.

The cheapest and best-suited vegetable oil for producing soap and margarine turned out to be coconut oil. It was already native to, and could be propagated easily in, the least-costly places in the world, Europe's Asian colonies. The Dutch firms Van den Berghs and Jurgens, which controlled the margarine industry, turned to their country's colony, Indonesia—to this day the world's second-largest coconut producer after the Philippines. British firm Lever Brothers turned to Sri Lanka and Malaysia for its coconut oil, and later merged with the Dutch firms to form Unilever, one of the world's biggest conglomerates. Not to be beaten, Unilever's arch-rival then and now, American soap manufacturer Procter & Gamble, turned to its nation's newest colony, the Philippines.

Vast tracts of land were planted with coconut trees with unprecedented speed. By 1930, about 450,000 hectares of land were planted to coconuts; the palm accounted for 13% of arable land, compared to just 5% less than 30 years before. In contrast, a more well-known export crop, sugarcane, took up only 180,000 hectares in the same period. Shipments of dried coconut meat—from which the precious oil is extracted—rose from less than 1% of total exports in 1890 to 20% in the 1930s, to become the Philippines' biggest export item.

The two world wars carved the coconut industry deeper into the Philippine economy. It wasn't only the surging demand for soap and cosmetics during World War I that increased demand for coconut oil. That war saw the first massive use of dynamite—whose main ingredient, nitroglycerine, could be made from coconut oil, which is 8% glycerine. The consumer boom in the U.S. after World War II, together with advances in chemistry, spurred the invention of other products that led to increased use of coconut oil, from cosmetics and pharmaceuticals to fuel additives and plastics.

By 1960, with nearly a million additional hectares planted with coconut trees after the war, coconut farms made up 15% of the country's total agricultural land. Until the 1970s, coconut oil continued as the country's biggest export product, accounting for 35% of the total. In the 1970s,

coconut hectarage increased to 3 million hectares, a quarter of the country's arable land. That figure increased even more, to a peak of 3.3 million hectares in 1986, after the U.S. and South America restricted the export of soybean and cottonseed oil—coconut substitutes—during the mid-1970s. (That figure has since dropped slightly.)

Foreign exchange from coconut exports made up a big chunk of the country's earnings throughout most of the century. However, there were huge costs in the expansion of the coconut industry. Coconut planting contributed to massive deforestation: In 1910, forests covered 66% of the Philippines' total land area; now, it's only 20%. Up to the late 1800s, Mindanao, for instance, was mostly virgin forest. By 1987, only a quarter of forest cover remained. A third of the arable lands carved out of the forests were planted to coconuts.

But perhaps the most destructive legacy of the West's demand for coconut oil is the Philippines' poverty and economic underdevelopment. Many were lured by the cash derived from an easily planted and tended crop. But in reality, industry profits went primarily to traders and exporters, not farmers. A third of the country's population now finds itself trapped on coconut farms.

Little can be done to increase a coconut's tree's productivity when it matures, five to six years after planting, further exacerbating the problem. Productivity declines dramatically after 20 years. The only solution is to uproot an old tree, and shift to other, more productive crops such as coffee or corn, or to new palm hybrids. The small farmers who dominate coconut agriculture, however, can't afford to do that. It means giving up for years their only source of income, as well as a considerable outlay for the new crops.

On top of declining productivity, the value of coconut oil has fallen in real terms through the decades, partly as a result of increased production of substitutes such as American and Chinese soybean and cottonseed oil as well as sunflower-seed oil form the former Soviet republics. The value of Philippine coconut-oil exports have remained static in the last two decades, even as a huge chunk of its productive lands are locked in the palm's production.

It's not coincidental that the poorest of the Philippines' poor are small coconut farmers and rural workers. It's also not coincidental that the country's most underdeveloped provinces—and the hotbeds of communist and Islamic insurgencies, such as Mindanao's Lanao del Sur—are its biggest coconut producers. Even now, the country faces tremendous problems that emerged a century ago, because of the West's cravings for soap and margarine.

Far Eastern Economic Review, August 31, 2000

Uncivil Society

Anger at the slow pace of decentralization is fuelling street protests, providing ammunition for a conservative backlash

By Shawn W. Crispin/
KHON KAEN PROVINCE AND BANGKOK

RURAL UNREST is rising in Thailand. For two months now, thousands of villagers have held protests in Bangkok against the impact on their livelihoods of the Pak Moon dam, a hydroelectric facility run by the state utility in northeastern Ubon Ratchathani province. Others have taken to the capital's streets to clamour for action to alleviate rural poverty. Behind this grassroots discontent is a problem that goes to the heart of how the country is governed: the failure to devolve more decision-making power to the local level.

Decentralization of power was promised in Thailand's new constitution, promulgated in 1997, but measures to implement it have been slow in coming, say many Thais. "It looks good on paper, but the reality is the old vested interests in Bangkok really don't want to relinquish power to the people," says Uaychai Watha, chairman of the Assembly of Small-Scale Northeastern Farmers, which represents farmers in 19 of the country's poorest provinces. "The only way real change will come is through more protests. Only then will the government understand this has become a political war—a war for people's democracy it cannot win."

Uaychai's battle cry is gathering resonance. The Bangkok protests are only the most visible of hundreds of disputes brewing upcountry over land rights, forests and dams—the legacy of decades of rural-development policies imposed from above without local input. In many cases, frustration with the Bangkok bureaucrats who handed down those policies is reaching breaking point. The concern among academics and others is that amid the country's current political uncertainties, more street protests might prompt a conservative or military backlash.

Yet more unrest is almost certainly in the pipeline. A recent survey conducted by Prasit Kunurat, associate professor of sociology at Khon Kaen University in northeastern Thailand, re-

corded over 500 conflicts similar to the Pak Moon dam case in nearby provinces, 81 of which have flared into violence over the past decade. "Society is changing very fast—many conflicts that have been quiet could break back into the open," says Prasit. "We desperately need local conflict-resolution mechanisms that create win-win solutions."

To date, though, the situation remains lose-lose. Decades of centralized rule have stifled the development of provincial institutions now needed to mediate conflicts between the state and local communities. That institutional vacuum is starting to matter as the new constitution opens a Pandora's Box of grievances over Bangkok's rural-development policies. "We are living in a vicious circle," says Surichai Wun' Gaeo, of the Centre for Social Development Studies at Bangkok's Chulalongkorn University. "We cannot have decentralization overnight because of the multilayered, centralized structures we created during our rapid development era. Yet people are demanding change now."

Just as worrying is the gulf in perception between the government and the rural population over how decentralization should be carried forward. For many villagers, the constitution's promise of local empowerment has opened political space to air grievances about past development projects—and in many cases demand financial redress. To the government, however, the constitution merely promises to grant local communities more say in *future* development projects, and according to strict regulations. Critics blame this officious insistence on the rule book for the government's failure to reach a settlement with the Pak Moon dam protesters during a nationally televised hearing on the dispute on August 17. "There is a dire need for a more differentiated government understanding of the rural poor's plight," says Surichai "Unfortunately the government still sees things in black and white."

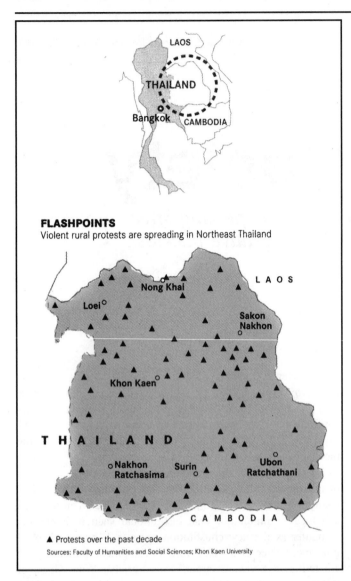

FLASHPOINTS
Violent rural protests are spreading in Northeast Thailand

▲ Protests over the past decade
Sources: Faculty of Humanities and Social Sciences; Khon Kaen University

Without the devolution of even basic decision-making authority, local governments are unable to mollify local protests. "Until more powers are given to the provinces, we cannot solve the people's problems," says Khon Kaen province's deputy governor, Thavat Satiennam. "How many more Isan [northeastern] protests can Bangkok handle?"

Social activists and NGO workers in Khon Kaen agree that further delay on decentralization will mean more rural anger erupting onto the streets of Bangkok. Many fear that escalating street protests could, in turn, provide a pretext for right-wingers or the military—whose interests have been eroded most by the new charter—to step in and impose order. Chulalongkorn University's Surichai and other academics are looking for ways to stave off a national crisis as rumours of military intervention circulate in Bangkok.

Jackboot intervention, though, would be out of step with the democratization sweeping Thailand. Since the 1980s, Thais have witnessed a burgeoning civil-society movement of NGOs and community organizations, usually led by a middle class demanding more democracy. Since then, the forces of old in the Thai bureaucracy have tried to guide the movement in directions that minimize the damage to their entrenched interests. In 1997, for example, the government announced a new "people-centred" approach to development. But in traditional top-down fashion, the Interior Ministry then directed district officials to "create" civil society. In Khon Kaen, local councils were instructed by the governor to disburse 10,000 baht ($245) each to approved civic organizations, paving the way for old-style patronage-based relationships. And in 1998, the government ordered the creation of "civic assemblies" in the provinces.

The Khon Kaen Civic Assembly has become "a tool for the government to co-opt and control civil society," says Ratana Boonmathya, a professor at Khon Kaen University and member of the assembly. The project has already "broken down," she adds. She believes Thai civil society is increasingly dominated by an emerging partnership between the government and a more empowered provincial middle class that still leaves the poor on the margins of local decision-making.

Wanchai Vatanasapt, former director of Khon Kaen University's Dispute Resolution Centre, is among those who see more conflict on the horizon. He believes the "government needs to look at such conflict constructively as a means of building representative democracy," but concedes that uncontrolled conflict "provides the background for a military coup."

History has a nasty habit of repeating itself in Thailand, where democratic flowerings have been nipped in the bud many times before. So long as Thailand's traditional government institutions fail to effectively mediate the democratic demands being unleashed across the country, grassroots frustration will mount. So too will the threat of a conservative backlash. The road towards political reform in Thailand looks set to get bumpier.

Indeed, many government officials continue to view rural protests with suspicion, either as politically motivated or as opportunistic attempts to wrest funds from state coffers. "Many of the so-called social reformers offer only unrest, not their partnership for finding systemic solutions to the problems we are facing," says Anek Laothamatas, who heads the committee charged with drafting a decentralization action-plan. "Most social activists have no grounding in the history of social change—they don't understand that an over-rapid transition will result in societal suicide." Other critics say Thai non-governmental organizations need to create "big bangs"—in the form of ever rowdier protests—to boost their negotiating positions for a piece of the foreign funding pie.

But many rural activists consider Bangkok's cautious approach to decentralization mere footdragging. "All the power is still in the hands of the Ministry of Interior," says Sanae Vichaiwong, a leader of the NGO Coordinating Committee on Rural Development in the Northeast of Thailand. "But we are finished with asking the government for help, we now demand it."

Article 29   *The Christian Science Monitor,* September 30, 1997

Little House on the Paddy

Life in a northern Vietnam village

By Cameron W. Barr
Staff writer of The Christian Science Monitor

MINH SON, VIETNAM

AT day's end there is a pretty view of rolling green hills from a porch in this village in northern Vietnam. The evening breeze shuffles the leaves on the eucalyptus trees and the sun's angled rays grace the rice paddies with a luminous sheen.

Life is good for farmer Hoang Van Quang, his wife, Nguyen Thi Hanh, their three children, and two in-laws—Mr. Quang's mother and his younger sister. The family's animals are just one sign of prosperity: A water buffalo glares from its paddock, four pigs grunt

and roll over in their sty, and underfoot, eight ducklings and 35 chickens peck their way toward table-readiness.

It is no thanks to socialism, but this family is in the middle of an economic great leap forward. Everyone is anticipating the day, a year or two from now, when they can trade in their wood house for the sturdy prestige of brick. The parents are proud that their two daughters do well in school and hope they will eventually attend a teachers' college in Hanoi, Vietnam's capital, about 45 miles to the south.

Not too long ago, everyone in this village farmed collectively. Whistles went off at 7 a.m. and then at 11 a.m. to mark the morning shift, and afternoon

work hours ran from 1 p.m. to 5 p.m. About five years ago—Quang doesn't remember exactly—the government gave him "land-usage certificates" that effectively granted the family ownership of several acres of paddy, as well as land to raise vegetables and fruit trees.

Technically, "the people" hold title to all the land in Vietnam. But Quang and his wife consider that they have "full autonomy" over their property. "If we work hard, we get more; if we work less, we get less," Quang says. And they set their own schedule.

The end of collective farming was part of a massive shift in Vietnam's economy that began in the late 1980s. The Communist planners in Hanoi de-

LEARNING ENGLISH: *Sisters Phuong and Huong struggle with pronunciations in their English-language textbook.*

THE FAMILY: *Squatting, from right: Nguyen Thi Hanh, daughters Huong and Phuong. At back from right: Grandmother Do Thi Bat, son Quyen, and Hoang Van Quang.*

ALL IN A DAY'S WORK: *Nguyen Thi Hanh carried just-washed sweet potato leaves from the stream running near the house to feed the family's four pigs. During the days of late summer, when the rice from their paddies has already been transplanted, the pace of life is slower for Ms. Hanh. 'Life goes on,' she says with a smile. 'If I don't finish something today, I can always do it tomorrow.'*

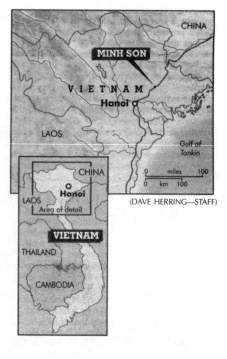

(DAVE HERRING—STAFF)

SIMPLE POSSESSIONS: *Their one-room house has a clay-tile roof, a dirt floor, and three large wooden beds where the seven family members sleep on thin, straw mats. They also own an electric fan, a television set, and a battery-powered clock.*

cided that state-controlled socialism—absent aid from the Soviet Union—would lead to disaster and embraced a policy of free-market reforms. The country is still poor, but many people are a lot less poor than they once were.

With the rice already transplanted and the children on vacation, the days of late summer aren't the most hectic in the year. Today, Ms. Hanh has washed some sweet potato leaves that will be fed to the pigs, done the laundry, and set some corncobs to dry in the sun. "Life goes on," she smiles. "If I don't finish something today, I can always do it tomorrow."

The one-room house has a clay-tile roof, a dirt floor, and three large wooden beds where the seven family members sleep on thin straw mats. An electric fan

hangs from the roof, along with bundles of mosquito netting over the beds. A battery-powered clock tells the time.

The kitchen is a separate, smaller building where Hanh cooks over an open fire. A cupboard holds a few metal pots, bowls, and plates. The meals mostly center on rice and animals the family raises. "If we want some meat, we kill a chicken," Quang says. Every two or three weeks they visit a nearby town to shop for fish sauce, salt, soap, and other necessities of life.

The money to buy these things comes from selling produce or slaughtering a pig. The family doesn't have a savings plan. "If I get some money, I usually spend it right away," Quang explains.

A tall bamboo pole at the edge of the wide, brick porch supports the television

antenna. Like parents everywhere, Quang and Hanh worry about how much time the children spend in front of the tube. "During the school year," the father says, fingering a remote control swathed in protective plastic wrap, "we try to keep it off so the kids keep studying."

Huong, the elder of the two daughters, doesn't seem to need much prodding. She brings out a worn English textbook to show two American visitors and begins trying out phrases.

A brand new straw hat with a bow hangs over one of the beds. The parents explain, beaming, that Huong received it for doing such a good job in school.

Article 30   *The Christian Science Monitor*, April 3, 1998

Vietnam's Communists Eye New Vices as Market Worries Rise

Warning against lavish weddings hints at party's concern over market effects on society

By Minh T. Vo

Special to The Christian Science Monitor

HANOI AND HO CHI MINH CITY, VIETNAM

To the young and hip in Ho Chi Minh City, the Dash Rush restaurant is more than a fast-food outlet serving novelties such as hamburgers: It is also a popular place to hold a wedding banquet.

In Hanoi, those looking for something more stylish can rent the ballroom of the five-star Hanoi Daewoo Hotel, where some 700 people attended a recent event.

Vietnam's young brides and grooms have more choices of large venues than ever. But not everyone is happy about this.

The government has launched a campaign against big wedding parties, using its press and brandishing unspecified punishments.

"Many families, including those of senior officials, have held big wedding parties or funerals to gain prestige or money," said the official Vietnam News. "These phenomena have become social ailments and degrade traditional morals

SIGNS OF THE TIMES: *Huge ads for foreign products tower over the streets of Ho Chi Minh City. The influx of Western goods came after Communist-run Vietnam adopted market reforms.*

MELANIE STETSON FREEMAN—STAFF

and the people's thrifty, modest, and fine way of living."

Socialism vs. markets

For a good part of the past decade, Vietnam's state-run media have periodically, and unsuccessfully, rallied to stamp out prostitution and illicit drugs. This latest addition to the government's list of social vices underscores the ruling Communist Party's predicament as it tries to reconcile socialist ideals with a burgeoning market economy.

Hanoi's leaders, watching how other Asian nations handle similar problems, see nearby Thailand as a mixed model. While Thailand allows private business to be dynamic, it also allows "too much of a widening gap between the rich and the poor, the cities and the rural areas," says government economic adviser Do Duc Dinh.

Opening in 1986

Ever since Vietnam embarked on its economic reforms known as *doi moi* in 1986, the gap between the rich and the poor has steadily grown. Moreover, some areas of the country have benefited more than others from the market economy. Ho Chi Minh City, the nearby Dong Nai province, and Hanoi alone attracted some 60 percent of the pledged foreign capital last year, yet they account for just 12 percent of the 75 million people in Vietnam. All this is not lost upon the Communist Party, which draws much of its power base from rural areas where some 80 percent of the population live.

Though some people complain that the pronouncement against big wedding parties is a case of excessive Big Brother intervention, the Vietnamese government sees it as a necessary step to defuse political unrest. Last year, a string of protests erupted in several provinces as villagers complained about low wages, tax increases, and corruption among local officials.

Vice worries conservatives

Some Western analysts believe that the uprisings fueled conservative elements within the party's ruling Politburo and slowed the pace of economic reforms. As a result, direct foreign investment in new projects fell by almost half last year compared with 1996. This was the first time that investment growth in Vietnam declined since *doi moi* was initiated.

"As the government courts international interest, it simultaneously constrains investors with tight restrictions on advertising spending and a long licensing process," says a Western businessman who declined to be identified.

The market has not grown as much as many international companies had expected, and some firms have moved out or scaled down operations. Dutch airline KLM Royal suspended its flights to Ho Chi Minh City in March.

This is in stark contrast to just four years ago, when the United States lifted its embargo against Vietnam and foreign investors rushed in, attracted to the country's relatively educated population, low-cost labor,

and opening economy. "The reforms in the early '90s were quite radical by any standards," says International Monetary Fund representative Erik Offerdal.

These days, investors are less enthusiastic, and they are increasingly concerned about the government's often-conflicting policies of freeing the market and tightening controls.

According to Daewoo's public relations manager in Hanoi, Phan Hong Nga, the South Korean conglomerate's apartment building adjacent to its hotel had a low occupancy rate in 1996 and early 1997 because companies were uncertain about the market. "Investors in Vietnam are quite cautious and unwilling to sign a long-term lease—which is understandable," she says. Ms. Nga adds that the building now allows three-month leases.

Government officials "haven't seen the need for further reforms in recent years," says Mr. Offerdal. "It has been several years since significant market reforms have been implemented."

There are signs, however, that Vietnam will kick-start its economy again. Last September, the Communist Party Central Committee changed its leadership by naming two market reformers to the posts of president and prime minister: Tran Duc Luong and Phan Van Khai.

New leader's direction

Yet Hanoi-based foreign businessmen caution that it is too early to tell what the new leadership will mean to the economy. In late December, conservative Army official Le Kha Phieu was elected Communist Party secretary general, traditionally the most powerful member of the trium-

virate. Mr. Phieu has a record of opposing a free-market system, and he is seen as a counterbalance to the new president and prime minister.

However, since assuming his new post, Phieu has said that he favors economic reforms. "We will continue to carry out renovation," the Nhan Dan daily reported him as saying.

Some businessmen warn that Phieu's recent statements in the official press cannot be taken at face value. "I doubt that he could have changed his ideology just because he has a new title," says one Ho Chi Minh City-based businessman who spoke on condition of anonymity. "A leopard can't change his spots."

Other analysts are more optimistic. "The reforms will definitely continue at a fast pace," says Mr. Dinh, the government economic adviser.

Offerdal agrees, pointing out that Vietnam's decisionmaking process is based on consensus: "I don't think that any one person can change Vietnam's commitment to market reform."

Vietnam's leadership overhaul came at a particularly difficult time. Vietnamese products' competitiveness fell as other Asian currencies plummeted. South Korea, Vietnam's largest investor last year, is tending to its problems at home. And investment from other East Asian nations, which together with South Korea accounted for some 70 percent of Vietnam's trade last year, may continue to decline as the region deals with its financial crisis.

Though the government forecasts GDP growth of 9 percent this year, many Western economists believe this figure is optimistic. "I don't think that it can be achieved given the regional situation," says Offerdal.

Control above reform

Ironically, the regional troubles and last year's disappointment may prove to be an "impetus for reforms," Offerdal says. "Over the past year, it's become apparent that new initiatives are needed."

However, party leader Phieu indicated that Communist control must still be incorporated in the government's economic plans.

"Maintaining political stability is the fundamental condition to carry out the renewal process," he told the official Nhan Dan daily newspaper recently.

Credits

Page 118 Article 1. Reprinted by permission of *Foreign Affairs,* July/August 1996, pp. 106-118. © 1996 by the Council on Foreign Relations, Inc.

Page 123 Article 2. From *USA Today Magazine,* May 1998, pp. 22-25. © 1998 by the Society for the Advancement of Education.

Page 127 Article 3. From *The Economist,* February 12, 2000. © 2000 by The Economist, Ltd. Distributed by The New York Times Special Features.

Page 134 Article 4. Reprinted with permission from *The National Interest,* Winter 1999-2000, pp. 105-108. © 1999 by National Affairs, Inc., Washington, DC.

Page 136 Article 5. From *Far Eastern Economic Review,* June 10, 1999, pp. 61-62. © 1999 by Review Publishing Company Limited. Reprinted by permission.

Page 138 Article 6. From *Discover,* June 1998, pp. 86-94 by Jared Diamond. © 1998. Reprinted with permission of *Discover* magazine.

Page 145 Article 7. Reprinted with permission from *Foreign Policy,* Summer 2000, pp. 74-88. © 2000 by the Carnegie Endowment of International Peace.

Page 151 Article 8. Reprinted by permission of *Foreign Affairs,* July/August 2000, pp. 26-39. © 2000 by the Council on Foreign Relations, Inc.

Page 156 Article 9. From *The Asian Wall Street Journal,* May 15-21, 2000. Reprinted by permission of *The Wall Street Journal.* © 2000 by Dow Jones & Company, Inc. All rights reserved.

Page 158 Article 10. From *World Policy Journal,* Fall 1998. © 1998 by the World Policy Institute. Reprinted by permission.

Page 162 Article 11. From *Look Japan,* August 2000. © 2000 by Look Japan, Ltd. Reprinted by permission.

Page 165 Article 12. From *Harvard International Review,* Fall 2000. © 2000 by Harvard International Review. Reprinted by permission.

Page 169 Article 13. From *Far Eastern Economic Review,* July 16, 1998, pp. 10-12. © 1998 by Review Publishing Company Limited. Reprinted by permission.

Page 171 Article 14. From *Look Japan,* June 1997. © 1997 by Look Japan, Ltd. Reprinted by permission.

Page 173 Article 15. Reprinted with permission from *The Christian Science Monitor,* December 8, 1997, pp. 1, 8. © 1997 by The Christian Science Publishing Society. All rights reserved.

Page 175 Article 16. From *Cultural Survival Quarterly,* Winter 1998, pp. 29-33. Reprinted courtesy of Cultural Survival, Inc. (www.cs.org.)

Page 179 Article 17. From *Parameters,* Winter 1999-2000, pp. 111-125. © 1999 by Ian James Storey. Reprinted by permission.

Page 187 Article 18. From *Time,* July 17, 2000. © 2000 by Time, Inc. Magazine Company. Reprinted by permission.

Page 188 Article 19. From *Far Eastern Economic Review,* June 10, 1999. © 1999 by Review Publishing Company Limited. Reprinted by permission.

Page 189 Article 20. From *Far Eastern Economic Review,* June 10, 1999. © 1999 by Review Publishing Company Limited. Reprinted by permission.

Page 191 Article 21. Reprinted with permission from *The New York Review of Books,* May 11, 2000. © 2000 by NYREV, Inc.

Page 196 Article 22. From *Far Eastern Economic Review,* September 21, 2000. © 2000 by Review Publishing Company Limited. Reprinted by permission.

Page 199 Article 23. From *CQ Researcher,* May 19, 2000. © 2000 by Congressional Quarterly, Inc. Reprinted by permission.

Page 203 Article 24. From *Vital Speeches of the Day,* July 1, 2000. © 2000 by City News Publishing Company, Inc. Reprinted by permission.

Page 206 Article 25. From *CQ Researcher,* May 19, 2000. © 2000 by Congressional Quarterly, Inc. Reprinted by permission.

Page 207 Article 26. From *The World Today,* February 2000. © 2000 by *The World Today,* a publication of the Royal Institute of International Affairs. Reprinted by permission.

Page 209 Article 27. From *Far Eastern Economic Review,* June 10, 1999. © 1999 by Review Publishing Company Limited. Reprinted by permission.

Page 211 Article 28. From *Far Eastern Economic Review,* August 31, 2000. © 2000 by Review Publishing Company Limited. Reprinted by permission.

Page 213 Article 29. Reprinted with permission from *The Christian Science Monitor,* September 30, 1997. © 1997 by The Christian Science Publishing Society. All rights reserved.

Page 215 Article 30. Reprinted with permission from *The Christian Science Monitor,* April 3, 1998. © 1998 by The Christian Science Publishing Society. All rights reserved.

Sources for Statistical Reports

U.S. State Department, *Background Notes* (2000)

CIA *World Factbook* (2000)

World Bank, *World Development Report* (2000/2001)

UN *Population and Vital Statistics Report* (January 2001)

UN *World Statistics in Brief* (2000)

The Statesman's Yearbook (2000)

Population Reference Bureau, *World Population Data Sheet* (2000)

The World Almanac (2000)

The Economist Intelligence Unit (1999)

Keizai Koho Center, *Japan 2000 An International Comparison* (1999)

Glossary of Terms and Abbreviations

Animism   The belief that all objects, including plants, animals, rocks, and other matter, contain spirits. This belief figures prominently in early Japanese religious thought and in the various indigenous religions of the South Pacific.

Anti-Fascist People's Freedom League (AFPFL)   An anti-Japanese resistance movement organized by Burmese students and intellectuals.

ANZUS   The name of a joint military-security agreement originally among Australia, New Zealand, and the United States. New Zealand is no longer a member.

Asia Pacific Economic Cooperation Council (APEC)   Organized in 1989, this body is becoming increasingly visible as a major forum for plans about regional economic cooperation and growth in the Pacific Rim.

Asian Development Bank (ADB)   With contributions from industrialized nations, the ADB provides loans to Pacific Rim countries in order to foster economic development.

Association of Southeast Asian Nations (ASEAN)   Established in 1967 to promote economic cooperation among the countries of Indonesia, Malaysia, the Philippines, Singapore, Thailand, and Brunei.

British Commonwealth of Nations   A voluntary association of nations formerly included in the British Empire. Officials meet regularly in member countries to discuss issues of common economic, military, and political concern.

Buddhism   A religious and ethical philosophy of life that originated in India in the fifth and sixth centuries B.C., partly in reaction to the caste system. Buddhism holds that people's souls are endlessly reborn and that one's standing with each rebirth depends on one's behavior in the previous life.

Capitalism   An economic system in which productive property is owned by individuals or corporations, rather than by the government, and the proceeds of which belong to the owner rather than to the workers or the state.

Chaebol   A Korean term for a large business conglomerate. Similar to the Japanese *keiretsu*.

Chinese Communist Party (CCP)   Founded in 1921 by Mao Zedong and others, the CCP became the ruling party of the People's Republic of China in 1949 upon the defeat of the Nationalist Party and the army of Chiang Kai-shek.

Cold War   The intense rivalry, short of direct "hot-war" military conflict, between the Soviet Union and the United States, which began at the end of World War II and continued until approximately 1990.

Communism   An economic system in which land and businesses are owned collectively by everyone in the society rather than by individuals. Modern communism is founded on the teachings of the German intellectuals Marx and Engels.

Confucianism   A system of ethical guidelines for managing one's personal relationships with others and with the state. Confucianism stresses filial piety and obligation to one's superiors. It is based on the teachings of the Chinese intellectuals Confucius and Mencius.

Cultural Revolution   A period between 1966 and 1976 in China when, urged on by Mao, students attempted to revive a revolutionary spirit in China. Intellectuals and even Chinese Communist Party leaders who were not zealously communist were violently attacked or purged from office.

Demilitarized Zone (DMZ)   A heavily guarded border zone separating North and South Korea.

European Union (EU)   An umbrella organization of numerous Western European nations working toward the establishment of a single economic and political European entity. Formerly known as the European Community (EC) and European Economic Community (EEC).

Extraterritoriality   The practice whereby the home country exercises jurisdiction over its diplomats and other citizens living in a foreign country, effectively freeing them from the authority of the host government.

Feudalism   A social and economic system of premodern Europe, Japan, China, and other countries, characterized by a strict division of the populace into social classes, an agricultural economy, and governance by lords controlling vast parcels of land and the people thereon.

Greater East Asia Co-Prosperity Sphere   The Japanese description of the empire they created in the 1940s by military conquest.

Gross Domestic Product (GDP)   A statistic describing the entire output of goods and services produced by a country in a year, less income earned on foreign investments.

Hinduism   A 5,000-year-old religion, especially of India, that advocates a social caste system but anticipates the eventual merging of all individuals into one universal world soul.

Indochina   The name of the colony in Southeast Asia controlled by France and consisting of the countries of Laos, Cambodia, and Vietnam. The colony ceased to exist after 1954, but the term still is often applied to the region.

International Monetary Fund (IMF)   An agency of the United Nations whose goal it is to promote freer world trade by assisting nations in economic development.

Islam   The religion founded by Mohammed and codified in the Koran. Believers, called Muslims, submit to Allah (Arabic for God) and venerate his name in daily prayer.

Keiretsu   A Japanese word for a large business conglomerate.

Khmer Rouge   The communist guerrilla army, led by Pol Pot, that controlled Cambodia in the 1970s and subsequently attempted to overthrow the UN–sanctioned government.

Kuomintang   The National People's Party (Nationalists), which, under Chiang Kai-shek, governed China until Mao Zedong's revolution in 1949; it continues to dominate politics in Taiwan.

Laogai   A Mandarin Chinese word for a prison or concentration camp where political prisoners are kept. It is similar in concept to the Russian word *gulag*.

Liberal Democratic Party (LDP) The conservative party that ruled Japan almost continuously between 1955 and 1993 and oversaw Japan's rapid economic development.

Martial Law The law applied to a territory by military authorities in a time of emergency when regular civilian authorities are unable to maintain order. Under martial law, residents are usually restricted in their movement and in their exercise of such rights as freedom of speech and of the press.

Meiji Restoration The restoration of the Japanese emperor to his throne in 1868. The period is important as the beginning of the modern era in Japan and the opening of Japan to the West after centuries of isolation.

Monsoons Winds that bring exceptionally heavy rainfall to parts of Southeast Asia and elsewhere. Monsoon rains are essential to the production of rice.

National League for Democracy An opposition party in Myanmar that was elected to head the government in 1990 but that has since been forbidden by the current military leaders to take office.

New Economic Policy (NEP) An economic plan advanced in the 1970s to restructure the Malaysian economy and foster industrialization and ethnic equality.

Newly Industrializing Country (NIC) A designation for those countries of the developing world, particularly Taiwan, South Korea, and other Asian nations, whose economies have undergone rapid growth; sometimes also referred to as newly industrialized countries.

Non-Aligned Movement A loose association of mostly non-Western developing nations, many of which had been colonies of Western powers but during the cold war chose to remain detached from either the U.S. or Soviet bloc. Initially Indonesia and India, among others, were enthusiastic promoters of the movement.

Opium Wars Conflicts between Britain and China in 1839–1842 and 1856–1866 in which England used China's destruction of opium shipments and other issues as a pretext to attack China and force the government to sign trade agreements.

Pacific War The name frequently used by the Japanese to refer to that portion of World War II in which they were involved and which took place in Asia and the Pacific.

Shintoism An ancient indigenous religion of Japan that stresses the role of *kami,* or supernatural gods, in the lives of people. For a time during the 1930s, Shinto was the state religion of Japan and the emperor was honored as its high priest.

Smokestack Industries Heavy industries such as steel mills that are basic to an economy but produce objectionable levels of air, water, or land pollution.

Socialism An economic system in which productive property is owned by the government as are the proceeds from the productive labor. Most socialist systems today are actually mixed economies in which individuals as well as the government own property.

South Pacific Forum An organization established by Australia and other South Pacific nations to provide a forum for discussion of common problems and opportunities in the region.

Southeast Asia Treaty Organization (SEATO) This is a collective-defense treaty signed by the United States and several European and Southeast Asian nations. It was dissolved in 1977.

Subsistence Farming Farming that meets the immediate needs of the farming family but that does not yield a surplus sufficient for export.

Taoism An ancient religion of China inspired by Lao-tze that stresses the need for mystical contemplation to free one from the desires and sensations of the materialistic and physical world.

Tiananmen Square Massacre The violent suppression by the Chinese Army of a prodemocracy movement that had been organized in Beijing by thousands of Chinese students in 1989 and that had become an international embarrassment to the Chinese regime.

United Nations (UN) An international organization established immediately after World War II to replace the League of Nations. The organization includes most of the countries of the world and works for international understanding and world peace.

World Health Organization (WHO) Established in 1948 as an advisory and technical-assistance organization to improve the health of peoples around the world.

World Trade Organization (WTO) Successor organization to the General Agreement on Trade and Tariffs (GATT) treaties. WTO attempts to standardize the rules of free trade throughout the world.

Bibliography

GENERAL WORKS

Mark Borthwick, *East Asian Civilizations: A Dialogue in Five Stages* (Cambridge, MA: Harvard University Press, 1988).
The development of philosophical and religious thought in China, Korea, Japan, and other regions of East Asia.

Richard Bowring and Peter Kornicki, *Encyclopedia of Japan* (New York: Cambridge University Press, 1993).

Barbara K. Bundy, Stephen D. Burns, and Kimberly V. Weichel, *The Future of the Pacific Rim: Scenarios for Regional Cooperation* (Westport, CT: Praeger, 1994).

Commission on U.S.–Japan Relations for the Twenty-First Century, *Preparing for a Pacific Century: Exploring the Potential for Pacific Basin Cooperation* (Washington, D.C.: November 1991).
Transcription of an international conference on the Pacific with commentary by representatives from the United States, Malaysia, Japan, Thailand, Indonesia, and others.

Susanna Cuyler, *A Companion to Japanese Literature, Culture, and Language* (Highland Park, NJ: B. Rugged, 1992).

William Theodore de Bary, *East Asian Civilizations: A Dialogue in Five Stages* (Cambridge, MA: Harvard University Press, 1988).
An examination of religions and philosophical thought in several regions of East Asia.

Syed N. Hossain, *Japan: Not in the West* (Boston: Vikas II, 1995).

James W. McGuire, ed., *Rethinking Development in East Asia and Latin America* (Los Angeles: Pacific Council on International Policy, 1997).

Charles E. Morrison, ed., *Asia Pacific Security Outlook 1997* (Honolulu: East-West Center, 1997).

Seijiu Naya and Stephen Browne, eds., *Development Challenges in Asia and the Pacific in the 1990s* (Honolulu: East-West Center, 1991).
A collection of speeches made at the 1990 Symposium on Cooperation in Asia and the Pacific. The articles cover development issues in East, Southeast, and South Asia and the Pacific.

Edwin O. Reischauer and Marius B. Jansen, *The Japanese Today: Change and Continuity* (Cambridge: Belknap Press, 1995).
A description of the basic geography and historical background of Japan.

NATIONAL HISTORIES AND ANALYSES

Australia

Boris Frankel, *From the Prophets Deserts Come: The Struggle to Reshape Australian Political Culture* (New York: Deakin University [St. Mut.], 1994).
Australia's government and political aspects are described in this essay.

Herman J. Hiery, *The Neglected War: The German South Pacific and the Influence of WW I* (Honolulu: University of Hawaii Press, 1995).

David Alistair Kemp, *Society and Electoral Behaviors in Australia: A Study of Three Decades* (St. Lucia: University of Queensland Press, 1978).

Elections, political parties, and social problems in Australia since 1945.

David Meredith and Barrie Dyster, *Australia in the International Economy in the Twentieth Century* (New York: Cambridge University Press, 1990).
Examines the international aspects of Australia's economy.

Brunei

Wendy Hutton, *East Malaysia and Brunei* (Berkeley, CA: Periplus, 1993).

Graham Saunders, *A History of Brunei* (New York: Oxford University Press, 1995).

Nicholas Tarling, *Britain, the Brookes, and Brunei* (Kuala Lumpur: Oxford University Press, 1971).
A history of the sultanate of Brunei and its neighbors.

Cambodia

David P. Chandler, *The Tragedy of Cambodian History, War, and Revolution since 1945* (New Haven, CT: Yale University Press, 1993).
A short history of Cambodia.

Michael W. Doyle, *UN Peacekeeping in Cambodia: UNTAC's Civil Mandate* (Boulder, CO: Lynne Rienner, 1995).
A review of the current status of Cambodia's government and political parties.

Craig Etcheson, *The Rise and Demise of Democratic Kampuchea* (Boulder, CO: Westview Press, 1984).
A history of the rise of the Communist government in Cambodia.

William Shawcross, *The Quality of Mercy: Cambodia, Holocaust, and Modern Conscience; with a report from Ethiopia* (New York: Simon & Schuster, 1985).
A report on political atrocities, relief programs, and refugees in Cambodia and Ethiopia.

Usha Welaratna, ed., *Beyond the Killing Fields: Voices of Nine Cambodian Survivors* (Stanford, CA: Stanford University Press, 1993).
A collection of nine narratives by Cambodian refugees in the United States and their adjustments into American society.

China

Julia F. Andrews, *Painters and Politics in the People's Republic of China, 1949–1979* (Berkeley, CA: University of California Press, 1994).
A fascinating presentation of the relationship between politics and art from the beginning of the Communist period until the eve of major liberalization in 1979.

Ma Bo, *Blood Red Sunset* (New York: Viking, 1995).
A compelling autobiographical account by a Red Guard during the Cultural Revolution.

Jung Chang, *Wild Swans: Three Daughters of China* (New York: Simon and Shuster, 1992).
An autobiographical/biographical account that illuminates what China was like for one family for three generations.

Kwang-chih Chang, *The Archaeology of China*, 4th ed. (New Haven, CT: Yale University Press, 1986).

___, *Shang Civilization* (New Haven, CT: Yale University Press, 1980).

Two works by an eminent archaeologist on the origins of Chinese civilization.

Nien Cheng, *Life and Death in Shanghai* (New York: Penguin Books, 1988). A view of the Cultural Revolution by one of its victims.

Qing Dai, *Yangtze! Yangtze!* (Toronto: Probe International, 1994).

Collection of documents concerning the debate over building the Three Gorges Dam on the upper Yangtze River in order to harness energy for China.

John King Fairbank, *China: A New History* (Cambridge, MA: Harvard University Press, 1992).

An examination of the motivating forces in China's history that define it as a coherent culture from its earliest recorded history to 1991.

David S. G. Goodman and Beverly Hooper, eds., *China's Quiet Revolution: New Interactions between State and Society* (New York: St. Martin's Press, 1994).

Articles examine the impact of economic reforms since the early 1980s on the social structure and society generally, with focus on changes in wealth, status, power, and newly emerging social forces.

Richard Madsen, *China and the American Dream: A Moral Inquiry* (Berkeley, CA: University of California Press, 1995).

A history on the emotional and unpredictable relationship the United States has had with China from the nineteenth century to the present.

Jim Mann, *Beijing Jeep: A Case Study of Western Business in China* (Boulder, CO: Westview Press, 1997).

A crisp view of what it takes for a Westerner to do business in China.

Suzanne Ogden, *China's Unresolved Issues: Politics, Development, and Culture* (Englewood Cliffs, NJ: Prentice-Hall, 1992).

A complete review of economic and cultural issues in modern China.

Li Zhisui, *The Private Life of Chairman Mao* (New York: Random House, 1994). Memoirs of Mao's personal physician.

Hong Kong

"Basic Law of Hong Kong Special Administrative Region of the People's Republic of China," *Beijing Review*, Vol. 33, No. 18 (April 30–May 6, 1990), supplement.

Ming K. Chan and Gerard A. Postiglione, *The Hong Kong Reader: Passage to Chinese Sovereignty* (Armonk, NY: M. E. Sharpe, 1996).

A collection of articles about the issues facing Hong Kong during the transition to Chinese rule after July 1, 1997.

Berry Hsu, ed., *The Common Law in Chinese Context* in the series entitled *Hong Kong Becoming China: The Transition to 1997* (Armonk, NY: M. E. Sharpe, Inc., 1992).

An examination of the common law aspects of the "Basic Law," the mini-constitution that will govern Hong Kong after 1997.

Walter Hatch and Kozo Yamamura, *Asia in Japan's Embrace: Building a Regional Production Alliance* (Cambridge: Cambridge University Press, 1996). Discusses the future likelihood of Japan building an exclusive trading zone in Asia.

Benjamin K. P. Leung, ed., *Social Issues in Hong Kong* (New York: Oxford University Press, 1990).

A collection of essays on select issues in Hong Kong, such as aging, poverty, women, pornography, and mental illness.

Jan Morris, *Hong Kong: Epilogue to an Empire* (New York: Vintage, 1997).

A detailed portrait of Hong Kong that gives the reader the sense of actually being on the scene in a vibrant Hong Kong.

Mark Roberti, *The Fall of Hong Kong: China's Triumph and Britain's Betrayal* (New York: John Wiley & Sons, Inc., 1994).

An account on the decisions Britain and China made about Hong Kong's fate since the early 1980s.

Frank Welsh, *A Borrowed Place: The History of Hong Kong* (New York: Kodansha International, 1996).

A presentation on Hong Kong's history from the time of the British East India Company in the eighteenth century through the Opium Wars of the nineteenth century to the present.

Indonesia

Amarendra Bhattacharya and Mari Pangestu, *Indonesia: Development, Transformation, and Public Policy* (Washington, D.C.: World Bank, 1993).

An examination of Indonesia's economic policy.

Frederica M. Bunge, *Indonesia: A Country Study* (Washington, D.C.: U.S. Government, 1983).

An excellent review of the outlines of Indonesian history and culture, including politics and national security.

East Asia Institute, *Indonesia in Transition* (New York: Columbia University, 2000).

An analysis of the revolution in Indonesian politics since the overthrow of Suharto.

Philip J. Eldridge, *Non-government Organizations and Political Participation in Indonesia* (New York: Oxford University Press, 1995).

An examination of Indonesia's nongovernment agencies (NGOs).

Audrey R. Kahin, ed., *Regional Dynamics of the Indonesian Revolution: Unity from Diversity* (Honolulu: University of Hawaii Press, 1985).

A history of Indonesia since the end of World War II, with separate chapters on selected islands.

Hamish McConald, *Suharto's Indonesia* (Australia: The Dominion Press, 1980).

The story of the rise of Suharto and the manner in which he controlled the political and military life of the country, beginning in 1965.

Susan Rodgers, ed., *Telling Lives, Telling Histories: Autobiography and Historical Immigration in Modern Indonesia* (Berkeley, CA: University of California Press, 1995).

Reviews the history of Indonesia's immigration.

David Wigg, *In a Class of Their Own: A Look at the Campaign against Female Illiteracy* (Washington, D.C.: World Bank, 1994).

Looks at the work that is being done by various groups to advance women's literacy in Indonesia.

Japan

David Arase, *Buying Power: The Political Economy of Japan's Foreign Aid* (Boulder CO: Lynne Rienner Publishers, Inc., 1995).
An attempt to explain the complexities of Japan foreign-aid programs.

Michael Barnhart, *Japan and the World since 1868* (New York: Routledge, Chapman, and Hall, 1994).
An essay that addresses commerce in Japan from 1868 to the present.

Marjorie Wall Bingham and Susan Hill Gross, *Women in Japan* (Minnesota: Glenhurst Publications, Inc., 1987).
An historical review of Japanese women's roles in Japan.

John Clammer, *Difference and Modernity: Social Theory and Contemporary Japanese Society* (New York: Routledge, Chapman, and Hall, 1995).

Dean W. Collinwood, "Japan," in Michael Sodaro, ed., *Comparative Politics* (New York: McGraw-Hill, 2000).
An analysis of Japan's government structure and history, electoral process, and some of the issues and pressure points affecting Japanese government.

Dennis J. Encarnation, *Rivals beyond Trade: America Versus Japan in Global Competition* (Ithaca NY: Cornell University Press, 1993).
Explains how the economic rivalry that was once bilateral has turned into an intense global competition.

Mark Gauthier, *Making It in Japan* (Upland, PA: Diane Publishers, 1994).
An examination of how success can be attained in Japan's marketplace.

Walter Hatch and Kozo Yamamura, *Asia in Japan's Embrace: Building a Regional Production Alliance* (Cambridge: Cambridge University Press, 1996).
Discusses the future likelihood of Japan building an exclusive trading zone in Asia.

Paul Herbig, *Innovation Japanese Style: A Cultural and Historical Perspective* (Glenview, IL: Greenwood, 1995).
A review of the implications for international competition.

Ronald J. Hrebenar, *Japan's New Party System* (Boulder, CO: Westview Press, 2000).
An analysis of the political structure in Japan since the end of complete LDP dominance.

Harold R. Kerbo and John McKinstry, *Who Rules Japan? The Inner-Circle of Economic and Political Power* (Glenview, IL: Greenwood, 1995).
The effect of Japan's politics on its economy is evaluated in this essay.

Hiroshi Komai, *Migrant Workers in Japan* (New York: Routledge, Chapman, and Hall, 1994).
An examination of the migrant labor supply in Japan.

Makoto Kumazawa, *Portraits of the Japanese Workplace: Labor Movements, Workers, and Managers* (Boulder, CO: Westview Press, 1996).
Translated into English from Japanese, the book includes reviews of the workplace lifestyle of bankers, women, steel workers, and others.

Solomon B. Levine and Koji Taira, eds., *Japan's External Economic Relations: Japanese Perspectives,* special issue of *The Annals of the American Academy of Political and Social Science* (January 1991).
An excellent overview of the origin and future of Japan's economic relations with the rest of the world, especially Asia.

E. Wayne Nafziger, *Learning from the Japanese: Japan's Pre-War Development and the Third World* (Armonk, NY: M. E. Sharpe, 1995).
Presents Japan as a model of "guided capitalism," and what it did by way of policies designed to promote and accelerate development.

Nippon Steel Corporation, *Nippon: The Land and Its People* (Japan: Gakuseisha Publishing, 1984).
An overview of modern Japan in both English and Japanese.

Asahi Simbun, *Japan Almanac 1998* (Tokyo: Asahi Shimbun Publishing Company, 1997).
Charts, maps, statistical data about Japan in both English and Japanese.

Patrick Smith, *Japan: A Reinterpretation* (New York: Pantheon Books, 1997).
A discussion of the rapidly changing Japanese national character.

Korea: North and South Korea

Chai-Sik Chung, *A Korean Confucian Encounter with the Modern World* (Berkeley, CA: IEAS, 1995).
Korea's history and the effectiveness of Confucianism are addressed.

Donald Clark et al., *U.S.–Korean Relations* (Farmingdale, NY: Regina Books, 1995).
A review on the history of Korea's relationship with the United States.

James Cotton, *Politics and Policy in the New Korean State: From Rah Tae-Woo to Kim Young-Sam* (New York: St. Martin's, 1995).
The power and influence of politics in Korea are examined.

James Hoare, *North Korea* (New York: Oxford University Press, 1995).
An essay that addresses commerce in Japan between 1868 and the present.

Dae-Jung Kim, *Mass Participatory Economy: Korea's Road to World Economic Power* (Landham, MD: University Press of America, 1995).

Korean Overseas Information Service, *A Handbook of Korea* (Seoul: Seoul International Publishing House, 1987).
A description of modern South Korea, including social welfare, foreign relations, and culture. The early history of the entire Korean Peninsula is also discussed.

___, *Korean Arts and Culture* (Seoul: Seoul International Publishing House, 1986).
A beautifully illustrated introduction to the rich cultural life of modern South Korea.

Callus A. MacDonald, *Korea: The War before Vietnam* (New York: The Free Press, 1986).
A detailed account of the military events in Korea between 1950 and 1953, including a careful analysis of the U.S. decision to send troops to the peninsula.

Christopher J. Sigur, ed., *Continuity and Change in Contemporary Korea* (New York: Carnegie Ethics and International Affairs, 1994).

A review of the numerous stages of change that Korea has experienced.

Joseph A. B. Winder, ed., *Korea's Economy 1999* (Washington, D.C.: Korea Economic Institute, 1999).

A review of the economic impact of the Asian financial crisis on South Korea.

Laos

Sucheng Chan, ed., *Hmong: Means Free Life in Laos and America* (Philadelphia: Temple University Press, 1994).

Arthur J. Dommen, *Laos: Keystone of Indochina* (Boulder, CO: Westview Press, 1985).

A short history and review of current events in Laos.

Joel M. Halpern, *The Natural Economy of Laos* (Christiansburg, VA: Dalley Book Service, 1990).

___, *Government, Politics, and South Structures of Laos: Study of Traditions and Innovations* (Christiansburg, VA: Dalley Book Service, 1990).

Macau

Charles Ralph Boxer, *The Portuguese Seaborne Empire, 1415–1825* (New York: A. A. Knopf, 1969).

A history of Portugal's colonies, including Macau.

W. G. Clarence-Smith, *The Third Portuguese Empire, 1825–1975* (Manchester: Manchester University Press, 1985).

A history of Portugal's colonies, including Macau.

Malaysia

Mohammed Ariff, *The Malaysian Economy: Pacific Connections* (New York: Oxford University Press, 1991).

The report on Malaysia examines Malaysia's development and its vulnerability in world trade.

Richard Clutterbuck, *Conflict and Violence in Singapore and Malaysia, 1945–1983* (Boulder, CO: Westview Press, 1985).

The Communist challenge to the stability of Singapore and Malaysia in the early years of their independence from Great Britain is presented.

K. S. Jomo, ed., *Japan and Malaysian Development: In the Shadow of the Rising Sun* (New York: Routledge, 1995).

A review of the relationship between Japan and Malaysia's economy.

Gordon Means, *Malaysian Politics: The Second Generation* (New York: Oxford University Press, 1991).

R. S. Milne, *Malaysia: Tradition, Modernity, and Islam* (Boulder, CO: Westview Press, 1986).

A general overview of the nature of modern Malaysian society.

Myanmar (Burma)

Michael Aung-Thwin, *Pagan: The Origins of Modern Burma* (Honolulu: University of Hawaii Press, 1985).

A treatment of the religious and political ideology of the Burmese people and the effect of ideology on the economy and politics of the modern state.

Aye Kyaw, *The Voice of Young Burma* (Ithaca, NY: Cornell SE Asia, 1993).

The political history of Burma is presented in this report.

Chi-Shad Liang, *Burma's Foreign Relations: Neutralism in Theory and Practice* (Glenview, IL: Greenwood, 1990).

Mya Maung, *The Burma Road to Poverty* (Glenview, IL: Greenwood, 1991).

New Zealand

Bev James and Kay Saville-Smith, *Gender, Culture, and Power: Challenging New Zealand's Gendered Culture* (New York: Oxford University Press, 1995).

Patrick Massey, *New Zealand: Market Liberalization in a Developed Economy* (New York: St. Martin, 1995).

Analyzes New Zealand's market-oriented reform programs since the Labour government came into power in 1984.

Stephen Rainbow, *Green Politics* (New York: Oxford University Press, 1994).

A review of current New Zealand politics.

Geoffrey W. Rice, *The Oxford History of New Zealand* (New York: Oxford University Press, 1993).

Papua New Guinea

Robert J. Gordon and Mervyn J. Meggitt, *Law and Order in the New Guinea Highlands: Encounters with Enga* (Hanover, NH: University Press of New England, 1985).

Tribal law and warfare in Papua New Guinea.

David Hyndman, *Ancestral Rainforests and the Mountain of Gold: Indigenous Peoples and Mining in New Guinea* (Boulder, CO: Westview Press, 1994).

Bruce W. Knauft, *South Coast New Guinea Cultures: History, Comparison, Dialectic* (New York: Cambridge University Press, 1993).

The Philippines

Frederica M. Bunge, ed., *Philippines: A Country Study* (Washington, D.C.: U.S. Government, 1984).

Description and analysis of the economic, security, political, and social systems of the Philippines, including maps, statistical charts, and reproduction of important documents. An extensive bibliography is included.

Manual B. Dy, *Values in Philippine Culture and Education* (Washington, D.C.: Council for Research in Values and Philosophy, 1994).

James F. Eder and Robert L. Youngblood, eds., *Patterns of Power and Politics in the Philippines: Implications for Development* (Tempe, AZ: ASU Program, SE Asian, 1994).

A review of the impact of politics and its power over development in the Philippines.

Singapore

Lai A. Eng, *Meanings of Multiethnicity: A Case Study of Ethnicity and Ethnic Relations in Singapore* (New York: Oxford University Press, 1995).

Paul Leppert, *Doing Business with Singapore* (Fremont, CA: Jain Publishing, 1995).

Singapore's economic status is examined in this report.

Hafiz Mirza, *Multinationals and the Growth of the Singapore Economy* (New York: St. Martin's Press, 1986).
An essay on foreign companies and their impact on modern Singapore.

Nilavu Mohdx et al., *New Place, Old Ways: Essays on Indian Society and Culture in Modern Singapore* (Columbia, MO: South Asia, 1994).

South Pacific

C. Beeby and N. Fyfe, "The South Pacific Nuclear Free Zone Treaty," Victoria University of Wellington *Law Review,* Vol. 17, No. 1, pp. 33–51 (February 1987).
A good review of nuclear issues in the Pacific.

William S. Livingston and William Roger Louis, eds., *Australia, New Zealand, and the Pacific Islands since the First World War* (Austin, TX: University of Texas Press, 1979).
An assessment of significant historical and political developments in Australia, New Zealand, and the Pacific Islands since 1917.

Taiwan

Joel Aberbach et al., eds., *The Role of the State in Taiwan's Development* (Armonk, NY: M. E. Sharpe, 1994).
Articles address technology, international trade, state policy toward the development of local industries, and the effect of economic development on society, including women and farmers.

Bih-er Chou, Clark Cal, and Janet Clark, *Women in Taiwan Politics: Overcoming Barriers to Women's Participation in a Modernizing Society* (Boulder, CO: Lynne Rienner, 1990).
Examines the political underrepresentation of women in Taiwan and how Chinese culture on the one hand and modernization and development on the other are affecting women's status.

Stevan Harrell and Chun-chieh Huang, eds., *Cultural Change in Postwar Taiwan* (Boulder, CO: Westview Press, 1994).
A collection of essays that analyzes the tensions in Taiwan's society as modernization erodes many of its old values and traditions.

Dennis Hickey, *United States–Taiwan Security Ties: From Cold War to beyond Containment* (Westport, CT: Praeger, 1994).
Examines U.S.–Taiwan security ties from the Cold War to the present and what Taiwan is doing to ensure its own military preparedness.

Chin-chuan Lee, "Sparking a Fire: The Press and the Ferment of Democratic Change in Taiwan," in Chin-chuan Lee, ed., *China's Media, Media China* (Boulder, CO: Westview Press, 1994), pp. 179–193.

Robert M. Marsh, *The Great Transformation: Social Change in Taipei, Taiwan, since the 1960s* (Armonk, NY: M. E. Sharpe, 1996).
An investigation of how Taiwan's society has changed since the 1960s when its economic transformation began.

Robert G. Sutter and William R. Johnson, *Taiwan in World Affairs* (Boulder, CO: Westview Press, 1994).
Articles give comprehensive coverage of Taiwan's involvement in foreign affairs.

Thailand

Medhi Krongkaew, *Thailand's Industrialization and Its Consequences* (New York: St. Martin, 1995).
A discussion of events surround the development of Thailand since the mid-1980s with a focus on the nature and characteristics of Thai industrialization.

Ross Prizzia, *Thailand in Transition: The Role of Oppositional Forces* (Honolulu: University of Hawaii Press, 1985).
Government management of political opposition in Thailand.

Susan Wells and Steve Van Beek, *A Day in the Life of Thailand* (San Francisco: Collins SF, 1995).

Vietnam

Chris Brazier, *The Price of Peace* (New York: Okfam Pubs. U.K. [St. Mut.], 1992).

Ronald J. Cima, ed., *Vietnam: A Country Study* (Washington, D.C.: U.S. Government, 1989).
An overview of modern Vietnam, with emphasis on the origins, values, and lifestyles of the Vietnamese people.

Chris Ellsbury et al., *Vietnam: Perspectives and Performance* (Cedar Falls, IA: Assn. Text Study, 1994).
A review of Vietnam's history.

D. R. SarDeSai, *Vietnam: The Struggle for National Identity* (Boulder, CO: Westview Press, 1992).
A good treatment of ethnicity in Vietnam and a national history up to the involvement in Cambodia.

PERIODICALS AND CURRENT EVENTS

The Annals of the American Academy of Political and Social Science
c/o Sage Publications, Inc.
2455 Teller Rd.
Newbury Park, CA 91320
Selected issues focus on the Pacific Rim; there is an extensive book-review section. Special issues are as follows:
"The Pacific Region: Challenges to Policy and Theory" (September 1989).
"China's Foreign Relations" (January 1992).
"Japan's External Economic Relations: Japanese Perspectives" (January 1991).

Asian Affairs: An American Review
Helen Dwight Reid Educational Foundation
1319 Eighteenth St., NW
Washington, D.C. 20036-1802
Publishes articles on political, economic, and security policy.

The Asian Wall Street Journal, Dow Jones & Company, Inc.
A daily business newspaper focusing on Asian markets.

Asia-Pacific Issues
East-West Center
1601 East-West Rd.
Burns Hall, Rm. 1079
Honolulu, HI 96848-1601
Each contains one article on an issue of the day in Asia and the Pacific.

The Asia-Pacific Magazine
Research School of Pacific and Asian Studies

The Australian National University
Canberra ACT 0200, Australia
General coverage of all of Asia and the Pacific, including book reviews and excellent color photographs.

Asia-Pacific Population Journal
Economic and Social Commission for Asia and the Pacific
United Nations Building
Rajdamnern Nok Ave.
Bangkok 10200, Thailand
A quarterly publication of the United Nations.

Australia Report
1601 Massachusetts Ave., NW
Washington, D.C. 20036
A monthly publication of the Embassy of Australia, Public Diplomacy Office, with a focus on U.S. relations.

Canada and Hong Kong Update
Joint Centre for Asia Pacific Studies
Suite 270, York Lanes
York University
4700 Keele St.
North York, Ontario M3J 1P3, Canada
A source of information about Hong Kong emigration.

Courier
The Stanley Foundation
209 Iowa Ave.
Muscatine, IA 52761
Published three times a year, the *Courier* carries interviews of leaders in Asian and other world conflicts.

Current History: A World Affairs Journal
Focuses on one country or region in each issue; the emphasis is on international and domestic politics.

The Economist
25 St. James's St.
London, England
A newsmagazine with insightful commentary on international issues affecting the Pacific Rim.

Education About Asia
1 Lane Hall
The University of Michigan
Ann Arbor, MI 48109
Published 3 times a year, it contains useful tips for teachers of Asian Studies. The Spring 1998 issue (Vol. 3, No. 1) focuses on teaching the geography of Asia.

Indochina Interchange
Suite 1801
220 West 42nd St.
New York, NY 10036
A publication of the U.S.–Indochina Reconciliation Project.

An excellent source of information about assistance programs for Laos, Cambodia, and Vietnam.

Japan Echo
Maruzen Co., Ltd.
P.O. Box 5050
Tokyo 100-3199, Japan
Bimonthly translation of selected articles from the Japanese press on culture, government, environment, and other topics.

The Japan Foundation Newsletter
The Japan Foundation
Park Building
3-6 Kioi-cho
Chiyoda-ku
Tokyo 102, Japan
A quarterly with research reports, book reviews, and announcements of interest to Japan specialists.

Japan Quarterly
Asahi Shimbun
5-3-2 Tsukiji
Chuo-ku
Tokyo 104, Japan
A quarterly journal, in English, covering political, cultural, and sociological aspects of modern Japanese life.

The Japan Times
The Japan Times Ltd.
C.P.O. Box 144
Tokyo 100-91, Japan
Excellent coverage, in English, of news reported in the Japanese press.

JEI Report
Japan Economic Institute
1000 Connecticut Ave., NW
Washington, D.C. 20036
A monthly analysis of Japanese government policies that affect the economy.

The Journal of Asian Studies
Association for Asian Studies
1 Lane Hall
University of Michigan
Ann Arbor, MI 48109
Formerly *The Far Eastern Quarterly;* scholarly articles on Asia, South Asia, and Southeast Asia.

Journal of Japanese Trade & Industry
11th Floor, Fukoku Seimei Bldg., 2-2-2 Uchisaiwai-cho
Chiyoda Ku
Tokyo 100-0011, Japan
A bimonthly publication of the Japan Economic Foundation, with a focus on trade but including articles on Japanese culture and other topics.

Journal of Southeast Asian Studies
Singapore University Press
Singapore
Formerly the *Journal of Southeast Asian History;* scholarly articles on all aspects of modern Southeast Asia.

Korea Economic Report
Yoido
P.O. Box 963
Seoul 150-609
South Korea
An economic magazine for people doing business in Korea.

The Korea Herald
2-12, 3-ga Hoehyon-dong
Chung-gu
Seoul, South Korea
World news coverage, in English, with focus on events affecting the Korean Peninsula.

The Korea Times
The Korea Times Hankook Ilbo
Seoul, South Korea
Coverage of world news, with emphasis on events affecting Asia and the Korean Peninsula.

Malaysia Industrial Digest
Malaysian Industrial Development Authority (MIDA)
6th Floor
Industrial Promotion Division
Wisma Damansara, Jalan Semantan
50490 Kuala Kumpur, Malaysia
A source of statistics on manufacturing in Malaysia; of interest to those wishing to become more knowledgeable in the business and industry of the Pacific Rim.

The New York Times
229 West 43rd St.
New York, NY 10036
A daily newspaper with excellent coverage of world events.

News From Japan
Embassy of Japan
Japan–U.S. News and Communication
Suite 520
900 17th St., NW
Washington, D.C. 20006
A twice-monthly newsletter with news briefs from the Embassy of Japan on issues affecting Japan–U.S. relations.

Newsweek
444 Madison Ave.
New York, NY 10022
A weekly magazine with news and commentary on national and world events.

The Oriental Economist
380 Lexington Ave.
New York, NY 10168
A monthly review of political and economic news in Japan by Toyo Keizai America Inc.

Pacific Affairs
The University of British Columbia
Vancouver, BC V6T 1W5
Canada
An international journal on Asia and the Pacific, including reviews of recent books about the region.

Pacific Basin Quarterly
c/o Thomas Y. Miracle
1421 Lakeview Dr.
Virginia Beach, VA 23455-4147
Newsletter of the Pacific Basin Center Foundation. Sometimes provides instructor's guides for included articles.

South China Morning Post
Tong Chong Street
Hong Kong
Daily coverage of world news, with emphasis on Hong Kong, China, Taiwan, and other Asian countries.

Time
Time-Life Building
Rockefeller Center
New York, NY 10020
A weekly newsmagazine with news and commentary on national and world events.

U.S. News & World Report
2400 N St., NW
Washington, D.C. 20037
A weekly newsmagazine with news and commentary on national and world events.

The US-Korea Review
950 Third Ave.
New York, NY 10022
Bimonthly magazine reviewing cultural, economic, political, and other activities of The Korea Society.

Vietnam Economic Times
175 Nguyen Thai Hoc
Hanoi, Vietnam
An English-language monthly publication of the Vienam Economic Association, with articles on business and culture.

The World & I: A Chronicle of Our Changing Era
2800 New York Ave., NE
Washington, D.C. 20002
A monthly review of current events plus excellent articles on various regions of the world.

Index